A History of Tin Mining and Smelting in Cornwall

D.B. Barton

CORNWALL BOOKS

First published in Great Britain in 1965 by D. Bradford Barton Ltd.

Revised edition 1969. Republished in 1989 by Cornwall Books.

British Library Cataloguing-in-Publication Data

Barton, D. B. (Denys Bradford)
A history of tin mining and smelting in Cornwall.
1. Cornwall. Tin mining industries, to 1988
I. Title
338.2′7453′094237

ISBN 1-871060-03-6
ISBN 1-871060-09-5 pbk

Printed and bound in Great Britain by BPCC Wheatons Ltd

CORNWALL BOOKS

An imprint of Wheaton Publishers Ltd.
A Member of Maxwell Pergamon Publishing Corporation plc

Wheaton Publishers Ltd
Hennock Road, Marsh Barton, Exeter, Devon EX2 8RP
Tel: 0392 74121; Telex 42794 (WHEATN G)

SALES

Direct sales enquiries to Cornwall Books at the address above.

to my father
and
the memory of my mother

ADDENDA

Page 251, line 2: *for* of the New Stamps at Carnkie, *read* to the New Stamps at Carnkie.

Pages 284–5: Output figures are gross, based on concentrates of approximately 70 per cent tin content. Those on page 286 [from The International Tin Council, *Statistical Supplement*, 1965] are to a higher and uniform grade.

'. . . *Adventuring for Tyn, which as a lottery hath a great many blanks to one prize.*'

Newton Papers, [Library of the Royal Mint].

CONTENTS

ILLUSTRATIONS

ILLUSTRATIONS IN THE TEXT

MAPS

FOREWORD

THAT NO HISTORY of Cornish tin mining has yet been written is a matter for some surprise when it is realised how important, interesting and singular an English industry it was. G. R. Lewis in *The Stannaries* covers its organization and development up to about the mid-eighteenth century but thereafter the subject has only been dealt with partially by any author and insufficiently so for the historical outline of the industry to be clear. This is particularly true of the period after 1873, to some extent the greatest yet most tragic years of Cornish tin. It is perhaps inevitable that the earlier period, with its romantic overtones of bal-maidens and grass captains in tall hats, of lavish count-house dinners and legendary profits, has been given the greater prominence in what has appeared to date, but in chronicling the full history of the industry there should also be told the more sombre tale in the later years when Cornishmen toiled in the lengthening shadows of the Straits and Bolivian tin giants. Then there were the grim realities of enforced emigration, of ankylostomiosis, and of such things as dangerously cheap dynamite in the battle for survival both of the industry and the men within it. The story of how Cornwall, from being the world's premier tin field in 1872, was reduced in half a century almost to extinction by foreign competition is not a happy one, although it will be found that these years of depression have an interest and a melancholy fascination all their own.

Nor can it be said that the history of Cornish mining during this time is at all well known. How many know, for instance, of the great tin rush of the early 1870's; of the West Basset scandal of 1878; of the financial disaster that was Pednandrea; or—to come to the present century—of the dramatic discovery of the new East Pool lode at Christmas 1915; of the bitter strike at Levant in 1918 when the pony in the 350 level had to be left to drown; of the disheartening negotiations with government for help in the crisis of 1921–2; of the crash of 1930 which brought Cornish mining once again to a temporary end; or of the long South Crofty strike in 1939 when a black-leg cargo of tin had to be smuggled to Holland from the port of Gweek? Even to those who can claim to recall some part of this happening, memory alone is a poor aid when one wants chapter and verse.

Even the history of Dolcoath in recent times is a closed book. Nothing in detail has ever been written on the Mayne forgeries of 1887; the notorious fine case of the following year; of the great runs of ground in the deep part of the mine in the 1890's in one of which seven men were killed and which was said to have crippled Dolcoath for ever; of its underground fires; or of the costly 'white elephant' that was Williams

Shaft. Indeed very much more has been written and re-written on the career of this premier Cornish mine during the period prior to 1850 than on its vastly more interesting career thereafter.

Inevitably it is more difficult to write the comprehensive history of an industry than it is to treat, piecemeal, a multitude of its constituent parts. In doing the latter it is possible to pick and choose; to indulge in recording that which is colourful or easy, whilst ignoring the dull or the difficult—the difference in fact between mining and fossicking. Moreover the problems are greater because one has to perceive and interpret the whole picture, and it is in the interpretation of the widely fluctuating fortunes of mining and smelting, the internal and external reasons for them, their immediate and their lasting effects, that most importance attaches in terms of economic history. This is doubly the case with Cornish tin mining in which the regularly repeated cycle of boom following slump, of feast following famine, is so marked. Distant and seemingly unrelated occurrences had far-reaching effects which superimposed themselves upon this rhythmic pattern and for this reason an attempt has been made to set the industry into a place in the wider context of the history of the world production of tin. To some extent this is essential, for the industry cannot fairly be considered in isolation: external factors, such as the development of the Australian tin diggings by emigrant Cornishmen in the early 1870's and the transition from silver to tin in the Bolivian mines in the 1890's, determined the pattern of Cornwall's industrial history from the mid-nineteenth century on. Few people realize the far reaching effects upon Cornwall and its premier industry of such diverse occurrences as, for example, the American Civil War, the feuds of Chinese secret societies in the Malayan states, or the building of a metre-gauge railway inland from Antofagasta on the Pacific coast of Chile.

Considering the basic pattern of Cornish tin mining in modern times, the industry has occupied two main phases, the one merging into the other in the middle years of the last century. The first was a period of tin mining proper which existed alongside but was very subsidiary to the copper mining which constituted most of Cornwall's industrial activity. The second was a continuation—almost a by-product—of the first, when copper mining declined and the production of tin took its place. The deep tin mines of Cornwall, comprising virtually all the post-1895 survivors, would never have come into being but for their previous careers for copper. This was true on two counts; firstly that the existence of tin in depth beneath copper deposits was a theory which arose only on account of practical results, and secondly that as tin mines alone only a bare handful could ever have paid their way opened solely for that metal.

Amongst them, Dolcoath is by far the best known, and the position she occupied in deep Cornish tin mining cannot be over-emphasised, for after tin came to prominence about 1860 Dolcoath was the yardstick by which other Cornish mines were measured. Her example was followed, for better or worse, by scores of other mines and a very great deal of money was lost in the process.

Two basic facts emerge that will be new to the great majority of readers. Firstly, that tin has been very much subsidiary to copper during the main period of Cornwall's mining history. The copper mines were more numerous, larger in size, and incomparably greater dividend payers: from 1800 up to the crash in copper mining in west Cornwall it is doubtful whether there was a year in which the richest single copper mine did not pay more by way of dividend than all the county's tin mines put together. Secondly, that tin mining has been, on balance, an unprofitable affair for those who found its capital. There were the very few rich exceptions such as Dolcoath, St. Ives Consols, East Pool, Tincroft and others, but these were a handful against the hundreds that never paid their way. A question which remains after any analysis of Cornish mining history is 'How it happened if the mines were not profitable, that so many persons were found adventuring in them?' The answer to this question, once posed by Pitt to an eminent Cornishman,[1] was—'For the same reason, Sir, that so many adventure in your lottery.' Certainly the innumerable attempts to re-work Cornish mines have shown time and again what a degree of financial risk attends every such venture. The penalties have been heavy and the prizes few, and a lot of money might not have been lost if greater attention had been paid to the value of historical research, for as Robert Hunt, Keeper of Mining Records, wrote in 1876, 'considerable loss of capital has frequently resulted from attempts to mine in localities where a little knowledge of what had been previously done would have shown there was but small prospect of success.' Fifty years later, moreover, this was still true.

How much has been lost and how much has been made, in terms of calls as against dividends, in Cornish tin mining since 1800 cannot be computed, but in this context it should be remembered that, contrary to popular belief, Cornish tin mining did not endure depression solely from 1895 onwards. It experienced it in greater or lesser degree throughout most of the century and after, a depression which grew steadily worse as a result of the expansion of tin output in Australia, Bolivia and south-east Asia. The whole nineteenth century was in fact one of borrasca for Cornish tin, relieved only by the short-lived bonanza of the early 1870's. Undoubtedly the great heyday of the industry came in earlier centuries, at a time when English tin had a monopoly all its own. Once the stanniferous muds of Banca and the surrounding lands had been discovered by the Dutch and worked by the industrious Chinese, Cornwall no longer ruled the world and the Banca sales set the standard by which the Cornish industry prospered, declined, or died.

[1] On the occasion of a deputation from Cornwall to the Prime Minister against the proposed admission of duty-free foreign ores. WB 17.8.38.

This is not to say, however, that Cornish tin mining was unprofitable for everyone, for this applied only to the outside investor. The main beneficiaries throughout its history have been the smelters, the merchants and the mineral lords. Theirs was the sure harvest, not only in the nineteenth century, but in every century that tin was raised in Cornwall, and what remained for dividend after the ministrations of this triumvirate was never very much.

One other interesting aspect that emerges from a detailed consideration of Cornish mining history is the great and abiding tenacity of the men engaged in it. There was—and, to a lesser degree, there still remains today—an inborn love of mining which alone can explain the continuance of the long and expensive rearguard action that was fought by Cornish mining against foreign tin in the half century after 1875. Other mining fields elsewhere in Britain and abroad under similar conditions ' burned out ' in a year or two and were abandoned for ever, but the Cornish clung to their mining heritage like men possessed. True, it was home they were defending but it was something more than this—in the shape of a tradition of centuries of mining perhaps—that sustained them during the last and worst thirty years after 1896.

If then it can be said that little enough has been recorded of the later history of Cornish tin mining, even less has been written on the subject of tin smelting. This is *terra incognita* indeed, for even those few lines hitherto written on the subject have been sketchily inaccurate. In print one will seek in vain for anything of substance on the great Cornish tin smelting houses of the past, on Carvedras or Treloweth, on Trereife or Trelissick. It was a quiet and devious trade, hidden in the hands of a few powerful men who ruled Cornish tin. In fact the extent to which the smelters fostered, promoted, and dominated the close circle of Cornish mining in the period up to about 1850 cannot be over-emphasised. Men such as Daniell, Daubuz, Fox, Williams, Batten, Bolitho, Michell, Harvey, Carne, James, Oxnam, Cornish, and Pascoe were the axis on which the universe of tin revolved. No mine of any consequence existed but they held a major share, so much so that the extent of smelters' holdings in a mine may now be taken as an acid test of its one-time value. It is not perhaps entirely without significance that it was in the later period when mine promotion and share-dealing became widespread—when the adventurer gave way to the shareholder—that so many worthless mines were set afoot and so much money was lost. Amongst these mines the names of the old smelters were conspicuous by their absence. And in this later era who can blame them for combining to secure continuing control of their industry during such a period of dilution. They were after all the hard-bitten and wily professionals amid a host of amateurs and dilettantes. After 1895, of course, all was changed, not only for miners but for the most hard-bitten of smelters. Tin had ceased to be a commodity of primarily Cornish concern, with horizons extending far beyond its former centres of Truro and

Penzance. Another twenty five years or so and even the second generation of Cornish smelters—Tregoning, Bolitho, Lanyon, Vigurs, Bain, Teague, Sampson, Strauss—had gone, and soon thereafter, even Cornish smelting itself.

Tin smelting is indeed of such importance in the county's industrial history that it deserves a monograph in itself. Unfortunately it is so inextricably bound up with mining that it is impossible satisfactorily to divorce the two. This necessity to treat two companion industries in a single volume has raised problems, not only in compressing into a few words that to which one would like to devote many lines, but in selecting and thus leaving out a great deal of subsidiary matter about individual mines and mining events that would have been of interest. It is intended that some of this detailed material will in due course form the basis for a collection of essays on various aspects of Cornish mining history.

Meanwhile it is hoped that this volume may be the means of bringing into focus a new awareness of an industry which looms large in Cornwall's not very distant past, and one which is still in being. It is a story told without compromise, omitting nothing of historical importance whether creditable or otherwise. At times the truth may seem harsh to some, but let no-one think that I have no affection for my subject.

D. B. BARTON

FEOCK,
CORNWALL

ACKNOWLEDGMENTS AND A NOTE ON SOURCES

THE THREE MAJOR printed sources which have been used for this history are the two county newspapers, the *Royal Cornwall Gazette* (RCG) from 1801 to 1945, the *West Briton* (WB) 1810 to 1880, and the *Mining Journal* (MJ) 1835 to 1922. In addition, use has been made of the *Mining World* (MW) 1920 to 1945 and the *Mining Magazine* (MM) 1910 to 1960. Except the latter, these are weekly publications and the task of extracting the relevant facts and figures from them can only be described as formidable. Most of this information has never been published before in book form and as some of it runs counter to what is popularly believed today about the history of Cornish tin mining the relevant references have been given in full.

The original records that have been used in addition to the foregoing comprise a great deal of material now preserved in a number of collections at County Record Office (CRO), Truro, or at the Royal Institution of Cornwall (RIC), the County Museum, Truro, consisting of mine cost books, ledgers, journals, reports, dues accounts, leases, ticketing lists, plans, etc., etc. These records, in all, comprise literally tens of thousands of individual items—a legacy of Cornwall's mining past which is in marked contrast to the records which have survived in other metal mining areas of Britain such as North Wales. There are, in fact, a dozen or more individual *parishes* in Cornwall that have more mining records than now remain of the whole latter area. The Stannary Court records, relating to some hundreds of individual mines wound up under its jurisdiction, are alone very extensive, and consist in some cases of a dozen or so large ledgers relating to single companies. Because of this great wealth of material, it has been thought advisable to relegate subsidiary material which is not of historical relevance to the main context of the book to foot-notes in an endeavour to keep the main historical theme of the industry's development clear.

By contrast, the surviving original records of tin smelting are few, and the only major source, apart from the remaining coinage books, is the Bolitho Collection at County Record Office. From the more or less interlocking nature of the trade, and the important position held by the Bolitho family in it, this includes notes and material which appertain to various other Cornish smelters, as well as to the Consolidated Tin Smelting Co. Ltd of 1891–1912. Records surviving in the possession of the companies still concerned in the trade are meagre, but Messrs H. J. Enthoven & Sons, and Messrs Vivian Younger & Bond Ltd., of London, were kind enough to supply what information they could. Unfortunately the historical records of other companies have not been made available.

It is a pleasure to record the author's grateful thanks to those who have assisted in one way or another in gathering material for this book: to my good friend H. L. Douch, Curator of the County Museum, Truro, to whom I am once again more than usually indebted not only for encouragement and assistance but for his kindness in making his own early newspaper and other research notes available; to P. L. Hull, County Archivist, and Miss C. Hawkridge, Assistant County Archivist, for their usual helpfulness; to J. H. Trounson, Redruth,—who has done more than any other man in the past twenty years, both personally and as chairman of the Cornish Mining Development Association, to further a lasting revival in Cornish tin; J. N. Rosewarne, Truro; Russell Bayles, Frodsham; T. J. W. Higgans, of Harvey & Co. Ltd., Hayle; E. J. Wigley, Zennor; W. L. Hart, Redruth; T. R. Harris, Camborne; F. Michell, Redruth; and W. J. Dickinson, Librarian, Truro City Library.

I

TIN DURING CORNWALL'S COPPER BOOM

THE FETTERS OF COINAGE: TO 1838

'*Surely the tin mines are not to be sacrificed in order to supply a revenue for the Heir Apparent?*'

West Briton, 7 September 1832

THE FIRST BEGINNINGS of the trade in Cornish tin, in pre-Christian times, are obscure and its later history thereafter, in the Dark Ages, is a matter of historical speculation. Of the late mediaeval period somewhat more is known and for this the standard work is G. Randall Lewis' *The Stannaries*.[1] In this is given the background of the rise of tin mining from the original alluvial moor- and stream-workings, to the primitive cliff and other early lode mines with their tiny adits and hand windlasses, and to the water-wheels and horse-whims of a later century. Of the period from the sixteenth to the eighteenth century, not enough is known in detail to enable any consecutive historical account to be written, although black tin[2] output is known from the coinage records with some accuracy.

From these figures (see Lewis, p. 257) it will be seen that the industry, although important both locally and as the primary world source of the metal, was small in size and that its output was more or less static. This is a noticeable feature when comparison is made with the progress of copper mining in Cornwall, which had increased by leaps and bounds from about 1720 onwards. Long before the end of the century copper was of far greater commercial importance to Cornwall, and when once the set-back of the great copper recession of the 1790's was over—the period when the Anglesey mines had flooded the copper markets of the world—the disparity in growth between Cornish tin and copper became even greater. It was copper that had brought the county to prominence as a mining field in the eighteenth century, and that had brought it steam pumping, steam hauling, tram-roads and the growth of the mining ports. True, a few of the bigger tin mines made use of these technical innovations in their

[1] Originally published 1908, reprinted 1966. Although written over half a century ago by an American scholar, this valuable treatise has been neither superseded nor supplemented by any other book.

[2] Tin concentrate, after separation from the tin-stone or ore by stamping and washing, in which state it was sold to the smelter. So called from its colour, dark brown or black. After smelting the metal was frequently referred to as white tin, and the two sides of the industry as black tinners and white tinners.

working, but the production of the two ores in Cornwall still remained distinct industries rather than two sides of the same industry, as later.

The differences between them were considerable and numerous, but all stemmed from the fact that tin mining was an ancient industry, governed and ruled by customs and regulations that had evolved with it, and to none of which the newly developed copper mining was subject. In view of this, not only to show the gulf that lay between the two, but also to provide some background to the events of the following decades it is necessary to consider—albeit briefly despite the complexity of the subject —the system of Stannary administration which governed almost all aspects of tin mining in Cornwall and Devon. The Stannaries[1] were created when the tinners of the two western counties came under the protection, by charter, of their own laws, much as in other mining fields in Britain, such as the Mendips, the Forest of Dean, the Peak District and elsewhere. In return for this they paid special taxes to the Crown or, after 1338, to the Prince of Wales as the Duke of Cornwall. A warden was appointed by the latter to administer the Stannaries, aided by a vice-warden and other lesser officials, and he held both civil and criminal jurisdiction over the miners. In the Stannary courts all offences, disputes and other legal matters appertaining to the mining of tin were heard, not only of labouring tinners but also of those who took shares in mines, lords of the soil who held mineral rights, as well as those who bought and sold black or white tin. Stannary parliaments were held at which laws were passed and amended and a whole judiciary system grew up, including the setting up of prisons at Lostwithiel and Lydford for the detention of offenders. Amongst the more important privileges granted to the tinners were their exemption from normal military service and from market tolls, and their right to divert any streams for their purpose and to cut fuel. They also enjoyed bounding rights, to search for and work freely for tin over areas of unenclosed common land. In return, the tin they produced was taxed after smelting through a system of coinage. A number of towns were designated for this purpose and at stated times during the year the ingots or blocks of tin were assayed and weighed by the Duchy officers.[2] Theoretically no tin could be sold until it had been coined, although needless to say a considerable quantity never saw the

[1] A stannary comprised one of the divisions of the two counties into separate mining regions. In Cornwall there were four; Foweymore (a name variously spelt), roughly corresponding to the present limits of Bodmin Moor; Blackmoor, between the rivers Fowey and Fal, centred on St. Austell; Tywarnhaile, roughly the triangular area from St. Agnes south to Truro and Redruth; and the united stannaries of Penwith and Kerrier, respectively west and south of the last named. In Devon, four stannary districts covered Dartmoor and its environs, whence all production came. For a full description of the organization of the Stannaries in all its aspects, see G. R. Lewis, *op. cit.*

[2] The actual process consisted of striking a corner from each block by hammer and chisel (hence the term coinage, from the French *coin*–corner) and after assaying this sample, weighing the block to assess the dues. On payment of these the metal was then stamped with the Duchy seal. The mediaeval coinage towns in Devon—where tin bore a lesser tax—were Ashburton, Chagford, Plympton and Tavistock. In Cornwall they were originally Bodmin, Lostwithiel, Liskeard, Truro and Helston but the increasing development of tin production westward was reflected in the disappearance of Bodmin and Lostwithiel from this list and the addition of Penzance.

beam at any coinage hall. When run into small bars and not into the regulation size $3\frac{1}{2}$ cwt. blocks that had to be presented for coinage, it was a commodity ideally suited to smuggling out of Cornish ports and creeks in small boats. A long, indented coast-line, a population already practised in the evasion of excise duties, and a waiting open market in Holland and other countries made it only too easy. The stannary laws designed to prevent trade in uncoined tin were numerous: no unstamped blocks to be moved except from smelting house to coinage town and even then by prescribed routes; no tin, black or white, to be conveyed by night[1]; all tin, once smelted, to bear the smelter's house-mark as well as marks to show the grade of purity, the owner and the actual workman who had cast it.[2] In addition all smelters had to swear on oath twice a year to abide by stannary laws, and that their books, showing the purchase and sale of all tin, would be open to Duchy inspection at all times.[3]

Thus it will be seen that the greatest difference between tin and copper in Cornwall lay in the smelting, the former being required to be produced solely within the county whilst the latter, for economic reasons, had become almost entirely the province of smelters situated on the South Wales coalfield. As a further result of the coinage system, in which tin sales were only held every three months, or in earlier times every six months, the black tinners had gradually become completely subservient to the white tinners. The latter, the smelter-tin merchants, were the financiers of the Stannaries, advancing money to the miners against the tin bills they themselves issued as receipts for the tin left with them for smelting. This was in effect a form of banking, and the close and lasting connection between smelting and banking was very marked.[4] The smelters ruled the roost firmly for various reasons; partly because they held shares in the most promising mines and stream works themselves and partly because by a mystique of assaying and measuring and the standard at which they would buy back their tin bills they overawed the small groups of tinners whose labours provided the great bulk of Cornish tin.[5]

[1] This rule was rigidly imposed and in the 1817 *Royal Cornwall Gazette* (1.11.17), for example, there was a warning notice to smelters and blowers of tin that '. . . no carrier of white tin, or other whatsoever, shall take and carry away any slab, block, or piece of tin, from any blowing-house, before the break of day nor after the setting of the sun, to any other place whatsoever . . . Sabbath has been frequently broken in upon, occasioned by the late hour at which Tin for Coinage has been received.'

[2] At this stage in the development of the industry, a great deal of the output was smelted on a commission basis, the smelter making his profit from that operation alone, and not, as became customary later, from buying black tin outright, and then smelting and selling it entirely on his own account.

[3] These, and the very many other laws laid down for the regulation of all aspects of the Stannaries, were published in *The Laws of the Stannaries*, a volume which ran to well over a hundred closely printed pages. This was regarded as virtually the tinners' Bible and was printed in something like a dozen editions in the period from 1750 to 1825.

[4] It is true to say that every Cornish tin smelter of any consequence in the nineteenth century was also engaged in banking activities. (See on, page 49).

[5] The calculations when selling black tin to a smelting-house were, designedly, so complicated—based on the smelter's own assay, the existing tin standard, and the price of tin bills per cwt.—that a publication entitled *The Tinner's Ready Reckoner or a Sure Guide at the Smelting House* was published in 1804. Nor was it only a web of mystique that the untutored seller of black tin had to contend with; ' some suspicions have arisen of frauds committed at some of the smelting-houses, by receiving black

In view of the importance of the smelters to tin mining an account of their distribution and status in Cornwall at the close of the eighteenth century is valuable. They were divided into two categories, the true smelting-houses (or melting houses as they were sometimes called) and the blowing-houses. The latter were much the older and were in common use by the fourteenth century at least: ' a rude structure, probably of rock and turf, with a thatched roof, the whole being so inexpensive that it was burned down in order to save the particles of tin which the blast had driven up into the thatch '.[1] The smelting equipment required was equally rudimentary: a granite furnace fired by charcoal and blown by a bellows worked by a small water-wheel, with the molten tin run into granite moulds. These blowing-houses were scattered thickly throughout Cornwall, probably three score or more in number,[2] serving in each locality the numerous small mine-workings and stream-works that then made up almost all Cornish output of the metal.

At the end of the seventeenth century the reverberatory furnace, first introduced about 1613, was used for tin smelting, employing the cheaper and more readily available coal for fuel instead of charcoal.[3] In this

tin by a larger weight than the white tin hath been delivered out by ' (*The Laws of the Stannaries*, 1824) and it was then laid down that official weights ' of Tower Measure ' should alone be used. ' Deceitful and untrue beams or scales ' were also not unknown in the early smelters' armoury and it was small wonder that the Cornish smelting-house proprietor of 1800 was regarded by his inferiors as a veritable Shylock.

[1] G. R. Lewis, *op. cit.*

[2] The sites of a great many early blowing-houses are now untraceable but evidence remains of the existence of a total of 28 active in the south-west in 1689. Apart from two in Devon, the 26 in Cornwall included one each in the parishes of Calstock, Linkinhorne, St. Neot, St. Blazey, St. Austell, St. Stephen-in-Brannel, St. Allen, Gwennap, Redruth, Illogan, Mawgan-in-Meneage, Breage, Ludgvan, Gulval, and at Bissoe. St. Agnes and Penryn—the latter an important centre for tin production where an unsuccessful application to become a coinage town had recently been made—had two each, and the parishes of Wendron and Kenwyn (immediately south-west of Truro) three each. (Calendar of Treasury Books 1685–89, BM, London, quoted in J. C. A. Whetter, *Cornwall in the Seventeenth Century: A Study in Economic History*).

This enumeration in the year 1689 by no means makes up the full number of blowing-houses that have existed in the county at other dates—their sites being perpetuated today in the place-name ' Blowing House ' in several parishes. Documentary evidence exists of them at Charlestown; Gweek; Trannack (near Helston)—a house probably synonymous with that of Gwavas, where part of the Old Blowing House was advertised for sale in 1833 (RCG 9.2.33); Treworder, in Kenwyn, in 1619; Penryn; Ponsanooth; at Nuland Praze, Cardinham; Nansdrissack, Sithney; another ' now a dwelling house ' at Kennall Vale, Stithians, in 1659; at Fursenuth, in the Manor of Bulland (west of St. Cleer) in 1664, and at Porth Towne [i.e., Porthtowan] in 1655 (RIC). In the following century there were others to supplement this list, such as one on Rashleigh's Manor of Lanyon, near Polgoon in Madron, which had been turned into a dwelling house (Coode–French Records, CRO, Truro) and the well-known Godolphin Blowing House, serving the Breage mines. Of this latter Hals records an interesting story about Godolphin and his steward Coke who both had their tin blown there, under the signs of a dolphin and a cat respectively. By degrees the latter, by its owner's astute adventuring and by some devious dealing, gradually ousted the former—until at last Godolphin woke to the facts—thus giving rise to the Cornish fable of the cat eating the dolphin. (See Wm. Hals, *Compleat History of Cornwall*, 1750).

The blowing-house listed in St. Allen parish was in Bishop's Wood, north of Idless (near Truro), a building which almost certainly became the Garras smelting-house, serving the nearby rich lead mine of that name. This smelting-house was advertised for sale in 1832 (WB 23.3.32) and some time thereafter the premises were used as mills for the manufacture of gunpowder.

[3] The introduction of reverberatory furnaces for tin smelting in Cornwall is usually ascribed to Robert Lydall, in conjunction with Moult & Co., who established the Newham smelting-house at Truro in 1705. Lydall was connected with copper smelting at Neath where furnaces of this type were in use prior to 1700. Lead had also been smelted in reverberatories elsewhere in Britain for many

furnace the ore was reduced to its molten state without contamination by the fuel, but solely by the application of heat. Apart from a radical increase in later years in the size of these furnaces, raising their charge capacity from a few hundredweights to as many tons, this mode of smelting was to continue to the end of the industry in Cornwall. Its initial introduction, however, gave rise to grave complaint about the quality of smelted as opposed to blown tin. In 1707 a petition was sent by the Company of Pewterers in London to Lord Godolphin as Lord High Treasurer complaining of the poor quality of tin smelted by new methods in Cornwall[1] and the following year the Officers of the Mint were called upon to examine this question. Although, with increasing skill, improvement was made in the quality of smelted tin, the blowing-houses were not superseded and the undoubtedly purer 'grain' tin[2] which they produced commanded a higher price. Stream tin, so-called because it was derived primarily from stream-works and other alluvial workings, was much purer than lode tin from mines, as a result of the accompanying sulphides and arsenides having been weathered out naturally in the process of time, and it became more and more the province of the blowing-houses alone. Thus two distinct branches of the tin trade, smelting and blowing, existed side by side after about 1700. As lode mining increased and streaming declined, so did the number of smelting works rise and that of blowing-houses fall. Detailed coinage figures exist for the years after 1800 and these enable the size and progress of the individual smelters and their houses during this time to be traced. Records prior to this date are fragmentary in the extreme and it is not possible to delineate the replacement of so many small scattered blowing-houses by relatively few much larger smelting establishments with, in one case at least, ten reverberatory furnaces.

By 1800 the principal smelting centres, as might be expected, lay close to the main coinage towns, Penzance and Truro. At the latter the oldest and probably the largest works was Calenick, a mile south of the town at

years and it is possible that Lydall's patent was for one particular design of reverberatory furnace and that other types were in use already. According to Thomas Tonkin (Mss. notes at RIC, Truro), 'This patent was an imposition on the Countrey for it was no new invention' and added weight is given to this by the evidence of Celia Fiennes (*Through England on a Side Saddle*, 1695) who states that Polgooth smelting-house was using reverberatory furnaces in 1695. In addition to this, W. C. Borlase (*A Historical Sketch of the Tin Trade in Cornwall*, 1874) states that the German metallurgist John Joachim Becher set up the first reverberatory furnaces at Treloweth smelting-house some years before 1700. (See also *Trans. Royal Geological Society of Cornwall*, vol. 4). As against this Lydall may have been the first to use 'raw' instead of 'charked' (i.e., coked) coal. The subject is a fascinating one and deserves further study. Adding to the factors also to be considered is that a form of small granite reverberatory furnace fired by furze was commonly used in the seventeenth century 'tin-kilns' or calcining-houses where the ore was roasted to drive off arsenical and other impurities. From the use of these it was but a short step, intentional or otherwise, to making use of them for running as well as roasting the ore. (See *Mineral Observations touching the Mines of Cornwall and Devon* . . . 1689, RIC, Min. 92, Truro).

1 DD/PW/46, CRO, Truro.

2 The name grain tin was derived from its quality of granulating when it was heated almost to melting point and then struck, for it then formed a loose mass of metallic nodules and grains. In this form it was used for coloured glazes, in glass-making and in acid solutions for dyeing fabrics and carpets, 'giving a degree of permanency and brilliance to several colours, not to be obtained by the use of any other mordaunts.' (Rev. D. Lardner, *A Treatise on . . . The Manufactures in Metal*, (vol. 3) 1834).

the navigable head of a creek branching westward from the Fal estuary. This was founded in 1711 and was the continuation of or successor to the Newham works established a few years before which has already been mentioned as possibly holding the honour of being the site of the first tin smelting in the county.[1] In 1794 Calenick smelting house had no less than ten furnaces[2] and was then occupied by Thomas Daniell who had succeeded to some of William Lemon's business interests in or about 1760. Farther up the same valley, at Treyew, there had also been another 'ancient blowing and smelting house,' the existence of which is recorded in a reference in the *Sherborne Mercury* in 1770.[3] Closer to Truro itself lay Carvedras, an important eight-furnace smelting house which was probably established only a short time after Calenick. Its first owner was Walter Rosewarne[4] followed by Henry Rosewarne, on whose death in 1783 the works was advertised for sale.[5] It was again advertised in 1792[6] at which date it was taken by a company of which L. C. Daubuz was the principal partner.[7] The Daubuz family were Huguenots who had come to England in an earlier century to escape persecution and settled originally in Falmouth, Theophilus Daubuz and his son Lewis Charles (senior, 1754–1839) establishing a name that was perhaps the foremost in early nineteenth century tin smelting. In 1807, 1809 and again in 1811 Carvedras was advertised to be let[8] but does not appear to have changed hands, for ultimately (in 1818) Daubuz acquired its freehold and for eighty years thereafter the Daubuz family carried on business there.

A mile upstream from Carvedras in the same valley lay Coosebean 'melting and blowing-house' which was advertised for sale in 1775 'with six melting furnaces and a complete blowing-house, the whole quite new built.'[9] The actual date of Coosebean's establishment is not known—the reference to the 'whole quite new built' being to the furnaces and not the works as a whole—but it was certainly presenting tin for coinage in 1771.[10]

1 Newham smelting-house was actually in production from 1705–1715, Calenick being started in 1711. A large collection of records relating to them, and the associated Angarrack smelting-house, are preserved at CRO, Truro (Bolitho Records).

2 W. G. Maton, *Observations on the Western Counties of England*, 1797.

3 31.12.70. Treyew Blowing-House was converted to a smelting-house in 1712 by Samuel Enys, in opposition to Moult & Co. at Calenick, and after five years was leased by him to Nicholas Donnithorne and partners. (Enys Accounts, CRO, Truro).

4 RCG 18.4.07. Walter Rosewarne the elder died in 1747, the younger died 1772.

5 SM 28.7.83. 6 SM 16.1.92 'with eight furnaces . . .'

7 Coode-French Records, 166/B14, CRO, Truro. John James, of Perranzabuloe, who died in 1819, was the other main partner.

8 RCG 18.4.07 '. . . where for many years the smelting business was carried on by Mr. Walter Rosewarne, and afterwards by Henry Rosewarne Esq., his son, and is now by L. C. Daubuz & Co. . . . spacious and uncommonly well supplied with water, and distant about half a mile from the town and navigable river of Truro': RCG 10.6.09: RCG 12.1.11 '. . . with offices . . . dwelling-house, assay office, and buildings, two grist mills, stable and crucible store; blacksmith's and carpenter's shop and store room with two gardens and five fields, containing more than six acres. For several years past in the occupation of Messrs Daubuz, James & Co.' On an undated plan of Truro (*circa* 1800, at CRO) Carvedras coal yard is shown as occupying a small quay on the Malpas Road.

9 SM 23.1.75. According to the Spry & Taunton mss (RIC, Truro) Coosebean smelting-house was under the direction of Francis Benallack 'but supported and built at the expense of another' and failed as a result of the former, who was an attorney in Truro, keeping no books of account.

10 Bolitho Records, Acc. No. DDX. 104, CRO, Truro.

The final advertisement of the premises as a smelting-house was in 1810 and thereafter the buildings were used for other industrial purposes.[1]

South of Truro, beside the then navigable waters of an arm of the Fal estuary, was the Perran smelting house. This was in operation in 1755 when there was a legal case involving a dispute over the inferior quality of the tin smelted there[2] and it was advertised to be let in 1774 'with five furnaces and room for as many more.'[3] In both cases its owner was Walter Reed of Stithians, whose son Thomas had control when it was once again advertised 'to be let or sold' in 1802.[4] There were then six furnaces in the smelting house, an assay office and dwelling-house, whilst a trade in coal and timber was also carried on. No new occupier was found and it was certainly derelict by 1811.[5]

With their focal point centred on the Penzance coinages were no less than five old established smelting-houses—Trereife (or Trereiffe) west of the town in the valley inland from Newlyn; Chyandour at its eastern end where a stream filtered through the sand into Mount's Bay; Ludgvan and Treloweth spaced across the neck of the Land's End peninsula; and Angarrack in the valley east of Hayle. The first-named of these was run by William, and later Uriah, Tonkin from 1732–46[6]—prominent Penzance merchants—and later by Ralph Hall & Co., in 1766.[7] By 1805 it had been occupied for some years by a company of which John Batten of Penzance was the chief partner. Of Chyandour smelting-house, the first recorded

[1] WB 5.10.10: RCG 6.10.10 '. . . with 100,000 Dinnabole scantle slates, 25000 bricks, 5000 fire bricks, 700 feet of moorstone, 30 block tin moulds, etc.'. The premises (which were also used at one time as a woollen mill) were used for papermaking, being sold up (with 16″ steam engine and two 14′ water-wheels) in 1847. (RCG 3.9.47).

[2] SM 13.10.55.

[3] SM 6.6.74.

[4] SM 4.1.02. Later, however, the Fox family of Falmouth (who promoted Perran Foundry) were connected with the smelting-house, for an advertisement occurs in 1806 of a fifty-ton lighter for sale, to be seen on application to George Fox at Perran Smelting House. (RCG 26.4.06).

[5] Christopher Wallis, *Journal*, 10 June 1811. In February 1813 the lease of the premises 'heretofore a tin smelting-house called Perran Smelting House . . . but now converted to the collecting of arsenic and other purposes' was advertised for sale (RCG 27.2.13). This in fact records the birth of arsenic production on a commercial scale not only in Cornwall but in England as a whole, for the parties concerned—Dr Richard Edwards ('a physician of Falmouth and a skilled chemist'), William Williams, and William Gregory—were the pioneers in this field, having taken the works on lease from Midsummer 1812. During a celebrated legal case later, in 1851, (see RCG 4.4.51) at a time when an action was brought against the owners of the works because of the deaths of cattle and other animals poisoned by eating herbage downwind from the stacks, some interesting detail on the subsequent history of this arsenic works is given. In 1830, after the death of Edwards, the principal partner, the works passed into the hands of others and was discontinued after a few years, at a time when the second arsenic works in the area (Conn's at Bissoe) was started. (WB 30.12.36). In 1844 Thomas Garland, of Redruth, acquired the works and four years later greatly extended it, making it by far the largest of the four arsenic works then active in the Carnon valley. It then had seven large chimneys and two very high stacks to help carry away the extremely poisonous and voluminous white smoke and the accompanying stench (said to be like garlic) produced in arsenic manufacture.

From this accumulated evidence it is possible to pinpoint the site of Perran smelting-house and its successors as being at Tarrandean at the head of Mellingey Creek, a then navigable inlet of Restronguet Creek. The blowing-house which preceded it is also recorded there on one seventeenth century map at least (RIC).

[6] Phillipps Mss. 46, Rental of Manor of Trewarveneth and Paul, RIC, Truro.

[7] SM 17.2.66. The works was again for sale 24.3.66, 25.2.71 and 6.7.72, variously named Trereife and Stable Hobba Smelting House. Hall was also a partner during this time in the smelting-house at Lostwithiel.

reference is in 1743 when it was leased by William Mackworth Praed to Richard and John Oliver of Gulval.[1] Fifteen years later it was advertised for sale together with its adjoining public house[2] and was then taken by the well known merchant of Marazion, George Blewett.[3] Although in the nineteenth century the name of the Bolitho family was to become closely associated with the works and with Cornish tin smelting as a whole, their first connection with Chyandour came relatively late, in 1781, when they leased lime pits for tanning purposes there. The family were originally tanners and merchants at Penryn but whilst it was common enough for tin smelters to be merchants as well, the Bolithos were something more—bankers, shipowners, importers and dealers in hemp and cordage and tallow, and, not least, the most important pilchard merchants and exporters in Cornwall. Even by 1805, however, they were only one of several partners in Chyandour smelting-house. The merchant Richard Oxnam was in fact the principal proprietor and with him as well as the Bolithos were William Cornish and James Pascoe.[4]

Ludgvan smelting-house is mentioned in a lease of 1736[5] and was for sale in 1774, then under the ownership of a Francis Thomas.[6] Four years later it was twice offered for sale: in March '. . . with smelting-house near 100 feet long and upwards of 40 feet wide, in perfect repair: five furnaces with floats, moulds, etc,' whilst in October it was further described as 'ever been esteemed the best and most reputable in the county.'[7] In 1798 it changed hands once again and some further detail of its size and capacity are recorded, for it then contained 'five capital furnaces . . . in the heart of a tin county' and was capable of smelting 600 blocks a quarter,[8] whilst in another contemporary advertisement it was stated to be 'not under any engagement to stock tin for any particular company at reduced prices.'[9]

Of the early history of Treloweth, a mile or so east of Ludgvan close by the main Penzance to Hayle road, rather less is known, but from the evidence of John Joachim Beccher this was the site of Cornwall's first smelting-house equipped with reverberatory furnaces.[10] It was for sale in

[1] BIR/3/5, RIC, Truro. As a blowing-house Chyandour had been owned in the seventeenth century by John and Henry Tremenheere, notable merchants in Penzance at this time. They were perhaps the greatest tin merchants of their day—apart from being notorious for their smuggling activities, and John Tremenheere was reputed to be sole owner of Chyandour in 1685. (J. C. A. Whetter, *op. cit.*)

[2] SM 14.8.58.

[3] BIR/3/10, RIC, Truro.

[4] See on, (page 46). Cornish, of Marazion, had been clerk to Oxnam's father, whilst Pascoe was a merchant in Hayle.

[5] Bolton Collection, Acc. No. D45, Dorset County Record Office. This was situated at Lower Quarter, Ludgvan, and was also known as Vellenoweth. Separate leases also exist of Ludgvan Blowing House.

[6] SM 5.12.74. [7] SM 19.3.78, 5.10.78. [8] SM 17.3.98.

[9] *Exeter Flying Post* 5.4.98. This reference was to the 'stocking' of tin for sale at special prices to the East India Company. (See on, page 28).

[10] Beccher, born in 1635, was a German philosopher and metallurgist from Speyer who came to Cornwall in 1680 and, with his assistant Loeffler, was connected with Treloweth. It was said that 'the discovery of the means of converting other substances into gold constituted the study of his whole life' apart from which he was in all probability the first to smelt tin in other than an air furnace. A few brief biographical notes on him, from which a paper in the *Trans. Royal Geological*

1782[1] and again in 1791[2] at which date, or shortly thereafter, it was acquired by L. C. Daubuz, who was already engaged in smelting at Carvedras in Truro.

Angarrack smelting-house was almost as old as Treloweth, being established by Francis Moult & Co. in 1704, with six of the new reverberatory furnaces, as a companion works to the one at Newham, Truro. By 1743 it was owned by William Lemon and was for sale in 1783, with William Tremayne then described as the sole proprietor.[3] In 1804 the four principal partners there were Richard Cunnack, William Carne, the Bolithos and James Pascoe.[4]

Apart from these major Cornish smelting-houses centred on Truro and Penzance, there had been various others within the county prior to 1800. These were situated at Gweek, Melancoyse near Helston, St. Blazey, Lostwithiel, St. Agnes, St. Austell, and perhaps elsewhere.[5] The St. Austell Smelting-House was the most important of these, run as part and parcel of the Old Blowing House which had been in existence at Tregorrick since 1690 at least. It was probably this ' new smelting house ' that was burned down at St. Austell in October 1782 only twelve months after it had been built.[6] In 1784 its owners were Sawle Fox & Co.,[7] who occupied the premises until 1809[8] when a new partnership, Foxes Rashleigh & Co. was formed.

Smelting apart, St. Austell by 1800 was the focal centre of much of the Cornish grain tin trade, and consequently of the few remaining blowing-houses. These drew their supplies of tin chiefly from the Hensbarrow district where the innumerable tin lodes in the decomposed granite were of a quality comparable to that of true stream tin. Old Blowing House,

Society of Cornwall (vol. 4) was compiled, exist in the Johnstone Records, Acc. No. 498/5, CRO, Truro. These are undated but are probably early 19th century, the Hawkins (Johnstone) family having connections in the time of Beecher with the manor on which Treloweth stood.

[1] SM 23.12.82. [2] SM 25.4.91.

[3] SM 20.10.83. Considerable correspondence relating to Angarrack smelting-house occurs in the Journal of Christopher Wallis, 1782–3. Several other persons (Daniell, Nowell, and Giddy) had held a proportion of the 400 shares into which the business was divided and it appears that Angarrack and Calenick were run jointly throughout the eighteenth century.

[4] Bolitho Records, DDRG 1/163, CRO, Truro. Richard Cunnack (or Connock) was a tanner and general merchant at Lescudjack, Penzance, who died in 1827.

[5] These were identifiable smelting- and not blowing-houses. Leases of the one at Gweek, which remained in operation until about 1794 are now at CRO, Truro (Coode-French Records) whilst the Melancoyse smelting-house, with which Christopher Wallis was concerned, was advertised to be let in 1799 (SM 4.3.99, 5.8.99). The Lostwithiel smelting-house was advertised in 1758 ' to be let, newly erected smelting-house called Middlesex House, with quay etc.,' (SM 13.2.58). It was probably then an important one, for Lostwithiel at this date was the administrative centre of the Stannaries but its site today is perpetuated only in the name of Smelting House Lane. In 1763 it was advertised as taken by Ralph Hall & Co., and described as ' under repair and soon to be at work ' (SM 8.8.63). In 1789 it was announced that it would shortly be re-opened with Messrs. Paget, Gordon, Chappell & Co., as proprietors (EFP 11.6.89) whilst two years later it was stated that John Paget had become sole partner (EFP 24.2.91). By 1805 according to coinage returns it was completely out of use. The St. Blazey Smelting House was for sale in 1757 (SM 31.10.57), whilst the St. Agnes house was to last until 1828 (see page 55).

[6] SM 21.10.82. [7] *Bailey's Directory*, 1784.

[8] RCG 6.5.09. To be sold, with lease 22 years unexpired ' the well accustomed tin blowing and tin smelting-houses in the parish of St. Austell in which the tin trade has been successfully carried on for many years past under the firm of Sawle Fox & Co.'

RCG 8.7.09 Sawle Fox & Co. partnership dissolved; now Foxes Rashleigh & Co,

which was probably established in mediaeval times, was for sale in 1775[1] and it was possibly then that Sawle Fox & Co. took it. In addition there was New Blowing House—deceptively named for it was probably also of almost equally early origin—and Higher Blowing House, the latter situated at Trethowel on the Bodmin road north of the town. This was also an establishment of some antiquity and was notable for being owned in the latter part of the eighteenth century by the partners in one of the earliest firms of tin-plate manufacturers in South Wales.[2] Nearby, on a now unidentifiable site near Biscovey, there had also been the St. Blazey New Blowing House,[3] although this was moribund by 1805 according to the evidence of surviving coinage records.

At this stage in the development of the industry it was usual for adventurers in mines of any size to receive their share of profits either in black tin or in block form, and thus it became common for almost every adventurer of any consequence to have shares in one or another smelting-house. Failing this, he would join forces on commission with the owners of one to produce and sell tin for him, with the metal either coined under his own name[4] or that of the smelter. This was particularly the case in that the profits in the industry came at the smelting rather than the mining stage, in the refining and merchanting of the tin rather than in its production as ore. Typical of the extent of the smelters' interest in a big mine was Ding Dong, in Madron, a mine reputed to be one of the oldest in Cornwall. Here in 1815, Daubuz held 5/16ths, Batten 4/16ths, Bolitho 3/16ths and eight others the remaining 4/16ths.[5] And if a smelter owned 5/16ths of a mine he took good care, as of right, that he handled at least 5/16ths of its produce. It was this widespread interest in smelting that was partly responsible for the divided ownership of many smelting-houses at this time.

The most important feature of the Cornish tin trade during the latter part of the eighteenth century was undoubtedly the institution of the East India Company contract. By this means the company, which had a monopoly of all official English trade with China, engaged to take a considerable quantity of tin each year at a special price. For some time before 1770 Cornwall had suffered a continuing depression in tin, a decline which was due primarily to that of the pewter trade 'which formerly exported vast quantities to America &c., exclusive of what was made use

[1] SM 20.3.75.

[2] Harford Getley & Co. (*Bailey's Directory*, 1784). James Harford and James Getley were Bristol iron merchants, owners of Melingriffith tin-plate works near Cardiff, established in 1759. From 1804 onwards Higher Blowing House was run by Harford & Co.

[3] Carlyon Records 5M/EO/7/21, CRO, Truro. This new house was active by 1729 at least, and presumably an Old St. Blazey Blowing House also existed.

[4] These were known as 'out-coiners,' a term which passed out of use—and has never been recorded by any other writer since—when this once widespread practice died out in the early eighteenth century. The companion term 'out-adventurers,' to describe those who had shares in a mine but had no share in the right to supply it with goods, survived very much longer.

[5] Bolitho Records, DDRG 1/74 CRO, Truro. Earlier, Richard Oxnam had been the chief adventurer in Ding Dong.

of in this country. This has been superseded by the general use of earthenware abroad and at home.'[1] Thus the opening up of a new market in China was seen as a way of lifting the tin trade back into something approaching prosperity. The East India Company had regularly shipped lead to Canton since 1672, almost since its trading inception, and in 1789 when it was suggested that tin should also be tried on a considerable scale, the Company was eager for new exports to counter the very adverse balance of its rapidly growing import trade in tea and silk. This was notwithstanding an abortive trial of the same metal some twenty years before, when its price was found to be too high for the Chinese market, and an equally unsuccessful attempt, also emanating from Cornwall, to ship out copper in the season of 1784–5.[2] The prime mover in promoting this trial consignment of 55 tons in 1789 was one George Unwin, then a supercargo in the service of the Company.[3] Two years earlier Unwin's ship had carried 200 tons of tin from Banca and had sold it at a profit in Canton, and knowing the depressed price of English tin, he proposed to the Court of Directors of his Company that this trial consignment might be made. In due course this metal was successfully disposed of, the demand being such that in the opinion of the Canton merchants ' 12,000 peculs, or even more, might be annually sent.'[4] Negotiations on behalf of the Cornish tin interest were led by Nicholas Donnithorne, a prominent adventurer in the St. Agnes area, who was Chairman of the Quarterly Tin Meeting in the County of Cornwall. This body later assumed the less cumbersome style of the Associated Tinners, a title which the smelters who dominated it perhaps thought might mask the fact that the miners, or true tinners, had little actual say in its affairs.

Even before the gratifying news of the Chinese demand for tin had reached the Company,[5] Donnithorne had prevailed upon it in September 1789 to take 600 tons then lying unsold in the smelting-houses of the county. This was offered at the low price of £72 per ton delivered on board, and if necessary Donnithorne was prepared to have it sent to Canton 'on account and risk of the County of Cornwall.'[6] Distress in the extreme south-west was particularly acute and lending urgency to the

[1] RCG 2.6.10.

[2] For these and all other details quoted herein see G. Unwin, *Letters, Remarks, etc. with a View to opening an Extensive Trade in the Article of Tin from the County of Cornwall to India, Persia and China*, (1790) and G. Unwin, *On Tin and Copper to India and China* (1813). This latter, an extremely scarce pamphlet, prepared in the year when the Company's monopoly was due for renewal, gives *verbatim* letters and reports, as well as statistics of the China trade in Cornish tin. A further source is H. B. Morse, *The East India Company trading to China 1635–1834*, (1926).

[3] Supercargoes were commercial agents of the Company, each East Indiaman carrying two or more in her complement. Together with the principal officers, they were allowed various privileges, including the right to a limited amount of private trading on their own account. By 1790 Unwin had left the Company's service and was then described as ' a purser in His Majesty's Navy.'

[4] Unwin, *op. cit.*, 1813. This quantity was about 800 tons. The only complaint was that the tin should be in more manageable form, in slabs of about 80 lbs. weight, similar to the form in which Banca tin was normally shipped.

[5] In those days, commercial intelligence from the East took well nigh a year unless it coincided with the return of a season's fleet. East Indiamen normally made the round voyage in two years.

[6] G. Unwin, *op. cit.*, 1790.

need for some plan to alleviate this state of affairs was that the 'Cornish miners have of late been very riotous and the Gentlemen of the County have been obliged to call in the Military.'[1] The extremely low price of tin and the very high price of corn were raising in Cornwall the spectre of the Revolution then raging across the Channel. The major obstacle standing in the way of the export of tin on this scale was the prevailing duty of 3s. 4d. per cwt. which ate considerably either into the East India Company's profits or into the price they would pay the Cornishmen, and a petition to Government for the removal of this duty was made from Cornwall in the January following. With the help of the Duchy of Cornwall officers, who inevitably took a more than passing interest in matters which bore beneficially or otherwise on their tin revenues, this was favourably received and an Act was duly passed in March taking off 'the Duties upon unwrought Tin exported to any of the Countries beyond the Cape of Good Hope.' The upshot of this was the decision by the East India Company to contract annually with Cornwall for a quantity mutually agreed upon, not to exceed 1200 tons in a year, at a price which was to be basically twenty per cent below that of tin in London. From Cornwall's point of view the certain sale of what amounted to one third of her tin output a year cleared up the problem of the perpetual surplus and at the same time removed any competition that might have existed between smelter and smelter as a result of over-production. The lower price they were to receive was compensated by the higher price which would rule, or would be made to rule, on the home and European markets, of which they still had a virtual monopoly despite the first tentative imports of Banca tin into Europe by the Dutch. Later, the even surer method of making up the deficit was to reduce *pro rata* the price paid for black tin which was intended for 'stocking' for the East India contract.

For twenty-five years and more the China market, where something like two thousand tons of tin a year were consumed in the form of tinfoil at religious ceremonies, was to prove of prime importance to Cornwall, sales to this quarter from 1790 onwards forming a considerable part of the whole output from the county.

ANNUAL QUANTITY OF TIN FROM CORNWALL SHIPPED BY THE EAST INDIA COMPANY 1789–1813

Season	Tons	Price	Total
1789–90	898	£68.13.4	£61,662.13.4
1790–91	1200	£68.13.4	£82,400. 0.0
1791–92	931	£71. 0.0	£66,101. 0.0
1792–93	800	£75. 0.0	£60,000. 0.0
	402	£68.13.4	£27,604. 0.0
1793–94	800	£75. 0.0	£60,000. 0.0
	430	£68.13.4	£29,526.13.4
1794–95	800	£75. 0.0	£60,000. 0.0
	402	£68.13.4	£27,604. 0.0

[1] G, Unwin, *op.-cit.*, 1813.

1795–96	800	£75. 0.0	£60,000. 0.0
	402	£68.13.4	£27,604. 0.0
1796–97	800	£75. 0.0	£60,000. 0.0
	261	£68.12.4	£17,922. 0.0
1797–98	800	£75. 0.0	£60,000. 0.0
	427	£68.13.4	£29,320.13.4
1798–99	802	£75. 0.0	£60,150, 0.0
1799–1800	686	£75. 0.0	£51,450. 0.0
1800–01	701	£75. 0.0	£52,575. 0.0
1801–02	302	£75. 0.0	£22,650. 0.0
1802–03	394	£75. 0.0	£29,550. 0.0
1803–04	615	£75. 0.0	£46,125. 0.0
1804–05	852	£75. 0.0	£63,900. 0.0
1805–06	375	£75. 0.0	£28,125. 0.0
1806–07	650	£75. 0.0	£48,750. 0.0
1807–08	700	£75. 0.0	£52,500. 0.0
1808–09	840	£75. 0.0	£60,300. 0.0
1809–10	390	£75. 0.0	£29,250. 0.0
1810–11	nil		
1811–12	449	£78. 0.0	£35,022. 0.0
1812–13	650	£80. 0.0	£52,000. 0.0
Total	18,559		£1,362,092. 0.0

During this period, the combination that existed between the smelters behind the façade of the Associated Tinners, was by no means absolute and there was occasional dissension amongst their ranks as to the continuing value of the system when once the worst of the depression was past. Even though there was provision for varying the amount to be set aside for stocking from year to year, there were still complaints at the money that was being lost during the years when tin prices were high and demand from home sources brisk. By 1800 too there was a growing demand in Britain from the burgeoning tin-plate trade and thenceforward the amount to be set aside for the China market was considerably reduced. During 1809 matters came to a head from a combination of circumstances —growing dissension amongst the smelters, a decline in tin output from the mines and an upsurge in demand from consumers in Britain—and the stocking system virtually broke down. The price of tin rose higher and yet higher, resulting in the allocations of stocked tin being raided by their proprietors to fill lucrative home orders. The Associated Tinners had jointly agreed to supply to the Company at 122s. per cwt. in London free of all charges but when delivery became due six months later the market price was well over 160s. on the quay at Truro,[1] and few if any smelters were prepared to forego this profit. In the year only about 2,000 tons of tin were smelted compared to almost 3,000 tons a few years before and the price of the metal rose to a figure which was entirely unprecedented and which was never again to be attained in the rest of the century.

[1] RCG 2.6.10.

Some valuable detail of the progress of this minor boom of 1810, the first of several like it in Cornish tin, exists in the journals kept during this time by the Helston attorney Christopher Wallis.[1] Even though these relate specifically to the Breage area, this was then one of the foremost tin producing districts in west Cornwall and the light they throw on tin mining and smelting are particularly valuable. 'Tin is now at a high price, and very much sought after; this metal has from the earliest days been the cause of much duplicity amongst the merchants trading therein, . . . the English East India Company have within these few years obtained the Tin Trade in the China market, and the Cornish miners sell the price of Tin supplied the East India Compy. from 20 to 30s. per hundredweight. This is called 'stocking,' and amounted at different quarters to 3/5th—2/3—1/2 of all the Tin produced—the Tin smelters sent these stockings to the East India Compy. in London and have constantly sent them a very impure quality, Iron, old stamp heads, base metal, etc, etc. Two smelting houses in the west of Cornwall, made a sort of Tin called Fine Tin and which they have been selling to Tin plate manufacturers at the same price as Grain Tin, and on a moderate computation have made thereby upwards of forty thousand pounds; of late a third smelting-house in the west of Cornwall has come in for a share of the plunder, the secret has just been opened generally, and the smelters are all in an uproar, and Tin bill price is now 110—and still looking up as there is a great competition amongst the smelters—I have been solicited by the first men of property and experience in the trade to join them in a blowing-house in this neighbourhood . . .'[2]

Revealed herein are the nefarious practices of the smelters in adulterating the tin[3] they had been supplying to the East India Company, for when tin could be sold upon the home market at £100 and upwards a ton, the temptation was strong to part with as little tin as possible into the requisite tonnage allocated to a smelter for his share of the China contract. Probably they realised, and certainly hoped, that some consumer in distant China many months later, at the end of a long chain of merchants, had little hope of redress against them as the original offenders.

Two months later Christopher Wallis wrote that tin was still advancing;

[1] Now at RIC, Truro. Wallis (1744–1826) had more enthusiasm for husbandry and his estates than for mining, and some of his strictures on the financial management and physical working of the latter make amusing reading. His own Trevarno estate, however, contained several promising lodes that could not be left fallow and he was drawn reluctantly into mining on his own account—much of it in driving long 'seeking' adits—as well as into the granting of setts to others.

[2] Christopher Wallis, *Journal*, 14 February 1810.

[3] The adulteration of tin—at least in a rather less outright way than by incorporating lumps of iron—had been a common practice for years by unscrupulous persons and one which the assaying part of the coinage system was intended to check. The usual stratagem was to mix a proportion of the very much cheaper lead in with the molten tin and except to an assayer a content up to about ten per cent of the latter was not easy to detect. In the early days, however, Cornwall also received some blame for adulteration of tin which had in fact emanated from the Continent, for in order to simulate English coined tin 'every tin founder in Holland has English stamps.' (P. W. Flower, *A History of the Trade in Tin*, 1880).

'owing to the competition amongst the smelters, Tin bills are now 120; 124; 128; 135; and most likely will soon be 140s.; the Tin trade is now opening to the county generally, owing to the disclosures by altercation between the Tin smelters, and thereby is discovered the methods taken to plunder the adventurers in Tin mines, the consumers of Tin, the foreign merchants trading in that article, and the large sums accumulated by individual smelters in Cornwall, and these bickerings will probably not subside until all the secrets of these iniquitous combinations are out, and then even then there is much danger of several of the smelters again connecting in fraudulent practices; the Penzance Coinage ended 21st instant, much was said by Smelter agst. Smelter, Merchants agst. Merchants, etc., etc.'[1]

It was an inevitable corollary of the rise in the price of tin that a number of mines would be re-opened and a week later Wallis wrote, 'Owing to the great advance in the price of Tin, the mines for that metal are now eagerly sought after and many are preparing to work, and 'tis probable that strangers to the county, manufacturers in Tin and others who consume that article, will become adventurers in the Tin mines, as not a block of Tin is in the market for sale, and Tin is much wanted; the Ordinance & other offices & compys. advertise to purchase.'[2]

In the area lying between Helston and Marazion, where Wallis was concerned not only as land and mineral owner, but also as attorney and adventurer, there were a number of important mines. Destined to be first and foremost amongst them was Wheal Vor, although at this date this was only one of half-a-dozen or more equally promising and well tried setts lying on or around Tregoning and Godolphin Hills—Wheal Vreah, Polladrass, the Bog, Godolphin Bridge Mine, Wheal Dolphin, Trevarno, Paul's Downs, Carn Meal and others. Wallis declined to take up any share in the proposed company for re-working Wheal Vor in January 1810, having only two days previously bemoaned his spending 'another unprofitable day about mines and mining matters' but on June 4 he was occupied 'making a copy for the present adventurers of the plan of her' [Wheal Vor]. The leading spirit behind the company was John Gundry, one of four brothers of Goldsithney who were then widely engaged in mining in the district. Concerned with him was Richard Tyacke, of Godolphin, another mine adventurer of some consequence in the area, who also re-started Polladrass and Paul's Downs in this year. But it was the smelter Daubuz who set afoot the re-working of what were then the most important mines in the area, Godolphin Bal[3] and Great Work. Both

[1] Christopher Wallis, *Journal*, 18 April 1810.
[2] Christopher Wallis, *Journal*, 25 April 1810.
[3] Bal is not a mine as we know it today but more a generic term for a group of small tin works occupied in surface and shallow lode mining. Borlase, writing in 1782, described it as 'a cluster of mines, where tin bounds are common' (A Survey of Tin Bounds, etc., B 16/1, RIC, Truro). The Cornish prefix Wheal may be taken to mean a true mine, although its origin was in the word 'work' which at first probably signified surface workings. Both terms are impossible to define precisely.

were re-started in June 1810: Great Work 'a very heavy concern and perhaps unproductive,' Godolphin Bal which 'Mr. Daubuz this day applied to put to work . . . will cost £16,000 to set her to work'.[1] These had been amongst the greatest of Cornish tin mines in mediaeval times, contributing largely to the fortunes of the Godolphin family. The latter had always been in the forefront of the technical development of early mining and it was here that Cornwall saw the first use of steam power for draining purposes as well as the introduction of gunpowder for 'shooting the rocks.' Another sett closely adjoining, specified only as 'an adventure in Godolphin Warren and Godolphin Hill, dues 1/10' was also taken up in March, with equal 1/4 shares held by Daubuz, Oxnam Bolitho & Co., Richard Tyacke and Wallis himself.

By mid-summer tin had risen to 'the greatest price that ever was known: now 155/- per cwt. for Tin bills and very eagerly sought after, the merchants and smelters bidding on each other daily for the black Tin and Tin bills, and there is a particular spirit at present in putting to work Tin mines.'[2]

Doubtless this great increase in mining activity had its parallels, unfortunately not recorded, in the other principal tin mining districts of the county and four months later, even though the price still kept high, a warning note crept into the Journal: 'Copper is fallen very low in price, and still falling, so as to become very alarming indeed to copper miners, yet Tin keeps up to a very high price, 155/- for Tin bills, and for good Tin 14 and 14½ for 20 and the smelters appear to be bidding on each other for fine Tin—however, many Tin mines have for some months past been preparing to work, and several are now nearly ready; Wheal Vor will likely fork in less than 14 days and be in full course of working within a month, and she will require a great number of men on Tin; Carn Meal is also nearly ready for to work a fire-engine, and Polladras steam-engine will be ready in the course of next month; Wheal Breage is in a course of working; all these are Tin mines and will likely take all the men that can be spared from the copper mines, but the copper mines when in full working employ a great number of women and children, who are useless about Tin, and I fear the want of usual earnings of women and children will be severely felt in large families . . .'[3]

Early the following month tin bills touched 157s. per cwt.,[4] but this was to be their peak for in the same month a re-action set in. This rapid reversal of fortune, from boom to incipient slump, was to be a phenomenon which recurred again and again throughout the rest of the century but initially, in the winter of 1810, it was brought about by a different cause: 'Tin

[1] Christopher Wallis, *Journal*, 26, 27, 30 June 1810. Both Godolphin Bal and Great Work had been stopped in 1793. (See *Journal*, 3 October 1795).

[2] Christopher Wallis, *Journal*, 16 June 1810.

[3] Christopher Wallis, *Journal*, 17 October 1810.

[4] 'Tin is in great demand, I adventure in Wheal Breage, she produces fine Tin for which the smelters give 14¼ for 20 and 157 for the Tin bill, perhaps the greatest price that ever was given for Tin . . .' Christopher Wallis, *Journal*, 10 November 1810.

bills, (fine Tin) having got up to 155, a price before never known, is now on the decline, owing to a very awkward circumstance, stocking was countenanced by a smelter and a few others for the East India company. Mr. Daubuz refused to stock, and having a considerable smelter, some adventurers in Tin mines have been prevailed upon to bring a vice-warden's order against him to refund to them the amount of such part of their black Tin sold him under an idea of his stocking—all the smelters west of Truro are alarmed at this proceeding and have fallen the price.'[1]

Some further light is thrown upon the start of this dispute with Daubuz, then the leading tin smelter in Cornwall, in the subsequently published autobiographical memoirs of one Tobias Martin, who was concerned in it.[2] Martin was managing agent at Trevenen, north of Helston, a mine that had been one of the richest in all Cornwall from 1801 onwards, and after a few years of service he discovered extensive fraud in the price being paid to the mine for its black tin, all of which was smelted by Daubuz—himself a shareholder in Trevenen—at Carvedras. Although blame for this was laid squarely at the door of the smelting-house manager, the Trevenen adventurers thereupon decided to commence producing their own tin and leased Ludgvan smelting-house for this purpose in 1806.[3] Daubuz then set about securing the dismissal of Martin. The latter was a man who had his faults, was far from a teetotaller and had the management of a deep mine troubled by frequent runs of ground: the adverse comments of those captains that Daubuz brought in to report on its working were sufficient for the purpose before many more months were out. Probably it was the Trevenen adventurers, from their own interest in the smelting trade, who brought to light the trick outlined by Wallis in 1810—buying black tin from mines at the lower 'stocking' price for the East India contract but subsequently selling the metal at normal market price—for Daubuz then turned his attention to the mine as he had done earlier to its agent. On October 15 1811 Wallis records the disposal of Daubuz' shares in Trevenen as a preliminary measure and in June 1813, after eighteen months of manoeuvres which are now unfortunately unrecorded, Wallis in his capacity as attorney 'attended Mr. Daubuz on his return from Trevenen Mine in Wendron which he caused to cease working this day . . .'[4] Two months later the sale of the lease of Ludgvan smelting-house was also in the attorney's hands.[5]

These disputes with Daubuz, and their outcome, showed only too clearly how powerful was the tin smelting interest in Cornwall and more

[1] Christopher Wallis, *Journal*, 25 November 1810.

[2] *The Remains of the late Tobias Martin, of Breage in Cornwall, Mine Agent*, 1856.

[3] This had been advertised for sale in 1806, the proprietor then being John Stevens, of St. Ives. (RCG 1.3.06).

[4] Christopher Wallis, *Journal*, 26 June 1813. The materials on the mine, comprising two (56″ and 30″) pumping engines and three steam whims (16″ 15″ 12″) were advertised for sale a month later (WB 22.7.13).

[5] Christopher Wallis, *Journal*, 17 August 1813. From Edward Rogers (purser of Trevenen Mine) to John Gundry for remainder of a ninety-nine year lease, for the consideration of £630, plus the materials as valued.

particularly the dominance of the smelter over the miner. Wallis himself, although distrustful of the smelting trade and its machinations, became embroiled in it.[1] On 25 April 1811 he was ' considering the propriety of becoming a partner in a Tin smelting-house, and the convenience and the inconvenience thereof, and writing Rev. Mr. Robinson, Bath, offering to purchase his smelting-house at Gweek.' A possible partner in this was John Gundry with whom he discussed a few days later ' building a smelting-house and therein smelting our own tin, etc., etc.' On 3 June he ' attended Mr. Reed of Trevales about putting to work his Tin smelting-house at Perran ' and a week later, with Reed and Tyacke, he viewed these premises and considered ' the repairs thereof, with a view to putting it to work.' Evidently this did not meet the case for on 19 June he, Tyacke and Richard Oxnam went to view the remains of the old blowing-house at Trannack, to consider ' whether twould be proper to re-build it and add thereto a smelting-house.' Meanwhile time was passing and the profit he and Tyacke might have been making on the smelting and sale of their share of the produce of the various Breage mines was slipping away. On 9 September, perhaps as an interim measure, the two men discussed the possibility of ' putting a furnace for melting tin in Wheal Dolphin[2] old engine-house ' but this was evidently not proceeded with for there was further consideration of ' building a tin smelting-house at or near Godolphin ' on 5 October. Finally on 1 February 1812, all this casting about, to Gweek, Perran, Trannack, Godolphin, and elsewhere, came to an end for Wallis then attended his friend Tyacke ' with an acct. of our first beginnings to smelt our Wheal Breage tin at Ludgvan smelting-house for a Penzance coinage which ends next Saturday.'

By this time in 1811 the boom had passed and, with tin bills down from above 150s. to a mere 90s., Wallis wrote sadly: ' mining materials very scarce and dear, and mining concerns spiritless. No market for Pilchards or Tin, and all the necessaries of life dear.'[3] In the following July it even became necessary for Tyacke to venture to Bristol and London himself to try for the sale of their block tin from Ludgvan, whilst on 12 September Wallis wrote despairingly: ' Tin very low and scarcely saleable so that very few mines can pay their cost, mining materials very dear, taxes very high, credit tottering . . .'

After the hiatus in supplies to the East India Co. in 1810, small increases in the price of stocked tin had been made in the two succeeding years but there were still dissidents amongst the smelters. At Ludgvan, the Trevenen adventurers had always held aloof but as they smelted solely their own ores, the Associated Tinners could do little to coerce them. Trevenen,

[1] Earlier, in 1795–9, Wallis had also had a share in a smelting- or blowing-house set up at Trannack (i.e., Melancoyse) near Helston.

[2] A sett lying ' to the east of Great Work ' according to Wallis. Operations had ceased here shortly before 1810 when the ' two large Boulton & Watt engines (one being of 63″) on Godolphin Adit and Wheal Dolphin ' were advertised for sale (RCG 9.6.10).

[3] Christopher Wallis, *Journal*, 25 September 1811.

moreover, was a declining mine by this time. When, however, a new smelting-house was started at Porth Towan early in 1811 by one Isaac Rogers of London,[1] and he flatly refused to co-operate, it was decided that pressure must be brought upon him, otherwise the stocking system would be invalidated. In February 1812, after long altercations between the two sides, William Reynolds and Captain William Davey were despatched to London to reason with the intransigent Rogers, and it was reported that meanwhile 'a resolution to send no tin to his smelting-house had been pretty generally adopted.'[2] A fortnight later Rogers had capitulated and at a meeting of the tin trade in Redruth it was decided that as Rogers had agreed to join, 'the stocking system be now universally renewed.'[3] A new contract was to be negotiated to supply the East India Company with a quantity not exceeding 1200 tons per annum at twenty per cent under the ruling London price, although after the Company's acceptance of this there were several adverse opinions voiced, by 'Adventuring Tinners' and others, querying the true value to the miner—as opposed to the smelter—of the East India sale at such low prices.[4] Adding to the uncertainty of the times was the pending renewal of the East India Company's monopoly of the China trade, and relief was prevalent, at least amidst the ranks of the smelters, when this was granted anew by Parliament. The contract was seen by them as an essential outlet for surplus metal during the years when European demand was slack, so much so that there were even suggestions that if the China trade had been thrown open to all, the Associated Tinners should set up their own factors in Canton and continue to ship tin there themselves.

The year 1814 saw a revival of demand for tin akin to that of 1810, but once again the high price was short-lived and it fell steadily thereafter despite the coming of peace in 1815. Although a pivotal date in European history, after a century of wars, the year of Waterloo did not mark any

[1] John James of Truro, who had a share in Carvedras, was also a partner in Porth Towan smelting-house in 1811 (Bolitho Records, DD RG 1/115, CRO, Truro).

[2] WB 14.2.12. The real weapon in the Associated Tinners armoury to coerce Rogers was that Lord de Dunstanville's lease of the ground on which the smelting-house stood—' at Porth Towan, near Wheal Lushington '—was now said to have been granted only for the purpose of Rogers smelting his own mine ores and no others. (AD 55/19, CRO, Truro). Reynolds was then steward to de Dunstanville.

The Associated Tinners, naturally, had great influence within the county and a resolution was passed in 1812 designed to prevent the leasing of sites suitable for smelting-houses by Cornish landowners, except to those who would conform to the Associated Tinners rules. This was a direct result of the establishment of the rebel Porth Towan house, and may well have accounted for the non-appearance of another smelting-house projected in 1812—when it was reported that ' several gentlemen of the Committee of the British Mining and Smelting Company are now at Redruth. A tin smelting-house will soon be erected there and some mines set at work in that neighbourhood ' (RCG 25.7.12). Of this venture, nothing more is unfortunately known.

[3] WB 28.2.12. ' For the trouble they have taken in waiting on the Proprietor of Porth Towan Smelting House ' Reynolds and Davey were voted '. . . one hundred guineas in addition to their expenses as a remuneration for their services.' This expense and trouble on the part of the Associated Tinners was in any case largely wasted, however, for smelting had ceased at Porth Towan by early 1814.

[4] WB 17.4.12. See also WB 8.5.12 for a long account of the East India trade in Cornish tin and the opposition to this renewal of the ' stocking ' contract.

turning point in the development of tin mining as an industry; despite this it is a convenient date at which a survey of the main tin mining districts of the county might be made. In so doing it is essential to rid oneself of the modern concept of the Redruth–Camborne area as the chief one. It was then almost entirely copper producing, and in terms of tin unimportant compared to Breage, Wendron or St. Agnes. Dolcoath, although a producer of small amounts of tin at shallow depth in the previous century, was now solely a copper producer, as were all the other big mines of this area. The entire run of mines north of Carn Brea—Cherry Garden, Stray Park, Dolcoath, Cook's Kitchen, Tincroft, Wheal Fanny, Wheal Vernon, Tregajorran and Wheal Druid—all figure regularly in the *West Briton* weekly ticketings as copper producers alone. A few, such as Tincroft, also produced tin but copper was regarded almost without exception as their *raison d'etre* at this time. Pednandrea in Redruth had been one of the most considerable of the tin mines in the area, although worked down to a mere 69 fathoms below adit, but like the adjoining Wheal Sparnon and its two satellites East and West Wheal Sparnon, this was standing idle in 1816. Gwennap was also regarded as the heart of Cornish copper mining and Poldice, together with the nearby Wheal Unity, Creegbroase, and Wheal Busy at Chacewater, alone produced tin as well as copper.[1]

The main centres of tin production lay scattered throughout the western part of the county: in Wendron, Breage, St. Just, St. Ives and St. Agnes, together with more scattered smaller areas around St. Austell and the Hensbarrow granite mass, and in east Cornwall near Caradon and eastward of Kit Hill. In Wendron lay perhaps a score of small mines which had been worked and re-worked throughout the eighteenth and earlier centuries, first and foremost amongst them having been Trevenen Mine, already mentioned. Trevenen was one of the deepest tin mines in all the county at this date, 177 fathoms, and worked by five engines up to its closing in 1813. During part of this time the mine was said to have given £80,000 in dues alone to its mineral lords[2] and also to have given a profit in one year, 1726, of £7,000.[3] By comparison Wheal Trumpet, later to rival Trevenen in

[1]Although almost entirely a copper producing area in modern times, Gwennap and district had given rise to the main centre of tin streaming in the county, along the length of the Carnon valley, whose constituent streams ramified throughout the parish. As early as 1700 W. Pryce (*Mineralogia Cornubiensis*, 1775) recorded 'some hundreds of poor men, women and children' washing tin out of 'the low lands and sands under Perran ar Worthall.' In 1788 John Williams of Scorrier took out a sett of the main Restronguet stream works from Sir William Lemon, at one/14th dues 'payable at the Smelting House in White Tin' and in order to extract the tin from an extensive area of ground below high water mark without the interference of the tides embankments were thrown up to enclose the whole upper part of the creek leaving channels at each side for the necessary navigation to Perran Wharf and Tarrandean on the west side and Devoran on the north. (Whitford-Lemon Records, Acc. 194, CRO, Truro). In 1816 this lower part of the valley, where a steam engine was also employed to keep down the surface water, was described as 'a desert of red sand, heaped into vast piles and hollows, the only herbiage visible being a few tufts of the sea daisy, while here and there in the trenches might be seen the tinners working knee deep in the water, and a few squalid half-clad boys wheeling the tin-ore to the stream head in barrows' (G. Henwood, *Four Lectures on Geology and Mining*, 1855).

[2] MJ 6.10.60. [3] MJ 13.7.61.

size and richness, was said only to have been 'worked for the last four years . . . a water-machine is now erecting . . . promises to be one of the best Tin Mines in Cornwall.'[1] Also active, adjoining Trumpet, were Wheal Ann, Wheal Nancy, Trenethick Wood, and Wheal Varnon, the latter sold up in 1815.[2] In Breage, the two most important mines were Great Work and Wheal Vor, both re-commenced in the boom of 1810. At the latter a 48″ engine was erected in the year and when the workings, then 90 fathoms deep, were cleared up and forked, sinking re-commenced. Success was far from immediate and in 1816, six years after the mine had been re-set to work, on the occasion of the advertisement for sale of a 1/29th share,[3] it was only described as 'a very promising adventure.' The adjoining sett of Wheal Vreah was also being worked, with a 45″ engine for pumping, whilst Polladrass Downs was offered for sale in March 1816, equipped with a 24″ steam engine and another water-pressure engine.[4]

In the St. Just district, a greater number of small mines existed than in Wendron and Breage combined, for this had probably been one of the most productive tin districts of eighteenth-century Cornwall, responsible for the institution of Penzance as a coinage town in 1663 and for its growing importance as such in the succeeding century. The physical conditions in and around St. Just were ideal for the early development of lode mining, permitting the opening up of an enormous extent of ground, to about 40 fathoms depth, by a series of short adits driven in from the cliffs. By the beginning of the nineteenth century the number of setts that had been granted in the mineralized ground extending from Priest's Cove to Pendeen Watch was legion. An advertisement of the sale of shares in a group of St. Just mines in 1815 lists some whose names are now well known, and others whose existence has hitherto gone unrecorded—Bosorne, Bernoal Ball Mine, Porth Just Mine, Wheal Spearne, Carnyorth Moor, Wheal Owles, Wheal Sanders, Truthwall Moor, and Botallack Cliff.[5] A few of the others in the immediate vicinity of St. Just whose contemporaneous existence can be traced included Wheal Buck, Wheal Vulcan, Wheal Owl—not to be confused with the better known Wheal Owles—Wheal Oak, Wheal St. Just, Wheal Bellan and Little Bounds. Of all these the best known was Botallack, a mine which was erroneously described by one contemporary writer as the richest tin mine in Cornwall in 1816. In fact Botallack, although it may have been an important producer of tin in previous years, was primarily a copper mine at this date, as original records of its output show. In 1817, 109 tons of copper ore were sold for £1358 and 123 cwts. of black tin for £372; in 1818, 292 tons of copper ore for £4074 and 288 cwts. of black tin for £644, whilst in 1819, perhaps because it was no longer worth raising and dressing black

[1] WB 22.4.14.

[2] In November 1811, Wallis records a legal dispute 'about Wheal Nancy adventurers having turned the water thereof into Wheal Trumpet mine and entirely filled her, and thereby occasioned a considerable damage.' (*Journal*, 6 November 1811).

[3] WB 9.2.16. [4] WB 1.3.16. [5] WB 27.1.15.

tin at the low prices of that year, none of the latter was sold at all, compared to 213 tons of copper ore, sold for £3115.[1] Around Trewellard and the slopes of Carn Kenidjack, were another ten or more distinct setts, all now long forgotten or recorded solely in the lode names of later big mines: Wheals London, Bal, Carn, Pens, Humfra, Widden, Chimney, Reeth, Sampson, Game and Stennack. The latter, subsequently forming part of North Levant, was perhaps the most considerable of these, employing about a hundred men in 1815[2] and pumped by a water-pressure engine, assisted by a steam whim adapted also for additional pumping. This last appears to have been the first steam whim in the area—although another with 15″ diameter cylinder was advertised for sale at Wheal Owles in 1815—whilst the first pumping engine was reputed to have been put in at Carnyorth in 1801. Pumping charges in the district, as well as in the rest of the Land's End granite, were light and the 36″ engine at Boscaswell Downs was the largest thereabouts. Surprisingly, in view of the apparent lack of surface water, the district was something of a stronghold of water-power, and the considerable winter rainfall on the uplands inland from St. Just village was made good use of in its short, steep course to the sea. The valleys north and south of the village each had a long succession of water-wheels used for either pumping, winding ore or working a set of stamps and even on open cliffs, seemingly remote from any direct supply of water, wheels of quite surprising size were put in. Considerable use was also made of water-pressure engines, the topography of the land and the numerous short sea-level adits being admirably suited to provide the requisite fall of water to operate such an engine and subsequently to discharge it.

In the St. Ives and Lelant area, amongst a considerable number of small adventures, one of the most important was Rosewall Hill Mine, which was said to have been worked without intermission from 1761 to 1811[3] and to a depth of 132 fathoms below adit. After the latter date it was continued by a new company, but unsuccessfully as the records of the Bolithos (who held 1/32 share in the mine) show, and when the sett was abandoned in 1817 the accumulated debt against the adventurers was £8,801.[4] St. Ives Consols, later to become famous as one of the richest

[1] Botallack Cost-Book 1815–20, RIC, Truro. In the succeeding forty years Botallack also continued to be more important as a copper than as a tin mine, although ultimately becoming exclusively a tin producer.

[2] WB 27.1.15. [3] MJ 1.1.59.

[4] Bolitho Records, DDRG 1/74, CRO, Truro.

	Cost Paid	Tin returned
to 1812	£38,199	£30,719
1813	£4,570	£4,484
1814	£3,501	£3,421
1815	£1,851	£1,142
1816	£953	£658
1817 (to Feb)	£181	£31
Balance (loss)		£8,801

The materials of the mine, comprising 30″ pumping engine, 12″ steam whim and 12″ pressure engine, were advertised for sale WB 21.2.17.

mines in the county, was only re-started in 1818, its two constituents being Wheal Mary and the old Wheal Ransom.[1] Close by were also Wheal Brothers, Balnoon, (with its constituents Halnoon, Vervis Downs, and Wheal Nicholas), Wheal Praed, and Wheal Speed, together with Wheal Providence at Carbis Bay which was sold up in 1810. In the Lelant area Wheal Reeth was by far the most important mine, and one which had possessed one of the first Newcomen engines in the far west of the county, being over 200 fathoms deep by 1820. Nearby also was Wheal Margaret, which was first started in 1782 and worked thereafter to 1804[2], being celebrated as one of the mines where Hornblower erected two of his patent two-cylinder engines in 1793,[3] one for pumping and another for winding.

In St. Agnes, geophysical conditions existed similar to those at St. Just, and for the same reasons this formed one of the early strongholds of cliff mining in the county. Carew's graphic description in the early seventeenth century admirably conveys not only the initial development of these mines but also the random nature of the innumerable lodes worked in them: 'Saint Agnes, one of the high hils, by his entrailes (like Prometheus) feedeth the Tynners' pecking, or picking bils . . .' Certainly the St. Agnes mines were proverbially rich in these early days and Polberro, a 'bal' comprising a host of old small setts and bounds which was pierced from the cliffs by three or more adits, was the most important of them. It was probably here that 'about the year 1720 a rich vein of tin was discovered in a cliff at St. Agnes for which the proprietor, Mr. Donithorne, was offered £100 a day for six months, which he refused.'[4] Seal Hole Mine was the most important constituent of Polberro and the first St. Agnes mine to possess steam power for drainage, having a Boulton & Watt engine in the 1780's.[5] Amongst a considerable number of other mines active in 1815 there were Wheal Coates, Wheal Harriet, Great Wheal Kine and Wheal Trevaunance,[6] whilst eastwards toward Perranporth, Blue Hills Mine had been re-started in 1813, working a number of old setts at the seaward end of Trevellas Coombe, as well as Penhalls on the cliff adjoining. Here a 58″ engine was put up, pumping from several shafts by means of flat-rods, with nine water stamping mills for dressing the tin.

In the rest of the county east of St. Agnes, the tin producing areas were scattered, and related primarily to the granite bosses of Hensbarrow, Caradon and the ridge formed by Kit Hill and Hingston Down. Somewhat outside this area, in mid-Cornwall, was Polgooth in the parish of

[1] WB 29.1.36. Part of what had once been Wheal Ransom was later worked again as a separate entity under that name (see WB 30.1.29) when it was stated that the levels of St. Ives Consols 'to the 57th (*sic*) inclusive are driven into the Ransom'. Subsequently this became part of the setts worked as Rosewall Hill & Ransom United.

[2] MJ 1.1.59.

[3] Advertised for sale RCG 8.12.04 comprising 27″ & 21″ pumping engine, plus another with two 10″ cylinders.

[4] RCG 22.6.72.

[5] This 40″ engine was advertised for sale RCG 15.6.05.

[6] Sale of shares (WB 31.3.15).

St. Mewan, a vast old mine or group of mines where Celia Fiennes, writing in 1695, recorded 'at least twenty mines all in sight which employs a great many people at work, almost night and day.'[1] Although an otherwise acute observer, this traveller records that in continuing her journey from there to Tregoney she passed by a further hundred mines and it would seem that each individual shaft fitted with windlass or other primitive head gear tackle was being counted as a separate mine. During most of the eighteenth century little is known of Polgooth's progress but one writer records that 'about 60 years ago [i.e. 1757] the mine stopped working, the receipts having exceeded the expenditure only £10 in ten years, although the cost during that time was £100,000.'[2] After some years Polgooth sett was granted in 1783 for a period of twenty-one years to a company of which Edward Fox of Wadebridge, and later Charles Rashleigh of Menabilly, were the principal partners. Working was on a considerable scale, with wages amounting to as much as £1,000 a month in 1802 for example.[3] The renewal or the relinquishment of the sett in 1804, on the expiry of the lease, became a legal matter between the adventurers and Lord Arundell, owner of the mineral rights, and during the dispute part of the workings were flooded by the actions of the two sides.[4] Thereafter, possession of the sett having been regained by Lord Arundell's men in June 1804, the mine was either idle or worked by tributers down to adit level. A scheme to recommence operations in 1812 by a company with a proposed capital of £20,000 came to nothing, although the prospectus and a historical account of the mine at this time[5] yield some interesting information on the cause of its stopping. This had been partly due to 'the owners being the proprietors of the smelting-house at St. Austell where all the tin raised from the mine was sold on terms not always satisfactory to those adventurers who had no interest in the smelting business'—a state of affairs that the new promoters proposed to rectify by erecting a smelting-house on the mine 'for the purpose of combining the profits of the smelter with those of the tinner, subject to the express condition of stocking tin for the East India Company as other smelters do.' It was alleged that the mine—'generally deemed the richest and most productive venture for Tin in the world'—had not stopped from any poverty of the lodes but from dissension amongst the adventurers, who had divided between them £110,000 in profit in twenty years. There was even some doubt as to whether, in fact, 'a *final* abandonment of the concern' had ever been agreed upon, for some of those who had an eye to securing a major interest in any new company that was formed had made sure that 'everything in the bottoms and engine shaft was amply

1 Celia Fiennes, *op. cit.*
2 J. Heard, *Gazeteer of the County of Cornwall*, 1817.
3 Edward Fox (of Wadebridge), Cash Book No. 3, CRO, Truro.
4 For this and further detail on the Polgooth dispute, see Christopher Wallis, *Journal*, 1804 *passim*. Wallis acted as attorney for Arundell in this matter.
5 WB 1.5.12, 12.5.12.

secured, at a great expense, for a future working.' It was also hinted that there were those in the old company who saw more to be gained by 'knacking' the mine rather than by continuing it: 'the estimated value of the tin leavings being very great, it became an object of desire.'[1] The response to this rather fulsome attempt to re-start the great Polgooth in 1812 failed, however, and it was not until seven years later that the mine was again taken up. This was by John Taylor, then prominent in the resuscitation of various old Cornish copper mines in Gwennap and elsewhere, and 80″ pumping and other engines were put in, chiefly on the newer, eastern part of the mine.

West of Polgooth were Dowgas and Great Hewas, the latter a mine which had been of only slightly lesser importance and value than Polgooth during part at least of the eighteenth century, but about which rather less is known. Two Newcomen engines were at work there in the 1770's and this in itself was a measure of the importance of Hewas, for the tin mines that could boast of one let alone two such engines were few and far between. In the big copper mines as many as five might be employed but if one sets aside those such as Relistian,[2] Chacewater and Poldice which produced both metals, the only tin mines with Newcomen engines were Great Work, Wheal Reeth, Polgooth and Ding Dong. Working ended at Hewas in 1774, but was resumed again for a time, using Boulton & Watt engines, in the 1790's. In 1813 the sett was again taken up by a company using London capital but promoted by two Cornishmen under the style of Malachy Lyle & Co., and a 58″ engine was put in.[3]

Elsewhere, around St. Austell, there were numerous small mines and stream-works on or around Hensbarrow Moor, a traditional stronghold of mediaeval 'moor' tin. The lodes in the soft and decomposed granite of this area produced an ore of exceptional purity, accounting for the pre-eminence of St. Austell as the one-time centre of the grain tin trade. The ground lent itself to being worked open to the day in long cuttings or quarry-like exposures, either for the tin alone or in conjunction with the china-clay which was later to provide such a bountiful future for Hensbarrow. Carclaze was the most celebrated of these early open-cast workings, a vast pit which was opened up and extended year by year to lay bare the numerous tiny lodes and stringers of tin disseminated throughout the ground. In 1816, 'Little Carclaize,' and six other adjacent 'grain tin works' were advertised for sale, having lately been held by R. A.

[1] This same report (WB 22.5.12) on Polgooth's proposed re-working stated that there was 'not perhaps a mine in the county of Cornwall so advantageously accommodated with numerous stamping mills, most of which are still working on the leavings belonging to the old concern, though the mine has been discontinued for upwards of five years.'

[2] Relistian Mine, at Gwinear, a 'great tynworke' whose early riches parallelled those of Poldice and Polgooth. It is recorded that in 1681 the mine was rendered unworkable for some time by an enormous fall of ground in the workings, in which 24 miners were killed—an accident that was to be the worst of its kind in Cornish mining history. (See J. C. A. Whetter, *Cornwall in the Seventeenth Century: A Study in Economic History*).

[3] It is possible that Hewas was the sett turned to by this company when the flotation of Polgooth failed the previous year.

Daniell and others.[1] Rather less well known but equally productive was Beam, otherwise Old Beam or Great Beam, which rivalled Carclaze in size and riches. Up to about 1810 moreover Beam was also worked like Carclaze entirely open to the day, the tin being ' scattered so promiscuously as to render it necessary to lay the mine open to a great extent and depth.' By 1817, however, at the time of this account,[2] Beam had shafts and levels and was 'beginning to assume a more regular appearance', whilst in 1824 it was reported that the mine had ' already produced a profit of £10,000 exclusively of the Steam-engines, Stamping Mills and other valuable materials now there-on, and one-19th part of that concern has been recently sold for one thousand guineas.'[3] Although no detailed record exists for the years thereafter, there is little reason to doubt the subsequent claim that Beam became ' for its extent, one of the most lucrative mines in the county ' and that ' upwards of £450,000 worth of tin ' was sold by its owners, Rashleigh Williams & Co.[4]

South and east of the Hensbarrow granite mass were also various true stream works, notably in the St. Austell river valley leading to Pentewan and on the alluvial areas near what was later to become the port of Par. The Porth stream works near the latter were noteworthy for being worked in a way which closely corresponded to that at Carnon further west. At first an area of ground was excavated and worked out by enclosing it within an embankment to keep out the sea at high tide, but in 1801 this was broken down and the whole inundated. A few years later operations were resumed further seaward by means of an iron shaft sunk through the sand about fifty fathoms out from the beach, this being drained—like the Wherry Mine at Penzance—by pumps worked from a line of flat-rods carried out from a water-wheel on the shore. About Christmas 1815 the sea broke in to this new ' mine ' from the original workings, fortunately without loss of life, and it had again to be abandoned.[5]

In eastern Cornwall, there had been surface working and mining for tin to shallow depths in many areas along the southern edge of Bodmin Moor, more particularly around Caradon Hill, but although these workings were quite extensive no major lodes had been opened up. Amongst a number of

1 WB 9.2.16. Under Daniell's ownership, Carclaze's output would doubtless be treated at his New Blowing House in St. Austell.

2 J. Heard, *Gazetteer of the County of Cornwall*, 1817.

3 WB 23.4.24. 4 WB 15.1.41.

5 F. Hitchins & S. Drew, *History of Cornwall*, 1824, Vol. 2. According to this same source, early in the 1820's workings at Porth were resumed and another iron shaft put down ' about fifty fathoms east of the old one, and about seventy fathoms from the shore.' The workings were about 19 fathoms deep and the shaft (6′ diameter) was sunk by the simple expedient of mooring a loaded barge across each successive cast iron segment at high tide, thus driving it down into the alluvium as the water level fell—a method used later at the Restronguet Creek Tin Mine in the early 1870's. The iron shaft was manufactured at Tavistock and the mine company headed by the ubiquitous John Taylor. Subsequently, perhaps in the same hands, the Tywardreath Mine was worked on Par Green up to about 1855, two iron shafts being sunk there by similar means, whilst ultimately no less than five iron shafts were put down in this area. (See R. Symons, *Alluvium in Par Valley*, Journal RIC, Truro, 1877).

tin works in the area, Stowes End Mine close by the Cheesewring had shown great promise in 1804 and was then described as 'one of the most promising Tin Mines in Cornwall.'[1] But this was brief-lived and the era of Caradon tin on any scale was still almost three quarters of a century away.

Farther east, shares were advertised for sale in 1817 in 'all that Tin Mine or adventure for Tin, called Kit Hill' but Drakewalls was perhaps the only mine of any consequence in the area. Here, where the eastern end of Hingston Down plunges to the Tamar, a tin lode or lodes had been worked for centuries, at first as an open cutting like Beam and Carclaze but later by levels and a shaft, making use of water-wheels for power.

In 1815, the amount of metal coined by the major smelting-houses in Cornwall was as follows:[2]

Angarrack	2726 blocks	Trereife	2087 blocks
Calenick	2336 blocks	St. Austell	1885 blocks
Treloweth	2329 blocks	(3 houses)	
Chyandour	2144 blocks	Portreath	1365 blocks
Ludgvan	2129 blocks	Carvedras	1241 blocks

Amongst these, the newcomer was the Portreath Smelting House, which had been set up at the end of 1813[3] at the onset of the brief boom of the following year. The principal partner in this was Collan Harvey of St. Day, a merchant who was already concerned in Portreath harbour in conjunction with the Williams family of Scorrier and the Fox family of Falmouth.

In 1816 two more smelters appeared, at Truro and at Great Hewas Mine. The first was set up by Reuben Magor and Edmund Turner, of Redruth and Truro respectively, who were engaged in banking, merchanting and brewing. This, the Truro Smelting Works, was entirely new and occupied a site immediately east of the present-day St. Austell Street, close beside the navigable head of the river Allen where it entered the Truro river.[4] The second, at Great Hewas, was established by the mine company three years after they had commenced operations.[5] This was far from an insignificant venture for in the following year 2361 blocks of tin were coined by the Great Hewas Smelting Company—a quantity almost as great

[1] RCG 14.7.04. Sale of shares in Stowes End Mine, Linkinhorne, 'in which are two capital Loads now working on tribute with great success, and from which eighteen men can raise Tin-stuff sufficient to employ ten or twelve stamping mills . . . The present Bottoms of Tin are nearly thirty-five fathoms long and still extending . . . In fact it is impossible to define in the space of an advertisement all the peculiar advantages which this Mine possesses over almost every other.'

[2] Bolitho Records, DDRG 1/163, CRO, Truro.

[3] RCG advertisement 1.1.14. 'Tin assay master wanted . . . capable of assaying tin and conducting a tin smelting-house . . . Apply to Collan Harvey, St. Day.'

The site chosen was on the south side of the valley running inland to Bridge. It later became the site of Glenfeadon house where the Bain family lived (see page 205).

[4] In February 1818 the Truro Smelting Co. advertised for tenders for building a quay on the 'Mopus' Road (WB 20.2.18), whilst later the same year it was reported that Sunny Corner Quay there was 'now occupied by Truro Smelting Co.' (WB 18.9.18).

[5] From entries in the *Sherborne Mercury* (5.4.73 and 6.9.73) it would appear that the adventurers at Great Hewas had also smelted their tin on the mine at this time.

as any of the big western smelting-houses. In 1818 this rose to 4388 blocks,[1] the largest output in the county, and forming a considerable proportion of the total number coined, 24851 blocks. In 1819 3222 blocks were coined (out of a total of 22006) and 3215 in 1820. It has subsequently been stated that the bulk of this came from other mines and that the principal up-country adventurers in the Hewas company were being deceived into thinking this came solely from their own mine. In fact, the smelting company did not commence purchasing ores from other mines until the end of 1817[2] and advertisements were then continued well into 1818 so that its shareholders were hardly being kept uninformed.[3] Thereafter some of the Hewas company's blocks were coined under the heading of Grain Tin—coinage records always differentiating between the two—and it is apparent that much if not all the outside tin bought by Hewas came from the mines of Hensbarrow and that a blowing-house had also been put in for this. In 1821 2813 blocks were coined, but many fewer in 1822 and in January 1823 the materials of the mine were for sale.[4]

As a result of the two successive booms of 1810 and 1814, and the number of mines that had then been set to work, the output of black tin increased considerably and in 1817, for the first time, it exceeded 4,000 tons. This was a quantity in excess of the existing demand, particularly as the amount that could be channelled off in the East India contract was diminishing year by year. To some extent the decline in demand and the receding price of tin after 1815 could be seen to go hand in hand with the diminution in the East India Company's Cornish purchases. After the hiatus in supplies in 1810–11 and perhaps also as a result of the indifferent metal that had then been supplied, the quantity of Cornish tin sold in China had greatly declined. In 1813, moreover, there was a portent of the way the tin trade was changing, for in that year the East India Company's ships carried 1700 Chinese labourers from Canton to the island of Banca to work the deposits there. Three years later, out of 550 tons of tin imported into China by the Company, 350 came from England and 200 'from Eastern ports.'[5] For Cornwall, what was worse was that permission was also 'given to commanders to take 157 tons of Banka tin as ballast from Canton to London.' Thus, for the first time, the trend of trade was ominously reversed, for this Eastern tin was sold in London at a handsome profit to the Company, at a figure below that of the highly priced

1 About 730 tons (metal), say 1100 tons of black tin.

2 WB 21.11.17.

3 Confirmation of this was given in 1856 at a time when Hewas was about to be re-worked. The books relating to the previous working were then shown, including the rates of tribute in the mine, to refute the allegation that sales had been mis-represented as bought from other mines. (MJ 2.2.56).

4 WB 10.1.23. With 58″ pumping engine, 22″ whim and 25″ stamps—the latter bought by the Polgooth adventurers.

The site of the Hewas smelting-house has not been traced and does not appear to have been separately leased as such, probably because it occupied one or other of the mine buildings on the sett. Coal was imported through Mevagissey (advertisement, RCG 28.3.18) whilst charcoal was purchased from the Bolithos of Penzance.

5 H. B. Morse, *op. cit.*

Cornish produce. The Chinese merchants whom the East India Company had supplied in the past were also soon to become concerned in working the Banca and other Straits[1] alluvial deposits, and before long they had almost entirely ousted Cornish tin from the China market. The East India Company's purchases from Cornwall did not end immediately, however, for there were certainly two subsequent ' stockings ' for contracts in 1819 and 1821. For the first, 500 tons were set aside whilst for the second 1/12th of three of the quarterly coinages of the year and 1/6th of the fourth coinage were set aside for the Company.[2] Thereafter the practice of stocking died a natural death, as a result of growing Chinese merchant influence, the rise of Straits tin, and the decline in the monopoly of the East India Company, finally abolished in 1834.

In Cornwall in the period after about 1815, the only constant demand was for the best qualities of grain tin to meet orders from the Welsh tin-plate works. These had expanded rapidly both in size and in number at the turn of the century, with exports quadrupling in the years from 1796–1802, despite the Napoleonic wars which affected the European markets. By 1800 nine works were in existence—a number that had grown to sixteen by 1825[3]—and the annual ' make ' of tin-plates was to rise from 4000 tons in 1805 to 18000 tons by 1837. Tin and iron formed ' the lightest, strongest, brightest, cleanest material that we have, which is beyond competition, and which above all produces the cheapest package which can be obtained '[4] and the demand for it grew insatiably during the first half of the nineteenth century. By 1815 the tin-plate works of South Wales and elsewhere were the Cornish tinners' most important outlet for their produce, as they were to be for many decades to come.

For plating iron, grain tin was far superior to smelter's common tin, not only on account of its finer colour and lustre but also because of its greater fluidity which made thin and more economical coating of the iron possible. A form of refined tin, made from the best selected mine ores which was re-melted and ' boiled ' by the smelters, had been developed in an endeavour to meet the tin-platers' needs but grain tin from blowing-houses still remained in the greatest demand, particularly for the better quality ' charcoal ' grade of tin-plate.[5] It was for this reason, in order to meet orders for grain tin, that the decision was taken at Chyandour in 1818 that ' a blowing-house be erected on the same premises [as the smelting-

[1] Straits in this context is a generic term intended to comprise not only the Straits Settlements themselves (Malacca, Singapore, and Penang, formed in 1826) but the independent native states of the Malayan peninsula, as well as the Dutch islands of Banca and Billiton (later known as Blitong) off Sumatra.

[2] Foxes Rashleigh & Co., Letter Book, (CRO, Truro) 4 May 1819. 2s. 6d. per cwt. to be deducted from all purchases of black tin in Cornwall, as agreed at the Quarterly Meeting and set aside to meet the lower price of this tin.

[3] P. W. Flower, *Origin and Progress of the Manufacture of Tinplates*, 1886.

[4] P. W. Flower, *A History of the Trade in Tin, 1880*.

[5] In October 1819, for example, the price per cwt. of the normal grades of tin was—Common 68s. in blocks, 70s. in bars: Refined 75s.: Grain 98s. 6d. There was also a superior and more expensive grade of grain tin, known as Dyer's, occasionally sold.

house] in equal shares agreeably to the new arrangement ',[1] whilst elsewhere it is probable that other similar air furnaces were put in at this time to supply the demand for grain tin. But in the tin trade as a whole, with supply generally in excess of the demand, the period from 1815 to 1819 was one of increasing depression. Despite a few flurries in the price from time to time, the trend was downward and by the latter year the metal stood at a figure half that of five years before. This was a serious matter for those who were deeply engaged in Cornish mining and to revert to the Breage district where contemporary comment survives in detail in Christopher Wallis' *Journal*, it is apparent that one of the major casualties of the slump was John Gundry of Wheal Vor. He was involved in many other mines[2] as well as in Ludgvan Smelting House and in banking at Goldsithney, all of which were commitments that suffered in the initial fall in metal prices after the peak of 1814. By October of the following year it was said that Gundry's bills ' are in general refused to be discounted, and apprehensions are entertained that there will be a run upon him . . .'[3] In November, Wallis paid off Gundry's share of Polladrass costs, amounting to £561[4]—probably in order that the mine should not be stopped—and on 14 December some of Gundry's shares in Wheal Vor were mortgaged for £3,000. His three brothers, James, William and Thomas Gundry, were then embroiled, pledging their own estates to help their brother and by the end of the year the joint affairs of two of the four at least were said to be ' very much deranged . . . they cannot *no how* make good their engagements.'[5] This state of pending insolvency was staved off for a matter of a few further months but in 1819, with the metal market settling lower and lower all the while, John Gundry and his brothers were declared bankrupt. Up to this time Wheal Vor had been worked on an extensive scale but was by no means rich, and the costs of continuing his share of the mine when calls were necessary probably contributed greatly to Gundry's troubles. Initially the mine had been worked by a 48″ engine, with a 45″ pumping on Wheal Vreah adjoining, which was held by the same adventurers. In 1815, to sink the mine deeper, the first of the famous compound pumping engines built by the celebrated engineer Arthur Woolf was also started (with 27″ and 53″ cylinders) and the following year a second set of steam stamps—amongst the earliest of their kind on any mine in the county—was put to work. Wheal Vor was thus on a very con-

[1] Bolitho Records, DDRG 1/75, CRO, Truro. On the bankruptcy of Richard Oxnam in 1818, the shares in Chyandour were re-apportioned as follows—T. & W. Bolitho 10/24, James Pascoe 7/24, William Cornish 7/24, whilst later in the year this was amended again to Bolitho 1/2, Pascoe and Cornish 1/4 each.

[2] Including almost a half share in Polladrass, major holdings in Wheal Oak, Wheal Margery and other Lelant mines, plus various others in the Perranuthnoe—Marazion area. Gundry had also lost heavily in the failure of Wheal Prosper at Marazion which had been worked for copper and went down in 1804 with a loss of about £30,000. Christopher Wallis, *Journal, passim.*

[3] Christopher Wallis, *Journal*, 10 October 1815.

[4] Christopher Wallis, *Journal*, 15 November 1815. The writer's own comment on his action was ' unwillingly; be cautious in future.'

[5] Christopher Wallis, *Journal*, 27 December 1815.

siderable scale by 1816, although it had not yet cut rich. When Gundry failed, it was determined to keep the mine going rather than surrender the lease, and his shares in it, together with those of his brother, were transferred or assigned to others.

ANNUAL OUTPUT OF TIN IN CORNWALL, AND AVERAGE PRICE, 1800–49[1]

	Metallic tin tons	Price per ton of metallic tin		Metallic tin tons	Price per ton of metallic tin
1800	2,522	£101	1825	4,358	£ 91
1801	2,328	£105	1826	4,603	£ 77
1802	2,627	£108	1827	5,555	£ 76
1803	2,914	£109	1828	4,931	£ 73
1804	2,993	£109	1829	4,434	£ 74
1805	2,742	£112	1830	4,444	£ 74
1806	2,855	£120	1831	4,300	£ 73
1807	2,426	£117	1832	4,323	£ 73
1808	2,330	£114	1833	4,065	£ 73
1809	2,508	£122	1834	3,989	£ 78
1810	2,006	£157	1835	4,228	£ 91
1811	2,384	£141	1836	4,054	£109
1812	2,373	£128	1837	4,796	£ 88
1813	2,324	£134	1838	4,900	£ 88
1814	2,611	£156	1839	5,200	£ 83
1815	2,941	£140	1840	6,250	£ 81
1816	3,348	£114	1841	6,050	£ 81
1817	4,120	£ 93	1842	6,150	£ 72
1818	4,066	£ 85	1843	5,820	£ 64
1819	3,315	£ 75	1844	5,060	£ 74
1820	2,990	£ 73	1845	5,250	£ 91
1821	3,373	£ 75	1846	6,060	£ 94
1822	3,278	£ 95	1847	—	£ 89
1823	4,213	£ 95	1848	—	£ 77
1824	5,005	£ 88	1849	—	£ 78

Another person who suffered a similar fate to Gundry during this same period was the unfortunate Richard Oxnam, of Penzance. He had been a highly respected figure in west Cornwall and High Sheriff of the county in 1810, but his speculations in tin, his extensive mining interests and his association in some ventures with the ill-fated Gundry—whom he may have tried to assist—brought him into financial difficulties by 1816 at least. The following year he was declared bankrupt and commenced a long confinement in the King's Bench prison for debt. His half-shares in Wheal Reeth and in other mines were acquired by James Halse of St. Ives, whilst his mansion at Rosehill was also sold and his holding in Chyandour smelting-house passed to his erstwhile junior partners.

[1] From Memoirs Geological Survey, *The Geology of Falmouth and Truro etc.*, 1906. Output figures for the period from the ending of coinage in 1838 to the commencement of official statistics in 1853 are imperfect, and are derived (1838–46) from the *Trans. R. Geol. Soc. Cwll.* Vol vi. In this table the output of Devon is also included from 1818 onwards.

In Wheal Vor the new shareholders were primarily Hayle merchants, to whom the supply of materials to the mine was a matter of greater commercial significance than it had been to Gundry, and in order that a due proportion of Wheal Vor's black tin could be smelted on his own account, one of them, Henry Harvey, commenced the building of tin smelting furnaces at Hayle Foundry in October 1816. This arrangement continued until 1820 when Harvey, although still remaining a major shareholder in Wheal Vor, ceased smelting.[1] It was at this date, in January 1820, that John and Thomas Gundry were finally adjudicated bankrupt, the former having held 591/1624 and the latter 203/1624 of the mine in 1819.[2] Some of these shares were then mortgaged and others relinquished, affairs being in a state of considerable confusion, more particularly as the Gundrys had made every effort to pledge and re-pledge every security they had in their continuing endeavours to keep solvent. On 18 February the assignees, whose duty it was to distribute the estate in dividends to the creditors, relinquished Gundrys' shares in Wheal Vor as being worthless, for the mine was then losing heavily. In June 1820 it was then bought for £18,000 by H. M. Grylls, on behalf of the new company, Grylls himself—one of Gundry's assignees—having at first 11/50 share in this.[3] Thereby was to start, a little later, one of the longest and most complex law suits in the history of Cornish mining.

The Ludgvan Smelting House, which had formed part of the ill-fated John Gundry's business estate, was acquired at the end of 1819 by Thomas and William Bolitho on a twenty-one year lease.[4] Shortly before this, an arrangement had evidently been contemplated between all the various partners in Chyandour, Angarrack and Trereife to take a lease of the Ludgvan premises either for joint use or perhaps with a view to closing it down and thus preventing the intrusion of any further smelters into the western area.[5] Under Bolithos' tenancy, whatever tin was produced after 1819 was not coined separately and Ludgvan as a name disappeared entirely from the coinage returns.

Gundry's bankruptcy, which brought Ludgvan onto the market in 1819 and Oxnam's similar troubles in 1817 which caused changes at Chyandour, were paralleled by the bankruptcy of Thomas Rawlings in 1820. Rawlings, a prominent and old-established merchant and banker at Padstow, was engaged in smelting at Carvedras and Higher Blowing House at St. Austell in partnership with Daubuz (as Rawlings Daubuz & Co.) and the latter acquired sole ownership of these two works in 1818[6] and 1820 respectively.

Another casualty in 1819 was the sole smelting-house in Devon, which

[1] Bolitho Records, DDRG 1/163, CRO, Truro. The actual period during which smelting took place at Hayle Foundry was Michaelmas 1816 to Ladyday 1820, output being 1807 blocks during that time. The smelting mark of Harvey & Co. on this tin was the Cornish arms (fifteen bezants in pile).

[2] RCG 27.2.46. [3] WB 19.3.41. [4] RCG 13.5.20.

[5] Bolitho Records, Memorandum Book DD RG1/75, CRO, Truro.

[6] WB 6.11.18. Advertisement to Smelters and Blowers of Tin that Carvedras will be opened on the 11th instant.

had been established at Tavistock by Gill & Co. four years before.[1] This was advertised for sale in June 1819[2] 'situate near the Iron Foundry . . . with one furnace now at work.' Also sold up during the worst of the depression was Old Blowing House at St. Austell where Foxes Rashleigh & Co. decided to quit the trade in July 1819,[3] advertising the house as a going concern in October. After some eighteen months of abortive negotiations with various possible buyers, stocks were gradually run down, the company's holdings in various local mines and stream-works disposed of, and the premises and materials advertised for sale in 1821.[4]

A new entrant to the trade in 1820, somewhat surprisingly in view of the existing state of depression, was that of George Grenfell of Penzance, a member of the family already famous for its connection with copper smelting in South Wales, who set up a works at Hayle. This was situated close beside the Penpoll stream[5] and was known as Trelissick smelting-house from the manor on which it stood. Four years later, finding the venture profitable and being in need of another establishment close to the easterly coinages, Grenfell decided to set up another works in Truro. He chose a site immediately beside the river alongside Garras Wharf, in a marshy area then known as Roper's Moor, and not far distant from the place where the Newham works had existed about a century before. The first stone of this works, 'Trethellan Tin Smelting House,' was ceremonially laid in April[6] with the first purchases of black tin recorded later in that year. Here, Grenfell was joined in partnership by Henry Samuel Boase of Penzance, a doctor and banker,[7] each holding a half-share in the new smelting-house.

By 1824, however, there had been a drastic change in the *status quo* of both tin mining and smelting in Cornwall, with the dramatic reversal of fortune at Wheal Vor. At the time that the new company had taken over in 1820 and in connection with the Gundrys' bankruptcy proceedings, a valuation of the mine was carried out and this serves to show how extensive a concern it then was. The total value of the machinery and materials was put at £15,643 17s. 3d., the engines comprising Pearce's

[1] Bolitho Records, DDRG 1/163 CRO, Truro.

[2] WB 18.6.19. John Gill was a banker and merchant, also a partner in the Tavistock Ironworks. According to the coinage records, the works must have been subsequently re-started, for Gill & Co., otherwise the Devon Smelting Co., continued coining until the late 1830's at least, at Morwellham, Calstock and Liskeard.

[3] Foxes Rashleigh & Co., Letter Book, CRO, Truro.

[4] WB 5.5.21. 'For sale all the implements and apparatus at St. Austell Blowing and Smelting House . . . Including four furnaces and stacks, assay-house, counting-house, etc.'

In the period from 1819 to 1823, during the winding up of the company, £18,700 was paid in dividends (or repayments of capital) to the members of the company, but matters were not finally wound up completely until 1830 due to the company holding various areas of woodland (for charcoal supply) on lease, and other lands subsequently taken in settlement of Rawlings' debts on his bankruptcy in 1820.

[5] On a site adjoining what is now the Post Office at Hayle, in Foundry Square.

[6] RCG 24.4.24.

[7] The ranks of the smelters were very closely inter-woven with those of the bankers of Cornwall—Batten, Carne, Turner, Bolitho, Boase, Williams, Enthoven, Grenfell, Rawlings, Oxnam, Teague, Bain, Willyams, Daniell, Magor, Davey, Daubuz. In fact, during the nineteenth century, there was not a tin smelter in Cornwall who was not a banker.

63", Wheal Vor Old 48", Woolf's 27" and 53" compound, for pumping; three steam whims (Western, Middle and Wheal Vrea); and two sets of steam stamps (Bratt's 24" and New Stamps 27"). No less than eleven sets of water stamps were also in use; Lower Chenhale, Lower Tuckingmill, Farmyard, Trevarnoe, Lowertown Lower, Lowertown Higher, Higher Bluebridge, Marks's, Higher, Mount Pascoe Higher and Lower Stamps.[1] That the mine was still an unprofitable one, the costs for 1820–22 show:[2]

		Costs	Sales	Balance	
1820	April—May	£ 6055	£ 3117	£2937	Loss
	June—July	£ 6119	£ 4468	£1651	Loss
	Aug.—Oct.	£10315	£10113	£ 202	Loss
	Nov.—Jan. 21	£ 8042	£ 7678	£ 363	Loss
1821	Feb.—April	£ 8935	£ 8143	£ 792	Loss
	May—July	£ 9649	£10105	£ 456	Profit
	Aug.—Oct.	£10013	£ 9580	£ 433	Loss
	Nov.—Jan. 22	£ 9210	£10612	£1402	Profit

The shareholders then comprised Trevenen, Carne & Wood 9/50; Harvey & Co. 8/50; H. M. Grylls 6/50; Christopher Wallis 6/50; Charles Trelawny 4/50; James Pascoe 4/50; and nine others 13/50 between them.[3] At this juncture the mine was on the brink of cutting rich, for in February 1823 part of the main lode was opened up in the 124 fathom level and was revealed as no less than nine feet wide and worth £150 a fathom—rich enough to be set on tribute, and readily taken, at a mere 6d. and 9d. in the £.[4] According to the *West Briton*, Wheal Vor 'recently made in three months a profit of £13,000, the value of its tin alone in that time amounting to £25,000—so rich a load [*sic*] of that metal has not, we are informed, been ever before worked in Cornwall—its width is said to be thirty-six feet.'[5] This veritable storehouse of treasure, which was to make the mine world famous was struck in Wheal Vor herself, but there were in fact four mines being jointly worked by the company along the length of the lode. These were Wheal Vreah, Carleen and Polladrass, with the first two being worked on a scale almost as large as the Wheal Vor section. The accounts of the tributers' tin, as sampled monthly, show the full extent of the richness of the workings at this time, as well as the numbers employed: Several interesting features emerge from this analysis of the tributers' produce at Wheal Vor during this period of its development. First the

[1] Account Book of Wheal Vor, 1820–24, RIC, Truro. This original information on what was Cornwall's premier tin mine in the first half of the nineteenth century is of considerable value as it has not been made use of by any other writer on Wheal Vor's history.

[2] *Ibid.*

[3] Of these, the first two groups named were merchants at Hayle, Grylls was an attorney in Helston, whilst Wallis and Trelawny were concerned as mineral lords of parts of the sett.

[4] i.e. the miners to receive this amount only out of every £ of ore raised. Broadly speaking, the lower the rate of tribute that the miner was paid, the more valuable the pitch being worked, and *vice versa*. Tribute at 6s.–10s. in the £ was an average figure.

[5] WB 26.8.23. It has hitherto gone unrecorded that there were also contemporary hopes of good copper ore sales from Wheal Vor to supplement those of tin. 723 tons of rich ore were sold in 1823, but this lode in the northern part of the sett petered out and only 479 tons were sold in 1824, 198 in 1825 and 226 in 1827.

WHEAL VOR: TRIBUTERS' TIN SAMPLED MONTHLY from November 1822[1]

	Wheal Vor	Men	Boys	Wheal Vreah	Men	Boys	Carleen	Men	Boys	Polladrass	Men	Boys	Total	Men	Boys
1822 Novr.	£2235	50	8	£2283	85	5	£ 721	42	2				£ 5244	177	15
Dec.	£2912	50	9	£2311	82	5	£ 556	60	2	£ 15			£ 5794	192	16
1823 Jan.	£2865	60	7	£2126	72	7	£ 578	82	—	£238	27	—	£ 5807	241	14
Feb.	£4214	46	8	£1706	58	7	£ 646	70	—	£253	14	—	£ 6818	188	15
Mar.	£3464	46	6	£2699	62	7	£ 870	60	5	£307	14	—	£ 7340	182	18
Apl.	£4034	38	7	£3402	76	8	£ 991	60	2	£464	18	1	£ 8899	192	18
May	£7176	47	8	£3600	80	9	£1076	56	1	£617	16	2	£12465	199	20
June	£7052	51	7	£2959	80	10	£ 894	52	—	£415	18	—	£11319	201	17
July	£7357	48	9	£2186	60	9	£1046	62	—	£507	18	—	£11096	188	18
Aug.	£4913	28	6	£3537	66	10	£1367	62	4	£471	18	—	£10287	174	20
Sept.	£2386	28	5	£1600	74	9	£ 672	61	2	£381	24	—	£ 5039	187	16
Oct.	£1977	24	6	£2282	66	7	£1063	85	3	£248	16	—	£ 5571	191	16
Nov.	£2906	32	4	£2337	—		£1429	59	3	£353	30	—	£ 7025		
Dec.	£2872	40	7	£1350	76	7	£1119	68	5	£341	30	—	£ 5682	214	19
1824 Jan.	£3587	41	11	£2561	66	6	£1411	85	5	£409	38	—	£ 7967	230	22
Feb.	£2912	58	8	£3100	56	5	£1174	84	9	£166	20	1	£ 7352	218	23
Mar.	£3316	58	12	£2275	52	6	£1293	84	10	£374	22	2	£ 7058	216	30
Apl.	£2264	52	11	£3368	60	7	£1290	80	10	£498	22	—	£ 7419	214	28
May	£3424	58	11	£3231	56	7	£1182	74	8	£627	28	—	£ 8464	216	26
June	£2735	52	15	£3441	44	5	£ 905	64	7	£426	24	2	£ 7508	184	29

[1] RIC, Truro.

numbers employed, for these figures include only the tributers, who represented about half those employed underground. In November, in addition to the 192 on tribute, there were 272 men and 24 boys on tutwork (84 and 11 Wheal Vor, 84 and 6 Wheal Vreah, 92 and 7 Carleen, 12 Polladrass) and by August 1823 this had risen to 309 men and 38 boys (116 and 9 Wheal Vor, 56 and 10 Wheal Vreah, 85 and 13 Carleen, 52 and 6 Polladrass) making a total of 541 men and boys then employed underground. Secondly, the relative importance of the Wheal Vreah section is shown, as also the fluctuating nature of the returns even after the first rich strike early in 1823.

Earlier, shortly before 1820, the adventurers had come to an agreement with the Copperhouse company of Trevenen, Carne & Wood (later Sandys Carne & Vivian) that a major part of the output should be smelted on the adventurers' behalf in the old furnaces remaining from the days when the old Cornish Copper Company had smelted copper at Hayle. Thereafter, however, when the mine cut rich, and when Sandys Carne & Vivian were themselves expanding into ironfounding and engineering at Copperhouse, different arrangements were needed—more particularly as bitter trade disputes between Harvey & Co. and the owners of Copperhouse had also sprung up. Accordingly it was decided to build a smelting-house on the mine itself. The great shoot of ore that had been opened up continued richer and yet richer, and when once the furnaces were in full commission[1] the quantity of tin presented by the Wheal Vor adventurers at the Penzance Christmas coinage in 1823 was so great that 'the officers could not coin it in the week—a Petition was sent for leave to coin about 340 blocks afterwards.'[2] Certainly the enormous output of the mine in the ensuing year must have struck something akin to dismay into others concerned in the tin trade of the county: at Ladyday, 1000 blocks coined at Helston and 856 at Penzance; at Midsummer, 1050 at Helston and 1010 at Penzance; at Michaelmas, 1500 at Helston and 499 at Penzance; and at Christmas, 1019 at Helston and 585 at Penzance. This was a total of 7269 blocks of tin, or about 1250 tons of metal, and probably represented an output of black tin from the mine of something like 2000 tons—a quantity very much more than double the output of any Cornish tin mine hitherto. Nor was it a flash in the pan, for it was said four years later that 'the more they extend and open the mines the richer and more productive they become, forming altogether an almost inexhaustible body of ore.'[3] Moreover, Wheal Vor continued to coin about one quarter of Cornish tin for another decade, the greatest riches of all coming between the 180 and 240 fathom levels where the lode was '100 fathoms long, 12 feet wide, and a small wheelbarrow of it was worth £3.'[4]

[1] Based on the output figures it seems probable that six or even eight furnaces were erected at Wheal Vor smelting-house. Its situation is now only recorded in the name Smelting House Shaft, about 200 yards east of Carleen village.

[2] Bolitho Records, DDRG 1/163, CRO, Truro.

[3] WB 16.3.27. [4] MJ 8.11.56.

The dominating position of Wheal Vor in terms of Cornish tin after 1823 is shown below, nor does there seem any doubt that it all came from Wheal Vor alone[1] and was not being smelted there for other local mines in which its adventurers had mutual interests. The Helston coinage, where most of the Wheal Vor blocks were regularly taken, had formerly been one of relative insignificance but it now assumed a new importance.

BLOCKS OF TIN (COMMON AND GRAIN)
COINED BY CORNISH SMELTERS, 1820–30[2]

	1820	1821	1822	1823	1824	1825	1826	1827	1828	1829	1830
Calenick	1645	1765	1334	1855	1156	1257	1196	1017	870	1919	1992
Carvedras	1382	1598	1956	1660	1719	1120	2687	4001	3740	2835	1826
Angarrack	1146	1417	1448	1948	1361	1604	1776	2816	2551	2176	2157
Portreath	1679	1778	2412	2157	1294	355	—	—	—	—	—
Gt. Hewas	2614	2676	1398	—	—	—	—	—	—	—	—
St. Austell [3]	1145	1792	1566	3206	4137	3204	3069	2707	1835	576	943
Chyandour	1515	1730	1665	1904	1789	1463	1540	2155	1798	1538	1730
Trereife	1676	2104	1801	1994	1855	1648	1587	2166	1777	1958	2275
Treloweth	1298	1860	1882	1765	1906	2142	2761	3100	3116	3090	2756
Trelissick	707	1174	1832	2073	1318	1100	981	2123	1526	1684	1850
Trethellan	—	—	—	—	1219	1428	3083	3281	2471	2013	2313
Whl. Vor	922	84	—	2677	7520	6454	6338	7099	7084	5822	5658
Whl. Reeth	—	—	—	147	1024	—	—	—	—	—	—
Truro S.Co.	417	1128	1262	2143	1231	1224	282	388	47	—	—
Devon S.Co.	258	167	115	148	314	325	218	876	898	927	564
Total	17084	19273	18671	23956	27481	24479	25518	31729	27713	24538	24064

On one occasion in 1827 no less than 1780 blocks of tin from the mine lay awaiting attention in Coinagehall and other adjoining streets, and coinage quantities of 1300 blocks were always common.

The sad thing about Wheal Vor's rise to prosperity was that the Gundrys, who had been the progenitors of its success by forming the company in 1810 to get the mine re-started, had no share in it. The mine's sudden but belated transformation from poverty to unbounding wealth had, however, occurred sufficiently close to the date of Gundry's insolvency for legal actions to be taken to recover the shares from those who had bought them from the assignee. The creditors also commenced an action to annul the sale of their shares and with the brothers' affairs being then in the Chancery Court—' a capital place for the lawyers '[4]—a field-day ensued for the latter. Cornish mining had always been a great legal battle ground, with a constant harvest reaped by the county's lawyers from litigation over setts and water-courses, stannary cases and patents, encroachments and the like, but the case of the Gundrys and Wheal Vor was exceptional. For twenty years and more the case was dragged out in disputed actions and counter-actions, not only by the assignees of the two men but also by the latters' son and daughter, in what was later described

[1] In 1824 the four hitherto separate mines were brought into one concern, known as Wheal Vor Consolidated Mines.

[2] From Coinage Book, 1820–34, RIC, Truro.

[3] Three houses (Old, New, Higher Blowing Houses).

[4] MJ 9.5.57.

as 'one of the most lengthened and tedious Chancery suits on record.'[1] The prize was certainly worth quarrelling over, for in the ten years from 1824–33 alone Wheal Vor Consolidated sold 60,342 blocks of tin (about 10,000 tons of metal) and was then far from exhausted. In Cornwall such a tin mine had not been known before and even though the divisible profits were not large by copper mining standards, this was largely because the price of black tin was only about £40 a ton throughout this period, and, more important still, the mine was chiefly held by merchants who were interested in the profits that came from trade turnover and not from shared dividends.

During the period of Wheal Vor's rise to pre-eminence in terms of Cornish tin, and also perhaps as a result of it, two successive mergers took place amongst the major smelters who suffered competition from its rise. The first of these commenced in 1825 when the Portreath Smelting Company decided to close its works and the last blocks were sent for coinage from there at Midsummer 1825.[2] The partners in the company at this date were Harvey, Davey, Williams & Co., whilst a few years earlier William Reynolds[3] had also had a small interest. The company had decided to join forces with Daubuz & Co. at Truro and elsewhere—possibly because they had found Portreath too distant from any coinage town—and thereafter for some years their joint output was to be coined under Daubuz' name alone. Harvey, Davey, Williams also acquired Old Blowing House at St. Austell—which had stood empty for about four years since Foxes Rashleigh & Co's. occupation—and an arrangement was entered into between them and Daubuz that all grain tin would be blown at his Higher Blowing House[4] and that Old Blowing House would be used solely for the production of common tin. This arrangement, of reviving the disused reverberatory furnaces at Old Blowing House and utilising it as a smelting works, was to last until 1839. In Truro the new partners decided to supplement Daubuz' existing capacity at Carvedras by acquiring the Truro smelting works which had been established eight years earlier by Magor & Turner, who traded as the Truro Smelting Company. This works was a large one and had produced more tin in 1823, for example, than either Chyandour or Calenick. In 1825 ownership of it passed to another company, styled the Cornish Smelting Co., in which Thomas Teague of Redruth was the principal.[5] This company was short-

[1] MJ 31.10.57.

[2] Tin Ticketings Book 1804–37, (Bolitho Records, CRO, Truro). Except where otherwise noted information on Cornish smelting works at this period is derived from this source.

[3] Of Trevenson, Pool. He was toller and steward to the Basset family of Tehidy.

[4] In 1814 Thomas Rawlings (of Padstow) had joined James Harford in partnership in Higher Blowing House, as Harford Rawlings & Co. Later the same year Daubuz acquired Harford's interest and it became Rawlings Daubuz & Co. On Rawlings' bankruptcy in 1820 Foster, a banker and merchant of Bodmin, joined Daubuz and the company then became known as Daubuz & Foster, to be replaced by Daubuz & Co. in 1825 with the commencement of the latter's sole ownership. In Daubuz' hands it was to remain active until 1863 at least. (Bolitho Records, DDRG 1/163, CRO, Truro).

[5] Teague was the chief shareholder in Poldice and a short time earlier, in 1823, tin had been smelted at the Truro works on contract for the mine.

lived, for Daubuz, on behalf of himself and his new partners, acquired the works in 1826 to run it as a subsidiary to Carvedras. His Treloweth works at St. Erth was also re-built and extended in 1825, and from this date onwards, almost to the end of the 1830's, the name of Daubuz—with which was coupled, unknown to the world at large, Harvey, Davey, Williams & Co.—occupied the leading position amongst Cornish smelters.

This grand alliance between Daubuz and his new partners thus had no less than five important works between them—Old and Higher Blowing Houses at St. Austell, Truro and Carvedras at Truro, and Treloweth at St. Erth. It was perhaps in an endeavour to combat the influence of this combine and Wheal Vor's dominance that the Bolithos, shareholders in Chyandour and Angarrack, and in Ludgvan if it was in fact being worked, spread their tentacles eastwards shortly afterwards by acquiring a major share in the Calenick works near Truro,[1] and through this same company of the associated New Blowing House at St. Austell. Both these latter were acquired from R. A. Daniell & Co. in 1828,[2] in conjunction with the Michell family who had been concerned with Calenick for many decades. In the new company formed, the Calenick Tin Smelting Co., the Bolithos held two-thirds, Robert and William Michell one-third.[3] At this same date the St. Agnes Smelting House (also known as the Primrose Smelting House) situated 'near St. Agnes Pier' was for sale[4] and it is possible that this was an outlying subsidiary works which had belonged to Daniell & Co.[5]

Another interesting but extremely short-lived newcomer to the list of smelters at the coinages during this time was Wheal Reeth. The adventurers here, perhaps following the example of Wheal Vor in the same year, erected their own smelting-house on the mine and in 1823 coined 147 blocks and 1024 in 1824. Thereafter, however, nothing further was

[1] The Bolithos already held an interest in a small 'melting-house' in Truro, which is recorded as having been set up, with a single air furnace, in part of the premises lately held by Whitbread and Withers, brewers (in 1817). This 'tin melting house' on Kenwyn Street was advertised for sale by the Bolithos in April 1838 (WB 27.4.38).

[2] This company had been formed in 1816 to acquire the smelting works from R. A. Daniell himself, who had found it necessary to take in partners to secure further capital. Of its capital of £12000, R. A. Daniell held £1500, the banker Humphrey Willyams £1500, Thomas Daniell £1500, Thomas John £1500, R. Michell £1200, John Giddy £1200, R. Michell Jnr. £1200, John Michell £1200, Wm. Michell £1200. All tin produced had to be sold together and not divided. The following year the company also took from R. A. Daniell the lease of the New Blowing House at St. Austell, one of the Michells being despatched there as manager. In this branch, R. A. Daniell retained a half share, all the others seven (R. Michell Senr. having died) 1/14 each. In addition the company owned Trevellas Stamps and Date's Stamps in St. Agnes in 1819 and Tolskithy Stamps in Illogan, and in 1820 took Michell's Stamps and Burning-House in Redruth. (Willyams Records, Articles of Agreement of R. A. Daniell & Co., 1816, CRO, Truro).

R. A. Daniell died in 1823 and his estate passed to his son Thomas. Neither was a man of business such as the founder of the family fortunes had been and by 1828 Thomas Daniell was in financial difficulties, being eventually driven to reside at Boulogne to escape the dreaded perils of bankruptcy.

[3] At this same time the Michell family, in conjunction with Penrose of Tregye, had acquired the Penpoll lead smelting works, a few miles south of Calenick, on the bankruptcy of John Swan who had established it about three years before.

[4] WB 1.2.28, RCG 2.2.28, 8.11.28.

[5] No coinages for the St. Agnes smelting-house are on record, a fact that would be accounted for by its being a branch works of which the output was coined under another name.

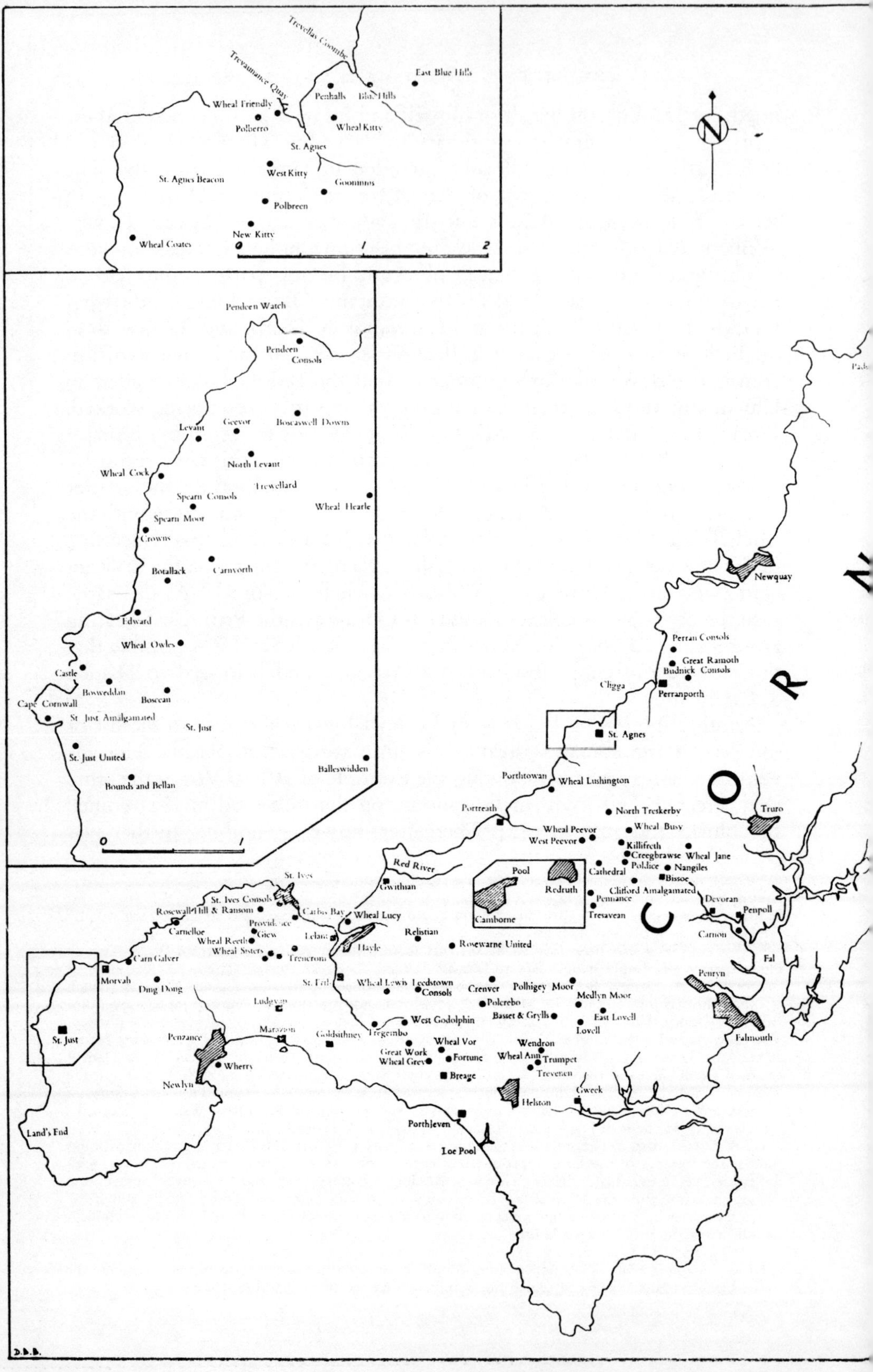

Trevellas Coombe
Trevaunance Quay
East Blue Hills
Penhalls
Blue Hills
Wheal Friendly
Polberro
Wheal Kitty
St. Agnes
West Kitty
St. Agnes Beacon
Goonmnos
Polbreen
New Kitty
Wheal Coates
0
1
2
N
Pendeen Watch
Pendeen Consols
Levant
Geevor
Boscaswell Downs
North Levant
Wheal Cock
Trewellard
Spearn Consols
Wheal Hearle
Spearn Moor
Crowns
Botallack
Carnyorth
Edward
Wheal Owles
Castle
Bosweddan
Boscean
Cape Cornwall
St. Just Amalgamated
St. Just
St. Just United
Balleswidden
Bounds and Bellan
0
1
Padst
Newquay
Perran Consols
Great Ramoth
Budnick Consols
Cligga
Perranporth
St. Agnes
Porthtowan
Wheal Lushington
Portreath
North Treskerby
Wheal Peevor
Wheal Busy
West Peevor
Killifreth
Creegbrawse
Wheal Jane
Poldice
Nangiles
Cathedral
Bissoe
Clifford Amalgamated
Pennance
Tresavean
Truro
Devoran
Penpoll
Carnon
Fal
Penryn
Falmouth
C O R N
Red River
Gwithian
Pool
Redruth
Camborne
St. Ives
St. Ives Consols
Rosewall Hill & Ransom
Carbis Bay
Providence
Wheal Lucy
Carnelloe
Giew
Lelant
Relistian
Wheal Reeth
Wheal Sisters
Trencrom
Hayle
Rosewarne United
Carn Galver
Morvah
Ding Dong
St. Erth
Wheal Lewis
Leedstown Consols
Crenver
Polhigey Moor
Medlyn Moor
Ludgvan
Polcrebo
Basset & Grylls
East Lovell
Lovell
St. Just
Penzance
Marazion
Goldsithney
Tregembo
West Godolphin
Wheal Vor
Wendron
Great Work
Fortune
Wheal Ann
Trumpet
Wheal Grey
Wherry
Breage
Trevenen
Newlyn
Gweek
Helston
Porthleven
Land's End
Loe Pool
D.B.B.

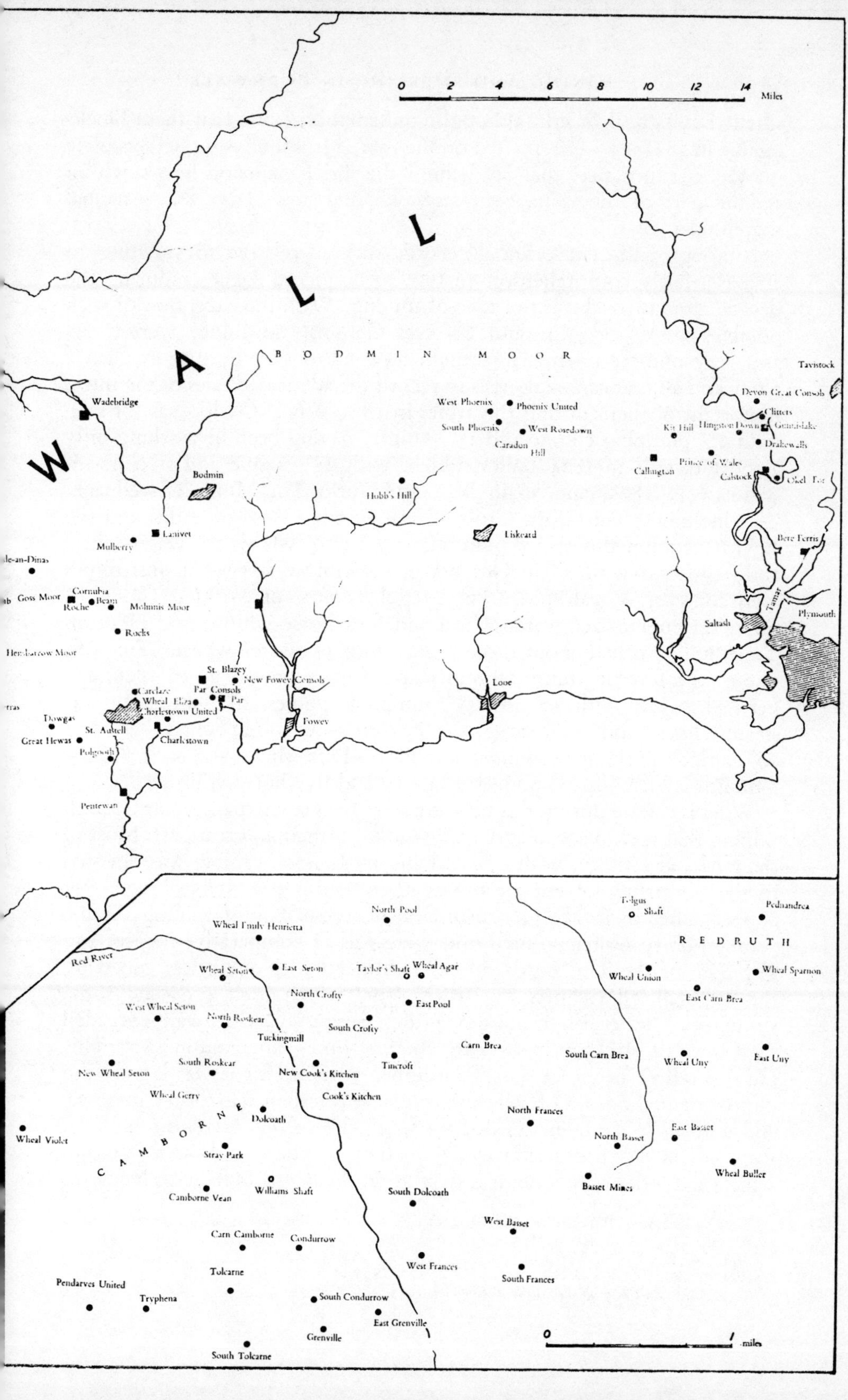

0 2 4 6 8 10 12 14 Miles
W A L L
B O D M I N M O O R
Wadebridge
Bodmin
Lanivet
Mulberry
Goss Moor
Cornubia
Roche
Beam
Molinnis Moor
Rocks
Hensbarrow Moor
St. Blazey
New Fowey Consols
Carclaze
Wheal Eliza
Par Consols
Par
Charlestown United
Dowgas
St. Austell
Great Hewas
Polgooth
Charlestown
Pentewan
Fowey
Looe
Liskeard
Hobb's Hill
West Phoenix
Phoenix United
South Phoenix
West Rosedown
Caradon Hill
Kit Hill
Callington
Prince of Wales
Hingston Down
Gunnislake
Clitters
Devon Great Consols
Tavistock
Drakewalls
Calstock
Okel Tor
Bere Ferris
Tamar
Saltash
Plymouth
North Pool
Tolgus Shaft
Pednandrea
R E D R U T H
Wheal Emily Henrietta
Red River
Wheal Seton
East Seton
Taylor's Shaft
Wheal Agar
Wheal Union
Wheal Sparnon
East Carn Brea
North Crofty
East Pool
West Wheal Seton
North Roskear
South Crofty
Tuckingmill
Carn Brea
South Carn Brea
Wheal Uny
East Uny
New Wheal Seton
South Roskear
New Cook's Kitchen
Tincroft
Cook's Kitchen
Wheal Gerry
C A M B O R N E
Dolcoath
North Frances
Wheal Violet
Stray Park
North Basset
East Basset
Wheal Buller
Basset Mines
Williams Shaft
Camborne Vean
South Dolcoath
West Basset
Carn Camborne
Condurrow
West Frances
South Frances
Tolcarne
Pendarves United
Tryphena
South Condurrow
East Grenville
Grenville
South Tolcarne
0 1 miles

smelted at Wheal Reeth, although confirmation exists that these blocks coined in 1823–24 were smelted on the mine.[1] It is interesting to speculate on the circumstances that lay behind this brief incursion into smelting on the part of the mine company, where James Halse was principal shareholder.

Although sufficient original records survive relative to smelting to enable a fairly comprehensive picture to be pieced together during the decade after 1820, this is not true of mining. With the exception of such notables as Wheal Vor and St. Ives Consols, tin mines were small and very numerous—many of them now known by little more than a name and all that can be done is to record the advertised sales of the more important of them during this time. In 1821 Wheal Owles was for sale, with 22″ pumping engine and 15″ whim, 'having been in working only a short time;'[2] in 1824 Balleswidden, with 28″ engine and two horse-whims:[3] in 1825, and again in 1827, Durlo Tin Mine, Towednack, '. . . including Tin Croft, Croft Wheal Reeth, Chylason Adit, and the Beyow,' equipped with 30″ pumping engine, 16″ whim, 30′ water-wheel and two horse-whims.[4] In October of the same year two St. Just mines were for sale, Wheal Spearn 'in a regular course of working' with 19″ pumping engine, 20′ water-wheel and four horse-whims, and Hermon Tin Mine 'worked about three years' with 32′ water-wheel.[5] Tincroft, one of the few tin mines around Carn Brea, was sold up in 1826, advertised in July with 70″ and 48″ pumping engines, 16″ stamps and 14″ steam whim;[6] and in the same year St. Agnes Consols in September, with 60″ engine and six horse-whims[7] and North Pink Mines, also in St. Agnes, a month later, with 48″ pumping engine and five horse-whims.[8]

What is certain during this time is that in the great expansion in Cornish mining that took place in and after 1820 tin played no great part. Most of the renewed activity, both by local and by London capital, was devoted to the re-starting of copper mines, there being a veritable 'rage' for copper—then by far the most profitable branch of English mining. As one contemporary writer pointed out, there was 'a comparative neglect and disrepute of the Tin, Lead and other Minerals.'[9] Nonetheless, aided by the unparalleled richness of Wheal Vor, which itself provided some stimulus to other tin mining ventures, output in Cornwall exceeded 5,000 tons of black tin in 1824 for the first time, and again in 1827. This did not reflect an increase in the number of mines for it was largely the unprecedented output of Wheal Vor that was responsible for the increase, and with it, to some extent, for the low price of tin. Adventurers in tin were not in fact prospering for, with the exception of those few mines which had lodes rich enough to pay good dividends even with black tin

[1] Bolitho Records, DDRG 1/163, CRO, Truro. According to this source, 342 blocks were coined by Wheal Reeth adventurers in 1823, 195 at Michaelmas and 147 at Christmas coinages.
[2] WB 1.6.21. [3] WB 17.12.24. [4] WB 7.10.25., 2.2.27. [5] WB 28.10.25.
[6] WB 7.7.26. [7] WB 15.9.26. [8] WB 27.10.26.
[9] G. Abbott, *An Essay on the Mines of England, 1833.*

at £40 per ton, the tin trade was going through what amounted to a time of prolonged, if not intense, depression.

The root cause of this could be traced to the loss to Cornwall of the East India contract—a market which had absorbed the European surplus of tin in previous years and which formed an outlet that the county could ill afford to lose. The value of the East India purchases to the tinners was self-evident in the price of metal during the relevant period. From 1783–88, the average price per cwt. had been 70s. 6d.; after the system of stocking had been instituted, it averaged from 1789 to 1816 89s. 3d. per cwt.; from 1817, when the China trade dwindled, up to 1831, the price was 75s. 9d., whilst in 1832 it was down to no less than 63s. 6d. per cwt.[1] By the early 1830's there was growing dissatisfaction, for the price was lower than it had been since 1750 and well might it be said that 'at present the adventurers in tin mines sadly want some addition to the miserable returns on capital outlaid.'[2] It was not, however, until the depression began to affect the smelters' trade as well as that of the true tinners that any public attention was focussed on the existing state of affairs. The chief occurrence threatening them and bringing about the sustained outcry at the low state of the tin trade was the rapid growth in the trade in metal from the East. In 1829 the tin brought to England for re-export had been 2674 cwts. whereas by 1832 this amount had increased ten-fold, to no less than 29,204 cwts. To make matters worse it was well known that 'a considerable part of this is foreign tin cast in Holland with our marks and brought here to be exported as English tin.'[3] The source of this competing tin was, of course, Banca and the Straits, where alluvial mining was making great progress. The trade from there had gradually ousted the East India Company's hold on the lucrative Chinese market and this became more and more marked as Chinese merchants took an interest in the Straits workings. By the end of the East India Company's monopoly of the trade in 1833, Cornish tin was a memory in China and the direction of the flow of trade had been reversed. The Dutch government assiduously fostered the Banca workings after 1816 and within a few years they had become a factor of growing importance in the realm of tin. By 1821 Banca output was 1250 tons and by 1846, using Chinese and convict labour to work the deposits, it had grown to 4400 tons—a considerable output by contemporary world standards.[4] In these figures one could read the genesis of the competition that was to afflict Cornwall and Cornish tin. The tin from there was marketed at Amsterdam and the biennial Banca sales became the yardstick of the entire trade. Cornwall had once had well-nigh the entire monopoly of the metal and had ruled the world, but by the early 1830's the course of the market in tin and its well-

[1] Anon., *The Tin Duties*, 1833 (pamphlet).
[2] *Ibid.* [3] WB 7.9.32.
[4] In 1856 the output of Banca rose to a peak of 6250 tons. During the 1870's it averaged about 4250 tons whilst by that date the nearby island of Billiton was producing almost as much.

being could no longer be determined in Truro at the smelters' meeting held after each quarter's coinage.

To meet the lower price of this foreign tin, produced from virgin alluvial deposits by what was virtually slave labour, Cornish tin would have to be sold more cheaply and with no other obvious economy in production possible—for wages were already at or near starvation level, and with the merchants so closely allied to the main adventurers the cost of materials sacrosanct—the target for most of the outcry in Cornwall became the coinage dues. These were in fact a duty then amounting to about 5s. 3d. per cwt.[1] or seven per cent upon the ruling price of tin. Feelings about the injustice of the coinage duty ran high; 'surely,' wrote one correspondent, 'the tin mines are not to be sacrificed in order to supply a revenue for the Heir Apparent.'[2] A pamphlet, anonymously compiled but doubtless emanating from the circle of smelters whose finances were threatened by the foreign tin, was published in 1833, setting forth the case against what it termed 'this grievous tax on an interest now struggling with great difficulties.' In this it was pointed out that the former concessions granted to the tinners in return for their payment of coinage dues—such as the immunities and privileges of being exempted from market tolls and military service, of being allowed to divert any streams and cut what fuel they needed—were no longer of any value: 'the advantages have long ceased and the burden remains without compensation.' Another anomaly was the assaying part of the coinage process, for this was no longer necessary now that the smelters presenting the blocks were the owners of them and not, as once, those for whom they had merely smelted the ore. But this was as nothing compared to the hindrance of having to regulate smelting capacity to the calendar of the coinages which took no account whatever of the fluctuating demands of the metal market. This might mean that part of a large order received by a Truro smelter a few days after a coinage had finished there would either have to wait three months until the next coinage before he could supply, or else blocks would have to be carted to say Penzance to catch the train of royal panoply.[3] In truth the whole coinage process had become an unmitigated nuisance to the smelters, not only because of such minor fatuities as the laid down scale of charges for such items as the oiling of the beam of the weighing scales, for 'adjourning' the coinage each time, and for the holding of candles during those in winter, but also in the carriage involved and the time

[1] 4s. per cwt. of 120 lbs. on all tin coined in Cornwall, plus 4d. for 'post-groats'—an extra duty that had been imposed when the coinages had been made quarterly instead of twice yearly. In Devonshire the basic duty was 1s. 8d., and not 4s. as in Cornwall. In addition, in both counties, there were fees 'for special coinages and what are called great pieces [blocks of size and weight greater than normal], fees to cutters, poisers, pilers, scalesmen; fees for the use of the beam; drink money, gift money, house money, and dinners to the officers at all the coinages.' Anon., *The Tin Duties*, 1833.

[2] WB 7.9.32.

[3] To minimise the cost and nuisance of coinage delays, smelters customarily worked together in supplying each other with locally held stocks so far as was possible—yet another instance of the harmony of those within the close circle of the trade. The necessity for common action over the East India contracts had already brought the main smelters together into a form of combination.

wasted. The expense and delay of transporting hundreds of blocks of tin each coinage day from the various smelting-houses to the coinage hall was a financial burden that was patently unnecessary. Why at least, the smelters argued, should the tin not be stamped and coined by the Stannary officers at the smelting-houses, as had once been the case up to the mid-seventeenth century? The nonsensical insistence on holding the coinage process in a central part of the town was apparent when one reflects upon the space occupied by three or four thousand blocks of tin, and the carts to carry them, quite apart from the physical labour and time involved in moving and re-moving them. At Penzance, for example, tin had to be brought in from as far afield as Angarrack, Wheal Vor and Treloweth. Chyandour and Trereife were relatively close but it would still have been more convenient and less costly had the officials gone to the tin and not *vice versa*. Yet another complaint was that blocks of other than the standard size involved special higher fees, even though bars and smaller ingots were much to be preferred by smelter, tin merchant, shipper and tin consumer alike. Once, of course, all this extra cost on the production of merchantable tin had fallen upon the latter, but with growing foreign competition, times were changing: 'when this country had nearly a monopoly of tin, the impost was lightly felt and the duty forming an addition to the price of the article was in fact paid by the consumer both at home and abroad.'

As a result of this agitation and, doubtless, by an unrecorded amount of lobbying by the Cornish M.P.s, the government agreed in March 1833 to amend the coinage but not to abolish it. The major alterations were the introduction of coinages every six weeks, twice as frequently as before, the addition of Calstock, Hayle and St. Austell as new coinage towns, plus the abolition of all the minor fees.[1] This was a welcome relief so far as it went, but at a meeting of the tin mining interest, advertised in the *West Briton* 19 April 1833, it was plainly stated that it was not enough. One of the smelters' main requirements, that the coinage officials should go to the smelting-houses and not have the tin carted to them, had not been met and during the following months dissatisfaction continued to be expressed at the continued levy.[2]

Nevertheless the more frequent coinages were a marked improvement and the designation of new towns for the purpose was of particular benefit

[1] WB 5.4.33.

[2] Relations between the Duchy of Cornwall and the mining interest were at their lowest ebb at this time, dissatisfaction on the part of the latter being due also to the existing system whereby the Duchy farmed out on lease its entire mineral rights to one person. Until this arrangement was abolished in the 1840's, all applications for setts etc. had to be made to this lessee, a system which was not conducive to the best interests of either the Duchy or Cornish mining.

The Duchy authorities, moreover, were probably the single largest mineral lords in the south-west, possessing—apart from those under the sea—the rights over very extensive tracts of ground in the chief tin producing districts, including Dartmoor, much of Hingston Down and Kit Hill, the Bodmin Moor area north of Caradon, part of St. Austell Moor and of the area around St. Agnes, plus most of that lying about Wendron. Duchy mines thus included such notables as Drakewalls, Phoenix United, Charlestown United, Polberro, Wheal Coates, Budnick Consols, Carclaze, Rocks & Treverbyn, St. Austell Hills, Trumpet and Wendron Consols, the Lovells, etc.

to various smelters. Gill & Co., of Tavistock—otherwise the Devon Smelting Co., as the concern had re-styled itself after 1819—had formerly

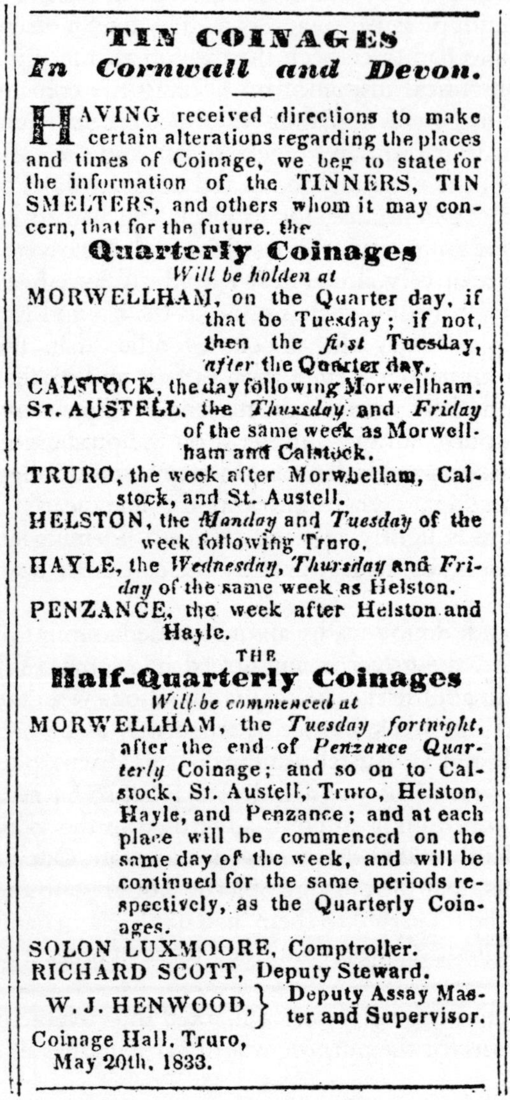

TIN COINAGES
In Cornwall and Devon.

HAVING received directions to make certain alterations regarding the places and times of Coinage, we beg to state for the information of the TINNERS, TIN SMELTERS, and others whom it may concern, that for the future, the

Quarterly Coinages
Will be holden at

MORWELLHAM, on the Quarter day, if that be Tuesday; if not, then the *first* Tuesday, *after* the Quarter day.
CALSTOCK, the day following Morwellham.
St. AUSTELL, the *Thursday* and *Friday* of the same week as Morwellham and Calstock.
TRURO, the week after Morwhellam, Calstock, and St. Austell.
HELSTON, the *Monday* and *Tuesday* of the week following Truro.
HAYLE, the *Wednesday, Thursday* and *Friday* of the same week as Helston.
PENZANCE, the week after Helston and Hayle.

THE
Half-Quarterly Coinages
Will be commenced at

MORWELLHAM, the *Tuesday fortnight*, after the end of *Penzance Quarterly* Coinage; and so on to Calstock, St. Austell, Truro, Helston, Hayle, and Penzance; and at each place will be commenced on the same day of the week, and will be continued for the same periods respectively, as the Quarterly Coinages.

SOLON LUXMOORE, Comptroller.
RICHARD SCOTT, Deputy Steward.
W. J. HENWOOD, } Deputy Assay Master and Supervisor.
Coinage Hall, Truro,
May 20th, 1833.

Notice of amended dates and places of coinage, 1833.
West Briton, 5.4.33

had to carry all their Cornish tin[1] to Liskeard for official attention, a circumstance which had necessitated its revival as a coinage town in 1820 after many decades during which it had been held in abeyance.[2]

[1] As tin produced in Devon bore a lesser coinage duty than that in Cornwall it is probable that Duchy regulations prevented the smelting of Cornish tin in Devon, or *vice versa*. Thus it is probable that the Devon Smelting Co. also had furnaces and a small smelting works at Calstock as well as at Tavistock.

[2] In 1827 it was reported that 'The Devon Tin Smelting Company coined the tin raised from

The blowing-houses at St. Austell, although of rapidly lessening importance at this time, had also hitherto been put to the expense of carting their produce to Truro as well as of the subsequent storage of it there until it was sold and shipped. The new regulations also benefitted the much more important three western houses at Treloweth, Trelissick and Angarrack which had henceforward only to carry their tin to the new coinage 'hall'[1] on Hayle wharves and not to distant Penzance. Hayle as a coinage town rapidly assumed an important position, for Wheal Vor tin was also usually consigned there and Helston declined to the position of the least important of all the western coinage towns.[2]

At St. Austell, after its establishment as a coinage town in 1834,[3] a new smelting company entered the list of coiners. This was the Charlestown Smelting House[4] set up by John Taylor, Sons & Co., which first commenced buying black tin early in the year and coined at the St. Austell Midsummer coinage. Initially purchases were made solely from Polgooth and Charlestown United—in both of which important mines Taylor was manager and shareholder—but before long ores were being bought from St. Agnes, Poldice, Great Beam, and from mines in the Tamar valley. Yet despite its link with two of Cornwall's most considerable tin mines, this smelting venture was not successful and in 1835 the number of blocks coined by the new house was only a small proportion of the whole:

BLOCKS OF TIN (COMMON AND GRAIN)
COINED BY CORNISH SMELTERS, 1835[5]

	Ladyday	Midsummer	Michaelmas	Christmas	Total
Devon Smelting Co.	265	155	64	61	545
Charlestown (Taylor)	300	155	300	32	787
Carvedras (Daubuz & HDW.)	1644	731	558	1146	4079
Trethellan (Grenfell & B.)	425	679	370	501	1975
Calenick (Bolitho & M.)	669	698	461	874	2702
Angarrack (Bolitho & Carne)	554	463	401	568	1986
Trelissick (Grenfell)	602	448	553	686	2289
Treloweth (Daubuz)	572	726	420	628	2346
Chyandour (Bolitho)	250	327	307	304	1188
Trereife (Batten)	407	745	472	739	2363
Whl. Vor	956	909	885	592	3342
Higher B. House, St. Austell	—	70	—	103	173

Drakewalls Mine, at Liskeard, on Tuesday last. This company now send their tin to market by the Liskeard and Looe Canal at much less than one half the former cost of conveying it to the place of shipment by land' (RCG 6.10.27).

[1] In a building rented from Harvey & Co.

[2] Typical figures, for the second part of the Christmas quarter of 1834, were—Morwellham 104, Calstock 102, St. Austell 214, Truro 1333, Helston 80, Hayle 1187, Penzance 1599 blocks.

[3] RCG 5.7.34. 'Much interest was excited at St. Austell on Friday and Saturday last in consequence of the coinage of tin taking place there for the first time. No hall having been built as yet, the coinage took place in the cow market.' At the following coinage there must have been more amusement than interest, for it was recorded that no tin whatever was brought 'although the officers were in attendance there' (RCG 26.9.34).

[4] This should not be confused with the blowing-house which had earlier been located at Charlestown. The smelting-house was then newly erected on a site to the north-west of the harbour whereas the blowing-house (some of the buildings of which still survive) was nearer St. Austell, at The Grove close to Mount Charles.

[5] Bolitho Records, DD RG 1/163, C.RO, Truro

By the end of that same year, after a venture into the trade lasting a mere eighteen months, smelting at Charlestown was stopped, a total of 1428 blocks having been produced and coined.[1]

In 1835 another outside group attempted to enter Cornish tin smelting, in the guise of the Cornwall Royal Tin Smelting Co. The prospectus of this company, with a projected capital of £100,000, was published on New Year's Day 1836[2] and in this it was stated that one Cornish smelting company had made fifty per cent profit per annum for the last twenty years. 'The tin smelting trade,' it continued, 'is one of the last monopolies remaining in this country to be opened to the public . . . from 15 to 20 per cent per annum is a profit which may be relied upon after allowing for competition.' Already, a week previously, a letter from the solicitors to His Majesty and to the Duchy of Cornwall had been published stating that despite the connotation 'Royal' in this smelting company's title there was no connection with either the Crown or the Duchy,[3] and further acrimonious correspondence on this subject took place between these parties during ensuing weeks. Evidently the royal prefix was subsequently dropped, for in a later report the Cornwall Tin Smelting Co. was referred to, stating that 'arrangements for this establishment will be finally completed forthwith. Tenders for land on the banks of the Fal near Truro will now be received and considered.'[4] In the event, however, no new works was built by the company, with or without royal patronage.[5]

It was in any event an inauspicious time to be entering the tin trade even on the smelting side, for the mid-1830's were times of depression. Admittedly the amelioration of the coinage process had 'removed an expense of not less than £3000 per annum from the tin trade' but this was of benefit to the smelters and not the mine adventurers. The latter had gained little or nothing thereby and the number of mines that were stopped and abandoned in the period are evidence enough of the price of tin. A typical year was 1832, when there were the sales of Wheal Fortune in Breage, with 20′ water-wheel used for pumping and stamping[6]; South Dolcoath, with 36″ engine[7]; Dowgas Consolidated Tin & Copper Mines—where no less than twelve water-wheels had been in use, one for pumping and

[1] According to one report (MJ 27.6.40) 'we believe Mr. Taylor and friends erected some tin smelting works . . . where some £60000 or £70000 were expended; but which project was not attended with success. It was perhaps found that 'Unity alone is strength' and that without the aid and support of the respective mines, no stand could be made against the united forces of the Williams, the Bolithos and others. . . .'

[2] WB 1.1.36. [3] WB 25.12.35. [4] WB 6.5.36.

[5] At this time there was a tendency to add the prefix 'Royal' to the names of various mines, on the basis that they were on Duchy land, on grants held directly from the Crown. Royal Polberro Consols in St. Agnes was a case in point, bearing this name certainly from 1836 onwards; also the Wendron Royal Mining Co., working the Wheal Whidden, Ruby and Garlidna setts north of Helston. This company was for sale in 1837 (RCG 17.11.37) with a 45″ pumping engine on the former and a 60″ on the latter. Wheal Whidden, famous for its wind-machine—a form of windmill adapted for pumping mine water—was advertised for sale WB 16.2.38.

[6] WB 20.1.32. This had only been a re-working of part of Wheal Fortune, a more important period of activity there having ended in 1828 when a 45″ pumping engine and six horse-whims were advertised for sale. (WB 15.8.28).

[7] WB 2.3.32. The mine, and engine, were also advertised on several subsequent occasions until 1837.

eleven for working seventy heads of stamps;[1] West Wheal Reeth, with 30′ water-wheel;[2] Wheal Adelaide, also in Lelant, with 22″ pumping engine;[3] and Carnelloe Mine, on the cliffs at Zennor, with a hydraulic pressure engine.[4]

A list exists of the chief tin mines active in 1836 and the numbers they employed:[5]

	Men	Women	Children	Total
Boscaswell Downs	182	5	59	246
Charlestown Utd.	431	120	263	814
Great Work	298	46	74	418
Wheal Kitty (Lelant)	118	52	88	258
Wheal Mary (Lelant)	118	14	31	163
Wheal Owles	110	—	21	131
Polgooth	141	10	51	202
Polberro Consols	362	34	82	478
Wheal Reeth	147	18	66	231
Carnon	144	2	66	212
St. Ives Consols	237	16	60	313
Wheal Vor	592	327	255	1174
Wheal Grey (Breage)				130
St. Austell Hills	72	13	24	109
Cwll. Great Utd.[6]	155	20	22	197

Amongst these major mines the predominance of Wheal Vor is readily apparent. By April 1831 a profit of £75750 had been cleared and in that year new setts were added to extend the company's ground.[7] The legal action still continued over the shares formerly held by the Gundrys, 7/100ths of these being sold by the assignees in November 1831 at a price of £565 per 100th share.[8] In 1834 no less than sixteen engines were at work on the mine as a whole, comprising six pumping engines (two 80″ Trelawny's and Borlase's, 53″ Woolf's, 45″ Carleen, 24″, and 48″ Old Engine), six stamping engines (two 24″, two 16″, 27″ and 14″) and four steam whims (two 18″ and two 16″). The amount of coal consumed by these engines was enormous, and the cost of its carriage, all brought by horse and cart from Hayle, was £1782 9s. 5d. in the twelve months of 1835.[9] Although still working on this gigantic scale, the main lode was falling off in value by 1834 but two years later it was reported that another great course of tin had been cut in the bottom level, restoring the mine to its former state of unparalleled richness.[10] Smelting on the mine had then reportedly been discontinued for some time but on account of the new discovery and the fall in the prices offered for black tin by the smelters, the Wheal Vor furnaces were brought into use again later in the year.[11]

1 WB 27.4.32. 2 WB 8.6.32. 3 WB 2.11.32. 4 WB 28.12.32.

5 H. T. De la Beche, *Report on the Geology of Cornwall, Devon and West Somerset*, 1839.

6 A London association working various mining setts in Cornwall for both copper and tin, the chief area for the latter being Caradon.

7 WB 9.4.31. Wheal Vor Consolidated thereafter comprised Wheal Vor, Wheal Vreah, Carleen, Wheal Zozen, Polladrass Downs and Penhale.

8 WB 25.11.31. 9 WB 20.5.36. 10 WB 6.5.36.

11 WB 4.11.36. According to the quarterly coinage returns, there was no intermission.

Amongst the other major mines, Great Work, which was 180 fathoms deep by this date, had given dividends of £8000 from 1834–36.[1] The company was not the same that had re-set the mine to work in 1810 but a new one in which the adventurers were 'with but few exceptions the same as in Wheal Vor,' and who had taken the lease in 1825.[2] It was in fact then an amalgamation of the old Great Work itself and the adjoining Wheal Breage, the whole being known as Great Work Consols and having two 60″ engines for pumping plus a steam whim and a set of steam stamps. Balleswidden, only 72 fathoms deep by 1836, even after many years of working, was very prosperous at this time, with three engines installed. 'The stamps which at present has 35 heads constantly at work is not sufficient for stamping one half of the tin stuff raised therein,' the main lode being described as 'more than ten feet big, exceedingly good and still improving.'[3] Wheal Reeth had five engines at work, including a 36″ for pumping, and according to a report in 1836 was stated still to have its own smelting-house on the sett.[4] James Halse of St. Ives owned half the mine—acquired from the estate of the ill-fated Richard Oxnam when the latter had failed—and profits in the previous fifteen years were said to have been not less than £60000. Halse had fared equally well with his other main venture, St. Ives Consols, which it was said had been started in 1818 primarily for electioneering purposes and 'commenced with but little hope of its becoming a valuable one.'[5] In 1822 a small 20″ engine was put in, pumping only from 51 fathoms depth, and something like 300–400 tons of black tin a year were raised during the subsequent decade or so, the mine becoming particularly renowned for its Great Carbona, a massive body of ore which was discovered in the early 1830's. By 1838 this 'had been worked for at least 10 fathoms in length, breadth and height. The scattered lights, the great number of miners in their soiled and torn working dress, the pillars and beams of wood which support the roof and walls, the rock lining this vast cavern, all dimly and at intervals discerned by flickering and uncertain gleams, produce a very striking and uncommon effect.'[6]

Further west Boscaswell was also a mine in which Halse was the prime adventurer.[7] Some details of its early history can be gleaned from an advertisement of 16/24 shares in it in 1822: 'a deep adit which was begun upwards of eighty years since has been nearly driven into this mine' enabling the deeper parts to be re-worked with much less expense than

[1] WB 2.11.38. [2] WB 16.3.27. [3] WB 12.8.36.

[4] WB 25.2.36. According to coinage records this was not in use, although it is possible that Wheal Reeth tin was smelted on the mine and coined under the name of one of the smelters who also had the marketing of the finished metal.

[5] R. Hunt, *British Mining*, 1887. This was a reference to Halse's aspiration to become the Member of Parliament for St. Ives and his endeavour to gain certain votes by increasing the number of house-holding miners in his employ. It is true that he built Halsetown to this end but the whole story is one that, like many others, may well have gathered spurious weight over the years.

Although given mention by various other writers, a full history of St. Ives Consols has yet to be written.

[6] WB 11.5.38. [7] MJ 17.1.52.

formerly.[1] A 28″ engine had been erected in 1827 and by 1836 the workings had been carried down to 166 fathoms below adit, making it then the deepest of all the St. Just mines.[2] In this same area, Wheal Cock—later to become part of Botallack—was sold up in August 1836, being then drained by a 10″ pressure engine working with a 40 fathom head of water,[3] whilst the nearby 'Wheal Caul & Boswidden' was also for sale in 1837, a mine notable for the size of its water-wheel, which at 65′ diameter holds pride of place as the largest on record ever to have worked on a Cornish mine.[4] Levant, like Botallack, was primarily a copper mine, although its two old constituents, Zawnbrinney and Boscriggan, doubtless produced quantities of tin in previous centuries before the red metal came to prominence. Re-started in 1820, it was said to have given £70,000 by 1836, its chief adventurers being Daubuz and his fellow smelter, Batten at Trereife.[5] At one period £4360 was paid in dividends on two months' working, to this date the most ever yielded by any St. Just mine and this on an outlay of only £400.[6]

Statistics of output from individual mines during this period are very imperfect and can be arrived at only from the published figures of the tin ticketings held at various centres in west Cornwall at weekly and fortnightly intervals. The following outputs are, however, recorded for 1837:[7] Charlestown United 583 tons, St. Ives Consols 397 tons, Polgooth 252 tons, Boscaswell Downs 239 tons, Great Work 225 tons, Wheal Reeth 192 tons, Wheal Budnick (Perranporth) 169 tons, Wheal Kitty 163 tons, Wheal Mary 161 tons, Polberro 149 tons, Carnon 122 tons, Tincroft 105 tons. To these had to be added the output of Wheal Vor, where the adventurers coined 2638 blocks of tin in the year, representing a mine output of about 800 tons of black tin.

Features of the tin trade which dwindled and almost disappeared at this time were tin bills, bounds, and the blowing-houses. The first, briefly described, were warrants given by the smelter to the owner of black tin, whereby the former undertook to deliver an agreed quantity of 'white' tin for it at the next coinage. The great majority of tinners being unable to wait for their money for this quite lengthy period usually immediately negotiated the tin bill at a discount to the smelter. The latter thus gained twice by his transaction, being smelter and money-lender—against unquestionable security—in one. In the early 1830's, however, the smelters discontinued this practice and began instead to purchase ore

[1] WB 5.7.22.

[2] At this date, the mines in the locality were comparatively shallow: Wheal Edward 65, Wheal Owles 54, Spearn Moor 45, East Levant (formerly The Stennack) 30, Wheal Maitland 33 (fathoms below adit).

[3] WB 5.8.36.

[4] RCG 6.10.37. By comparison with this wheel, which was only 7′ smaller in diameter than the now world-famous one at Laxey Mine in the Isle of Man, there were several 60′ ones in Cornwall and numerous others of about 50′ diameter.

[5] WB 29.1.36. John Batten was also principal adventurer in Wheal Owles and was known locally at this time as the 'King of St. Just.' (RCG 2.1.68).

[6] RCG 21.3.67.

[7] H. T. De la Beche, *op. cit.*

outright at a certain price per cwt., thus obviating the tin bill entirely. Little if any tin was now being smelted under the old system, i.e. for the miner who owned it, as the smelters had developed into being metal merchants as well—a trade which more than made up for the loss in the discounting of tin bills.

The second item to pass into limbo at this time, albeit more gradually, was the old practice of bounding—a valuable privilege under Stannary law of the early tinners, whereby they laid legal claim to areas of unenclosed or common land where they thereafter held the rights to search and dig for tin, paying to the owner of the soil toll-tin or dues varying from one-tenth to one-twentieth of the tin yielded. The bounds themselves were marked out within limits defined by the placing of poles and furze bushes, or more often by piles of stones or turf at each corner. Tin bounds were small in extent and quite innumerable, existing in their hundreds in many tin producing parishes. In 1817, for example, shares in no less than 237 separate tin bounds were advertised on the occasion of the death of their owner, 112 of these being in St. Agnes parish alone. Every one had its own distinctive name, a collection as diverse and whimsical as could be imagined: Prevention, Space Bean, Pennall's Spar, Great Dribble, Wheal Tesgentle, East (and West) Wheal Saturday, Wheal Touch-Pipe, Pitslewren White Work (part of Chacewater Mine), Black Nag and Little Drink-if-you-can, (both part of Baldhu sett, south of Chacewater), West St. Gracious, Wheal Rag, Two Pair Goodlucks, Clear Diamond—to name but a few.[1] A single big mine might include a score or more of separate bounds, as witness Polgooth in 1824 where 23 were named whose owners had then to come forward and prove their claim.[2] These, however, were not mineral rights as we know them today and could be extinguished or might lapse from a variety of causes—the most frequent being the failure to work them or renew them each year. By the 1830's the rights of bounders had become suspect—partly because of the constant whittling down of common land by enclosure, legal and otherwise, and partly because they could seldom afford the law costs to combat encroachment. It was a system becoming increasingly ill suited to the changing times, as was instanced in one legal case when it was pointed out that Helston parish cemetery actually lay within bounds and that the graves there could be turned over for tin with impunity if the bounders chose to do so. By 1849 it was a mild under-statement to refer to the 'supposed unsettled state of the law as to tin bounds'[3] and shortly thereafter, their demise hastened by the disappearance of small scale mining and streaming

[1] WB 14.3.17.

[2] WB 29.10.24 comprising Casualty, Mary-Gold, Wind Sheet, Blowing-House, West Stitch, Bunnys, Vanvean Step, North Yeo, and fifteen others. Adjoining, in the same parish, were the following bounds—Small Hopes, Good Friday, Shillings-go-by, Great Groan, Ployden's Misfortune, Welcome-in, Justification, Rusty Hammer, New Bounders Folly, Fear Nothing, Caffler's, and a score more. WB 17.3.26.

[3] H. English, *Mining Almanac*, 1849.

in most areas, they were either extinct or considered worthless.[1] This cause, and the rise of Banca alluvial tin which was coming into Europe in increasing quantities, was also responsible for the disappearance of much of the Cornish trade in grain tin in the 1830's, although on a small scale the blowing-houses at St. Austell were to linger on until the 1860's (see page 129). From about 1815 the number of blocks of grain as against common tin had fallen steadily (for example 2377 of grain, 18405 of common in 1816: 1440 of grain, 24247 of common in 1826) and this could largely be ascribed to the advance that had been made in tin-plating methods in South Wales. It had become standard practice to use only common tin for the first and heaviest dipping of the plates—a process known as tinning—and the more expensive grain tin only for the second and final coating, known in the trade as 'washing' and this greatly reduced the consumption of the latter grade, or of its substitute, 'refined'. Improved tin-plating techniques generally were also rendering thinner coats of tin possible.

After a brief-lived rise in price in the previous year, 1837 was a particularly unprofitable one for Cornish tin mining as a whole, and another attempt was made to secure the abolition of the entire coinage system. This was precipitated by the death of George III, for it had become realised in Cornwall that the only time when there was a hope of the dues being abolished altogether was at the beginning of a new reign, when a new Civil List was being prepared and when an amendment of the Duchy officials—and of their stipends—could be constitutionally arranged. This, it was generally believed, was the prime reason why the abolition had been refused in 1833.[2] In November 1837 there was a meeting in Truro to frame the requisite petition to Parliament and the new Queen[3] but two months later it appeared that the success of this was perhaps not as assured as had been presumed. For Cornwall the matter was made more imperative by the prospect of a long reign before Her Majesty, during which any amendment of the Civil List was unlikely. Moreover a similar petition to the House of Commons by the tin-plate manufacturers of South Wales had also been made with a view to the abolition of the import duty on foreign tin—hitherto at the prohibitive rate of £50 per ton. To assist the Cornish members in putting their case, a deputation of smelters and

[1] One of the last big sets of tin bounds advertised for sale was in the parish of Perranzabuloe in 1853 when about one hundred were named, including Wheal Dobrean in the Sands, Early-up and Peep of Day, Wheal Eagles, Goodmorrow Neighbour in Perran Coombe, Smutty Wheal Gordan, Wheal Popery in the Sands, Quick Come-by, Quick Return, Wheal Slip or Catch Snibs. (WB 8.7.53).

[2] Since this date there appear to have been obstructive tactics at times on the part of the smelters in order to show up the shortcomings of the amended coinage system: at the Michaelmas coinage at Penzance in 1835 it was reported that 'owing to a disposition on the part of some of the smelters to keep as large a portion of their tin as possible until the last day of the coinage, the streets near the hall presented a scene of most disgraceful confusion, which obstructed the thorofare, and certainly required the interference of the authorities.' This moreover, was despite 'the scarcity of water at the stamping mills' which had rendered the amount to be coined 'smaller than for many years past' (RCG 30.10.35).

[3] WB 19.1.38.

adventurers agreed to go to London and a levy of 1d. per cwt. on all black tin smelted was agreed upon to meet their expenses. To reinforce the case with facts and figures, a return of the financial state of the tin mines in Cornwall in 1837 was prepared for the information of the government, setting out the loss then being made—figures which certainly showed the unprofitable balance-sheet of the industry:[1]

Loss on 58 mines in 1837		£111517.0.0
Less estimated increase in value and property in mines	£31000	
„ profit on 10 mines	£20358	51358.0.0
	Net loss	£60159.0.0

Apart from these statistics, which told their own tale, it was pointed out that the coinage dues were the sole surviving British tax in which the proceeds did not go to the Exchequer. It was said they constituted a direct tax of 5 per cent, or 15 per cent when carriage to a coinage town, delay, and other expenses were reckoned in, an impost that was 'oppressive and vexatious, and entirely contrary to all sound modern principles of free trade.'[2]

This deputation, and its lobbying, had the desired effect, and in May 1838 it became known, albeit unofficially, that the Duchy dues would probably be abolished but that, in conjunction with it, the duty on foreign tin would be reduced to £15 per ton.[3] In August the question was formally debated in Parliament and approved, the Tin Duties Abolition Act being passed on the 16th of that month, ending the centuries old system of coinage and coinage dues entirely. Compensation was paid by the Exchequer from the Consolidated Fund to the Duchy of Cornwall, the latter's annual revenue from this source having been an important one in the past few decades. In addition, it was necessary for compensation for loss of office to be paid to the various Duchy officials and employees concerned.[4]

The only drawback in Cornish eyes was the fact that the duty on imports of foreign tin had also been reduced but even this was as nothing compared to their new found freedom from the bondage of 'coining.' This system had become, long before 1838, an anachronistic survival of mediaeval times and one which was a positive burden to a growing industry facing the threat of foreign competition. Its ending in 1838—the

[1] MJ 25.10.51. [2] WB 19.1.38. [3] WB 4.5.38, 18.5.38.

[4] Compensations granted on the repeal of the Coinage Duties on Tin, Parliamentary Accounts and Papers, xxx 1839, State Papers Room, B. M., London. The emoluments of the Duchy officials had been as follows—The Comptroller, Assay Master and Deputy Steward £200 p.a. plus £85 travelling expenses; one Poiser £36 p.a.; two Cutters £24 each p.a.; one Hammerman £24 p.a. and one assistant ditto £8 p.a.; eight Porters £24 each p.a. In the period from 1828 to 1837 the average annual expenses and charges of the coinages had been £861 14s. 8d. whilst the gross amount of the income derived by the Duchy in this same time had averaged £16430 15s. 10d. per year.

Contrary to much that had been said during the agitation to abolish the coinage system, the greater part of this revenue was thus not wasted on administration or sinecures.

final coinage taking place in September and the abolition coming into effect on 10 October[1]—may be said to mark the belated entry of Cornish tin mining into modern times.

COINAGE ABOLISHED: 1839–1849

'In his conflict with this competitor the Cornish miner asks not favour, but a fair field for his exertions ... and that his industry should come unshackled to the fight'
West Briton, 15 March 1833

THE ABOLITION of coinage in 1838 inevitably had more immediate effect upon the smelting rather than the mining side of the industry and this was reflected in a period during which there was a considerable upheaval in the *status quo* of the Cornish tin smelters. Already, in 1837, an important new company had appeared as buyers of black tin, comprising several of the partners in the old Harvey, Davey, Williams & Co. that had joined forces with Daubuz twelve years before. The latter had terminated this agreement to smelt and coin jointly and instead his partners formed a separate company known for a brief time as Williams & Harvey, but shortly thereafter as Williams Harvey & Co.[2] In February of that year they purchased Trethellan smelting-house at Truro, with all its furnaces, machinery, stores, etc., and the remaining part of its 99-year lease, from Grenfell and Boase for £2250. At the same time they acquired land at Hayle from Harvey & Co.[3] for the building of another new smelting works at Mellanear. This was on a site then known as Rope Walk Moor and the company was granted the right to the use of Harvey's quay at special rates for the landing of coal for the works as well as for the shipment of tin. The first purchase of black tin for Mellanear was recorded in April 1837 and thereafter, with two big works in operation, Williams Harvey became one of the leading smelters in the county and were to retain a well-nigh dominating position into the twentieth century. At first in fact they had three works, until 1839 when Old Blowing House at St. Austell was disposed of.[4]

After their establishment in 1837, Williams Harvey's rivals in terms of size were the Bolithos who had held an interest in no less than five works during the preceding decade—Chyandour, Ludgvan, Angarrack,[5] Calenick and New Blowing House at St. Austell. Of these the Ludgvan premises

[1] WB 7.9.38. Needless to say, the bare minimum of tin was then presented by the smelters.

[2] Harvey Records, DD 852/11, CRO, Truro. The partners were John Williams Junior, Michael Williams and William Williams, together with Richard Harvey (son of Collan Harvey) of St. Day.

[3] This company had no financial connection with Williams Harvey & Co. at this date. Fifty years later, in 1888, their principal partner (Frank Harvey) acquired a considerable interest in it.

[4] This works was then described as fitted up 'for the blowing of Grain Tin Ores and the smelting of other Tin Ores.' (WB 24.5.39, 12.7.39). No new occupier was found and in 1845 the premises were let to John F. Hodge to become part of the adjoining St. Austell Foundry. Coode-French Records, DDCF/5, CRO, Truro.

[5] In 1838 Angarrack smelting-house was advertised for sale (RCG 13.4.38) being then 'in the occupation of Joseph Carne esq. [who died in 1858] as tenant from year to year.'

had been advertised for sale in 1829 and again in 1835,[1] and the last-named in 1833.[2] The withdrawal of Daubuz' erstwhile partners of course meant a considerable reduction in the amount of tin coined under this name and from June 1837 to June 1838 the following number of blocks were coined by the various smelters as a whole:

Bolitho	6194 blocks	Grenfell	2424 blocks
Williams Harvey	5580 blocks	Carne	1255 blocks
Daubuz	4938 blocks	Devon S. Co.	1149 blocks
Batten	3967 blocks	Michell	634 blocks
Wheal Vor	2669 blocks	Cornish (Wm.)	511 blocks

The four works run by Daubuz—Treloweth, Truro, Carvedras and Higher Blowing House (St. Austell)—were more than were needed and in January 1839 it was decided to sell the Truro one which was no longer required as a subsidiary to Carvedras.[3] Shortly thereafter, two members of the family died within a few months of each other[4] and by the 1840's the name of Daubuz had moved from the front to the third rank in the tin smelting hierarchy of Cornwall.

These were merely new dispositions amongst Cornishmen already engaged in the trade. What was about to occur was not only an attempt by outsiders to enter it but, a year or so later, to oust them entirely. The first new company was the Bissoe Bridge Smelting Co. which made its initial purchases of black tin in September 1838.[5] From an advertisement at this time it is evident that one C. K. Vigurs was the principal shareholder.[6] He was the son of John Vigurs of Penzance, who was concerned in tin-plate works, collieries and iron-works in South Wales, as well as the

[1] RCG 12.12.29, WB 12.11.35. For sale, with residue of 99 year lease, Ludgvan smelting-house '. . . consisting of a Smelting House 75 feet long, 30 feet wide and 25 feet high; a Blowing House 42 feet long, 21 feet wide, and 20 feet high; Charcoal House 42 feet long, 21 feet wide, and 20 feet high; and a Linhay 42 feet long and 12 feet wide, with a large Yard attached thereto . . . these premises until lately were used for Smelting Tin but are at present unoccupied . . . also a dwelling house, four cottages, stamping mill, etc . . . lately in the occupation of Bolitho & Co. or their under-tenants . . .'

[2] RCG 2.3.33.

[3] WB 18.1.39. Also advertised for sale was Daubuz' Mansion House together with his gardens—the latter being the site of Treseder's Nurseries in Truro today, in an area still known as Daubuz' Moors.

[4] L. C. Daubuz Senior died in December 1839, aged 85, and his son, bearing the same names, in an accident in 1840, aged 39. The surviving son William died in 1854, aged 49, and the business was continued thereafter by J. C. Daubuz. L. C. Daubuz Senior had been the *eminence grise* behind Cornish tin smelting, his ability to command aided because he ' so remarkably resembled the Duke of Wellington ' (RCG 20.12.39).

[5] Bolitho Ticketings Book, Oats Collection, RIC, Truro. It is not known whether this was, as seems probable, a companion or associated venture of the Bissoe Bridge Mining Co. which was active at this same time. This latter, a London company, was sold up (with 50″ pumping engine WB 29.11.44) in 1844 and sometime thereafter the name Bissoe Smelting Co. became usual. The works was possibly the old Bissow (lead) Smelting House which was advertised for sale in 1830 (WB 31.12.30) and 1832 (WB 3.8.32 RCG 6.10.32) adapted for tin smelting. Alternatively it may have been a continuation of ' a new smelting work lately erected on Tregothnan Consols Mines [i.e. Baldhu and Tregothnan Mines, see WB 27.4.38] for smelting gossan and other ores, by a company of gentlemen from London, which has started working on a plan laid down by the late Fredk. Hall ' (WB 15.1.36). This same company, known as the Tregothnan Consols Smelting Co., had advertised the previous year to buy lead and silver ores from other mines (WB 14.8.35).

[6] WB 9.11.38.

important Cwm Avon (Port Talbot) copper smelting works.[1] In 1839 another member of the same family, William Revell Vigers,[2] who was in business as a metal merchant in London, acquired the Truro works which Daubuz had occupied. Here he was joined by C. K. Vigurs and the Bissoe Bridge works was stopped. As may well be imagined this new smelting venture was subjected to cut-throat competition from the existing ring, not only as regards the buying of black tin but in the sale of the smelted metal, and with the Vigurs' interests in tin mines limited primarily to a major share in Polberro Consols, they suffered accordingly. After some months, to escape from this web of competition, Vigurs formulated a plan to turn his works into a co-operative smelting establishment where miners could have ore smelted on contract terms. Such a plan obviously required widespread, almost total, support from the miners if it were to succeed, and a meeting of the tin mining interest was held in May 1840 at Helston. This was convened by a committee formed and headed by Vigurs, the meeting being for the express purpose of 'considering the present depressed state of the tin mines of Cornwall' and to 'protect the adventurers in tin mines and obtain a fair and reasonable price . . . and adopt some measures to destroy the ruinous system at present and for many years pursued by certain parties engaged in smelting.' Over 150 mine agents and other adventurers attended—necessitating adjournment to the more commodious Town Hall from the hotel that had been arranged—and here the plan outlined by Vigurs was enthusiastically received: that 'two houses should unite for the purpose of smelting your tin, selling it, and acting as your agents—you to have all the benefit arising from that operation, with the exception of a fair remuneration in the shape of commission and guarantee to those houses which so act for you.'[3] It was pointed out that under this system the tin trade in Cornwall could prosper uniformly and be free from the depression that dogged it. In the past few years there had been many who spoke out against what was considered the oppression of the smelters, for mining itself had received no gain from the coinage abolition, even though it was this side of the trade and not the smelters that the Duchy

[1] Early in his business career John Vigurs had been the London agent of the Associated Tinners. In company with others he established the Cwm Avon Ironworks in 1819 and a tin-plate works in 1825. In the 1830's Vigurs & Co. owned and worked the Ladock Iron Mine (and others), near Truro, to supply Cwm Avon with ore (see WB 12.7.39), whilst by 1849 their various Welsh enterprises employed about 1500 hands.

[2] There is confusion in all contemporary reports of the name Vigurs and Vigers, the two spelling variations being frequently used. So far as can be traced, all were of the same family.

[3] WB 5.6.40. Vigurs detailed how the other smelters 'made it a rule to endeavour to find out when a transaction was in progress of negociation or about to be completed. They would then go underhand and offer to the same person who was in treaty with our agents a shilling a cwt. less in order to get their orders for the supply . . . If these smelters by their combination (though they are few in number) and the ridiculous system of ticketing, had not the mines to fall back upon, to remunerate themselves for the sacrifices they may make, they would not be such fools as to throw money away.' Vigurs knew but did not name 'which house has generally been the leader in a reduction of prices' but he knew that 'other houses which have been many years in the trade and which have a list of customers that they have secured by their own industry and serving them well, are obliged to follow the leader, or consent to see their customers taken from them and their trade destroyed.'

and most of those who had agitated for abolition had intended to be the main beneficiaries. During Vigurs' long speech, there were frequent and sustained cheers at his outspoken defiance of the Cornish smelters who, after he had ' taken the liberty of entering their *sanctum sanctorum* and prying into the mysteries of the trade ' had tried, as a result, to ruin him. The upshot of this meeting, and a further one held at Redruth a week later, was that the scheme should be brought into operation at once, allowing ' subsist ' for the time being of 75s. for every cwt. of grain tin brought in and 70s. for common. By July it was reported that a large number of mines were sending their produce for smelting on the new plan to Trereife, Truro and Charlestown[1] whilst the *Mining Journal*, a particularly enthusiastic exponent of the scheme, jubilantly recorded the next month that ' four-fifths of the tin mines of the county of Cornwall, whether as adventurers, agents, or others, have determined on throwing off an incubus to which they have been subjected—to release themselves from a thralldom which has been so injurious, and for once to evince a spirit of independence.'[2]

One of the prerequisites for the success of the new plan was the existence of sufficient established works dispersed throughout the county ready to co-operate, plus a considerable amount of capital in their hands to enable advances to be made to the mines until such time as metal sales commenced. To this end, Batten & Son at Trereife were joined by Trevenen James, a tin merchant and broker in London,[3] and others, whilst the Vigurs family were joined at Truro by another wealthy merchant and shareholder in Polberro, one Fermin de Tastet. In addition there was at first the works formerly owned by John Taylor's company at Charlestown as well, for Richard Taylor (son of John Taylor and his manager in Cornwall) had supported the miners new plan at the outset. Unfortunately this decision to make the disused Charlestown furnaces available to the new company for smelting was later annulled—a serious setback to the plan, for Taylor also controlled various important tin mines including Polgooth and the extremely productive Charlestown United. For reasons never made clear, before many weeks had passed, the ores from these mines were being sold openly to Williams Harvey, Bolitho, Daubuz and Grenfell. Nevertheless in November it was stated that ' the results of the new measure are highly gratifying . . . but for the defection of the Charlestown Adventurers the new measure would have been more successful in its results as to the average price obtained for tin.'[4]

The apparent success of the scheme to this juncture then brought forth an even more ambitious plan—one perhaps that had been formulating as long if not longer than the original—for the smelting of copper on a

1 WB 17.7.40. 2 MJ 8.8.40.

3 James, originally from Goldsithney, had earlier been concerned in banking at Penzance with the smelter Boase.

4 WB 13.11.40. A rise of about £5 per ton had taken place in the price of black tin during the preceding six months due to increased competition for supplies.

similar basis, using the Cwm Avon smelting works of Vigurs & Co. This venture, variously referred to as the Free Miners Copper Co. or the Miners Smelting Co., was a sister project to the tin smelting venture already under way and was sponsored by much the same persons. It was, however, of far greater importance for it carried the war out of the limited field of the tin trade into the vastly more far-reaching and lucrative one of copper. Against it there was the inevitable opposition of the smelting giants such as Williams Foster, Vivian & Sons and Pascoe Grenfell & Sons who between them dominated the industry. Despite this formidable opposition it was decided by Vigurs and his friends to proceed—their hand strengthened by their controlling interests in Carn Brea Mines which were currently the most productive for copper in the county—and early in 1841 the old moribund joint stock company known as The Governor and Company of Copper Miners in England,[1] that had originally been incorporated by royal charter in 1691, was acquired by the Miners Co. and in March smelting of copper ore on contract was commenced at Cwm Avon. This company also henceforward assumed control of the tin smelting scheme already in operation, meeting as it did so even stiffer opposition from the existing independent smelters' 'ring.' A tolerably open and above board system of tin ticketings had come into operation in the years after 1830, made use of by a considerable number of mines,[2] and with published summaries in the *West Briton*, but after 1840 these disappeared when once black tin started going to one side or other of the two warring factions. Neither side wished it to be known from whom and at what price black tin was being bought, secrecy being particularly vital to the smelters' ring for it enabled them to cripple the Miners Co. by buying at special rates entirely by private contract.

So far things had gone well enough for the new company but the acid test of its ability to sell the finished metal on the open market in the face of such determined competition was still to come. In May 1841 there was a secret meeting at Camborne between Trevenen James for the Copper Miners Co. on the one hand and Bolitho, Williams and Daubuz on the other, at which he endeavoured to sell all the smelted tin *en bloc* to the latter at a special price.[3] At this meeting, however, the two sides were unable to come to terms but whether it occurred as a prelude to anticipated marketing difficulties, or as a consequence of them, is not known. Certainly by the autumn of 1841 the Miners Co. had accumulated an enormous quantity of metal[4] and in an endeavour to clear some of this, one thousand

[1] Commonly known as the Copper Miners Co. Amongst the Cornish committee were John Batten, Louis Vigurs, John Vigurs and Trevenen James.

[2] These ticketings were similar to those held for copper, although collaboration between the purchasing smelters in comparing assays, and in fixing the standard at which to buy, was known to take place and the element of competition in the true sense of the word was conspicuous by its absence. The meetings for this purpose were held weekly or fortnightly at various places in west Cornwall, usually at Redruth or Treloweth, but occasionally at Truro, Marazion and Halsetown.

[3] Bolitho Records, DDX 104, 31M/BPR/11, CRO, Truro.

[4] Evidence of this came at Balleswidden where, in August, the adventurers 'resolved to offer samples to other smelters in view of the refusal of the Governor and Company of Copper Miners in

tons 'of British Tin, the property of British Mining Adventurers, and lying at the Miners' Smelting Works at Trereife House, near Penzance, and at Truro House, in Truro' was advertised in October.[1] Yet few tin merchants or consumers would buy, afraid of their business connections and of engendering hostility on one side or another, as well as being wary of a market that was, from such cut-throat competition, in an unpredictably fluid state. In January 1842 another attempt was made to sell a similar quantity by public auction in London. At this only 130 tons were disposed of, and this at a reduced price in a private sale after the auction had ended with nothing sold.[2] This was, with a vengeance, 'civil war amongst the metallic potentates of England.'[3] The Cornish tin and the Welsh copper smelters were out to crush once and for all time the idea of any national miners' smelting combines, and at a second adjourned auction held in February not a single bid was made for the tin on offer. The only person to speak was the auctioneer: 'we never witnessed so determined and dogged a silence as that observed throughout the proceedings.'[4] Shortly afterwards it was reported that 'Messrs Enthoven & Co. have bought the very heavy stock of tin said to be upwards of 1000 tons in the hands of the Miners Company. It is said, however, that this Company will endeavour still to carry out its views by taking the black tin from the miner . . .'[5] Despite this latter reassurance, a few months later, in July, W. R. Vigurs and Trevenen James withdrew from the Copper Miners Company entirely and the works at Trereife and Truro commenced buying black tin and smelting it on their own account. This could only mean the end of the Company as regards its co-operative tin smelting venture and, despite further suggestions in later decades, the idea of miners themselves combining to smelt tin on a co-operative or contract basis was never tried again, the whole business recalling in many respects the gigantic and equally abortive attempt of the similar Cornish Metal Company (inspired by Matthew Boulton) in the 1780's. It was all too apparent that the Cornish tin trade was in determined and united hands and that breaking this ring was not easy. Anyone could smelt tin—for it required neither a great works, great skill, nor great capital—but selling it was another matter. The trade in the metal was almost entirely in Cornish hands, with brokers far from occupying the position of importance that they later assumed, whilst the smelters, by and large, were their own miners and their own merchants, and with a lucrative trade shared harmoniously amongst them competition was non-existent. Their numbers were few and many of them—like so many Cornishmen—were cousins. Into this tightly knit band no outsider could penetrate and outside it no large smelter could survive.

England to advance more than two-thirds of the value of the tin deposited with them' (Oats Collection, RIC, Truro).

[1] WB 8.10.41. The 1000 tons comprised: 20 tons Best Granulated; 90 tons Tin Plate Grain; 370 tons Refined; 370 tons Common; 50 tons Ingots, assorted sizes; 100 tons Bar, in barrels of 4 cwt.

[2] MJ 29.1.42 WB 4.2.42. [3] MJ 5.2.42. [4] WB 18.2.42. [5] RCG 11.3.42.

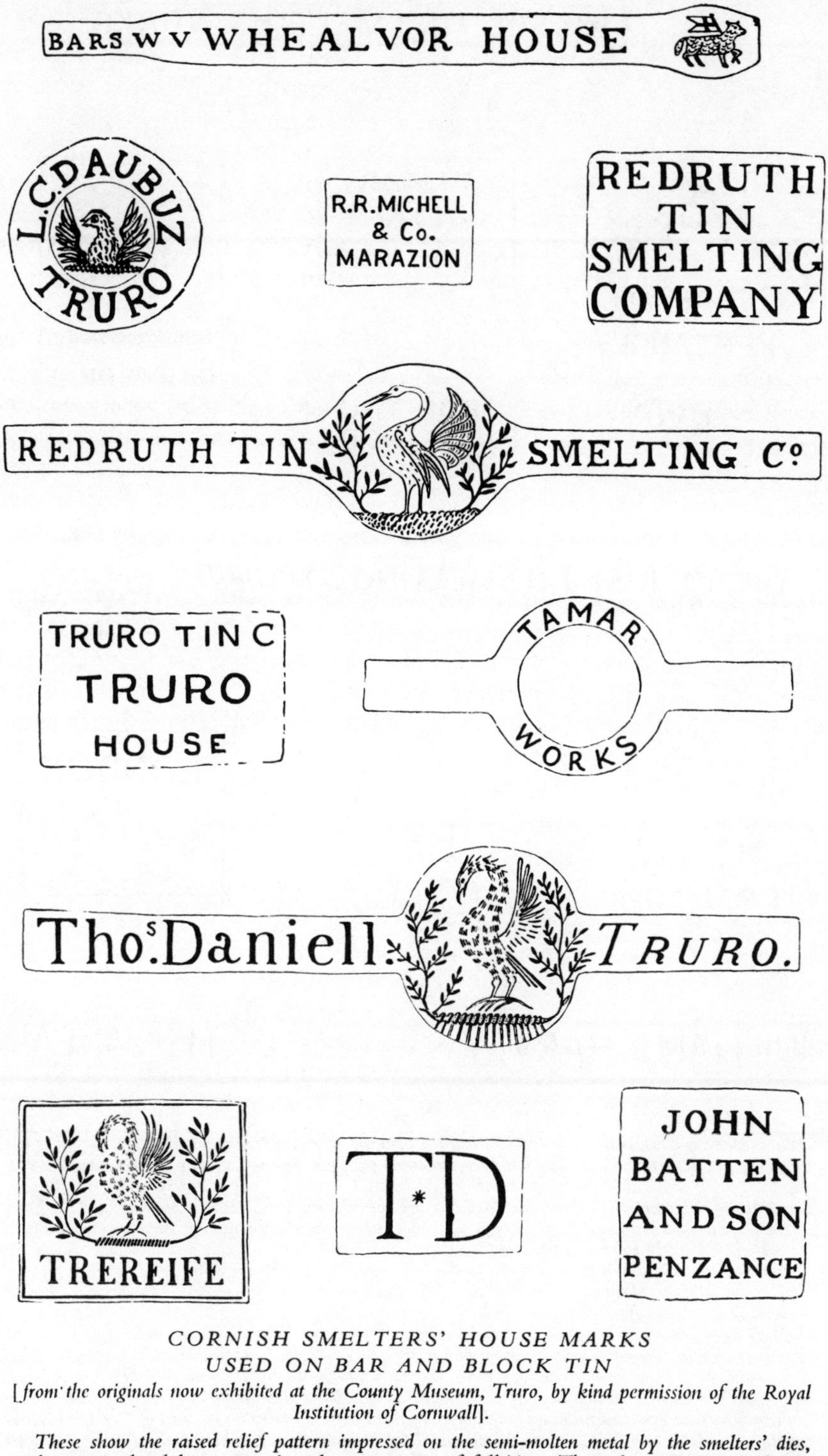

CORNISH SMELTERS' HOUSE MARKS
USED ON BAR AND BLOCK TIN

[*from the originals now exhibited at the County Museum, Truro, by kind permission of the Royal Institution of Cornwall*].

These show the raised relief pattern impressed on the semi-molten metal by the smelters' dies, and are reproduced here approximately one quarter of full size. The only one not immediately identifiable [P.S.C.] is thought to be Portreath Smelting Company. The Lamb and Flag, and the

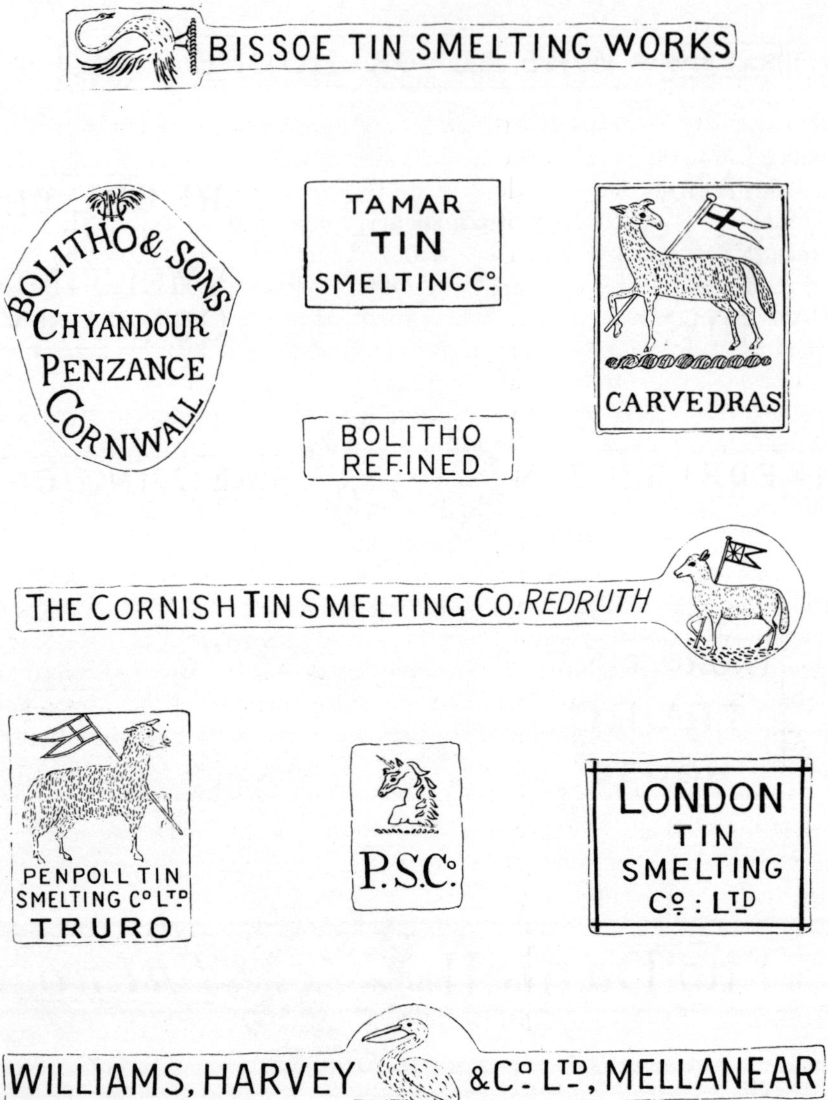

Pelican, are regarded as the standard Cornish tin smelting marks and although their true origin is lost in obscurity, a variety of theories has been advanced for their use in this way. One writer (W. C. Borlase, A Historical Sketch of the Tin Trade in Cornwall, 1874) *specifies that the former, symbolising innocence and purity, was used to mark tin destined for Christian countries, and the latter for Mohammedan, with both originating in the badges of the princely families of Italian merchants who controlled the early European tin trade. Certainly the Lamb and Flag was a device widely used by merchants in mediaeval times, having its origin in the wool trade which then formed England's greatest industry. On blocks of tin exported, the lamb and its banner of St. George bespoke England, a national symbol in the days before the Union Jack.*

Both signs, however, were not universally used in Cornwall. Other devices included a phoenix and a variety of other heraldic birds, a stag, a plume of feathers, fifteen balls (emblem of Cornwall today), a unicorn, etc. Two or more designs were also sometimes used by the same smelter concurrently. It should be remembered that they were not used as individual trade marks as we know these today in registered and protected form, although in 1908 there was an outcry on the part of Cornish smelters against the deliberate piracy of the Lamb and Flag insignia by a Chilean company in 1908. (*See* M.J. 7.3.08). *Even as late as 1936 'Lamb and Flag' was the brand name advertised by a Dutch tin smelter at Arnhem.* (*See MW* 5.9.36).

Little is known in detail of the progress of the Copper Miners Co. during the ensuing years when they were engaged in combat with the other smelters in Wales but the company quickly found itself in growing financial difficulties. By 1846 it was necessary to raise considerable sums on loan, with the issue of debentures to the Bank of England, and two years later complete control passed for some time to this body.[1] The connection with Cornish tin smelting was finally severed about this time, and the Trereife, Bissoe, Truro (and possibly Charlestown) works resumed entirely independent careers. The first named was valued in 1846 at a time when its two chief partners, Frederick Ricketts and Trevenen James, became bankrupt[2] and so far as is known it then reverted to the Batten family of Penzance, who had owned it before the new scheme had been started. The Truro smelting works was shut down in 1846 by Vigurs & Co.[3] but the fate of the Charlestown and Bissoe works at this time is not known. The latter works may have remained in the possession of C. K. Vigurs, but by December 1848 it had passed entirely into new hands.

During this time of turmoil in smelting, Cornish tin mining had fared little better in financial terms than it had done before 1838. The year 1839 was typical of the depression that still showed no signs of lifting. In March there was the sale of Great Wheal Prosper, at Roche, with 36″ pumping engine and two horse-whims,[4] and the announcement of the stopping of Polgine, Newton Moor and Condurrow, as well as the once famous Relistian at Gwinear.[5] In April Great Wheal Venture was sold up at St. Dennis,[6] and Wheal Uny at Redruth—the latter having a 36″ pumping engine installed and being described as 'in a good course of working to the 44 fathom level under adit.'[7] In August West Poldice in Lelant was for sale, with a 56″ engine and 24″ stamps working 36 heads[8] and also the little North Godolphin Mine, worked by means of a water-pressure engine ($11\frac{1}{4}$″ diameter) and two horse-whims.[9] The following month, Wheal Spearne at St. Just was sold up, with 24″ pumping engine,[10] together with Gwallon Tin & Copper Mine at St. Hilary, with two (63″ and 36″) pumping engines, 17″ whim and four horse-whims,[11] whilst during the remainder of the year, Parbola Mine (Gwinear), South Polgooth, East Ding Dong and part of Wheal Trumpet were 'knacked.'[12]

[1] After reconstruction in or about 1854, the Company had a more successful career.

[2] Ricketts and James, together with various other merchants (including Edward Turner the son of one of the original partners in the Truro Smelting Works, his brother-in-law James Oliver Mason of London, and H. J. Enthoven), were also concerned in banking (Ricketts, Enthoven & Co.—later taken over by Bolithos), in the West Cornwall Railway, in Carn Brea and other mines, as well as the ill-fated Copper Miners Co.

[3] MJ 3.4.47. [4] WB 22.3.39. [5] WB 29.3.39.

[6] WB 26.4.39. With 50″ pumping engine, 27″ whim, two horse-whims and two 16′ water-wheels working 12 heads of stamps.

[7] WB 19.4.39. [8] WB 9.8.39. [9] WB 23.8.39.

[10] WB 20.9.39. [11] WB 27.9.39.

[12] Parbola, with 30″ engine, (WB 8.11.39): South Polgooth, with 36″ engine for pumping and stamping (24 heads) plus four horse-whims, (WB 8.11.39): East Ding Dong, with a single 19′ water-wheel, (WB 6.12.39): Wheal Trumpet, with 10″ steam whim and 17′ water-wheel, (possibly only spare materials) (WB 13.12.39).

In the following year there was the sale of Tregavara Downs in January, with 36″ pumping engine, 18′ water-wheel working nine heads and four horse-whims;[1] Wheal Gilbert at St. Erth, in February, with 60″ pumping engine and three water-wheels working 24 heads;[2] Lelant Consolidated, East Levant, and the important Godolphin Mines, in March[3]—the latter a most important mine but worked during this time primarily for copper, and having two 80″ engines installed for pumping; Balnoon, near St. Ives, in May, with 30″ pumping and 24″ stamps engines[4]; Dartmoor United in July, with two 24′ diameter water-wheels for pumping and working stamps[5]; Morvah & Zennor Mines in August, with 40″ pumping engine, 60′ diameter water-wheel and four horse-whims[6]; North Trevenen with 18″ pumping engine and two horse-whims; Wheal Richmond with 30″ pumping engine in Wendron parish in October[7]; and in November, Great Ramoth, near Perranporth.[8] Succeeding years followed the same pattern, with an increasing number of mines of importance closed as the price of tin crept ever lower. Seven major mines were stopped in 1842[9] including Polgooth, Wheal Kitty, Rosewall Hill and Polbreen. Also sold up during 1842, although perhaps of less importance as regards production, was the little Wherry Mine, celebrated for its unique position on the tidal rocks two hundred yards off-shore at Penzance. This had previously been worked in the 1790's[10] and in 1835 another attempt was made to re-open it by a London company.[11] Eight hundred shares were

[1] WB 31.1.40. [2] WB 14.2.40. [3] WB 6.3.40, 13.3.40. [4] WB 8.5.40.
[5] WB 5.7.40. [6] WB 7.8.40. [7] WB 16.10.40. [8] WB 6.11.40.

[9] Polgooth, with 66″ pumping engine, 26″ stamps, a small water-pressure engine and twelve water stamping mills (WB 11.2.42): Wheal Kitty, St. Agnes, with 40″ pumping engine, 32″ stamps with 72 heads, and 16″ whim (MJ 28.5.42); Polbreen, with 50″ pumping engine, 40′ stamps water-wheel, and five horse-whims (MJ 11.6.42); Tincroft Consols, Towednack, with 30″ and 19″ pumping engines and four horse-whims (WB 2.9.42); Wheal Cock, St. Just, with 16″ steam whim (WB 2.9.42); Rosewall Hill, with 36″ pumping engine, 25″/48″ stamps and 20″ whim (MJ 15.10.42), Wheal Henry, Wendron, with 50″ pumping engine and three horse-whims. (WB 11.11.42).

[10] Some valuable and hitherto unpublished information on the Wherry Mine exists in the Diary of Charles Hatchett (*penes* Dr. Arthur Raistrick, Linton, Yorks), who visited it in 1796. Two marine shafts existed; the 'old,' 20 fathoms deep and 114 fathoms out from the shore, which was destroyed and the mine 'filled by the sea during the great storm in January 1796,' prior to which occurrence 80 men were employed; and the new shaft, 70 fathoms out, intended to be sunk to 24 fathoms. It was presumably the staging and headgear at this latter shaft which was carried away in the accident in 1798 when, as has so often been described, a sailing vessel collided with it during a storm.

It is interesting to speculate on the derivation of the mine's name. A wherry, in addition to being a well known East Anglian class of vessel, was also a craft of a very different type which was used in former centuries off the coasts of south-west England and Ireland. In 1860 there is a specific reference to a wherry as a type of boat then in use in Mount's Bay (RCG 10.2.60) and the original name of the mine may well have come from the boat which must have played an important part in its first finding and working.

[11] WB 24.4.35. 'A party of gentlemen from London . . . came down last week, in order to view the mine, previously to commencing the undertaking . . .' There had earlier been another attempt to re-start the Wherry in 1822, under the name of the New Wherry Mine, and shares in the venture were advertised early in January 1823 (WB 3.1.23) with 'work to be started in March.' This occurred shortly before the death of the man who had first discovered the lode, one Thomas Curtis, a working miner of Breage, who died in 1828 aged 76. (See obituary WB 14.11.28).

This re-working of the Wherry Mine had a parallel in the very similar but far less well known Wheal Mexico nearby 'about half way between the towns of Marazion and Penzance' (i.e. at Long Rock where its site is perpetuated in the name of a public house). Here a shaft with a wooden framework protecting it from the sea had been put down about fifty fathoms from the shore and in September 1834, it was reported that 'men are busily preparing for the erecting of a steam engine on the beach to drain off the water from the mine . . .' (RCG 13.9.34).

issued and a 40″ engine erected on the shore, as on the previous occasion, to work the pumps in the shaft by a long run of flat-rods. The house for this was reported to be complete and the engine ready for work in a fortnight, in May 1837,[1] whilst in March 1839 the Wherry was described as 'a mine of great promise and giving the Lords a considerable profit.'[2] The latter were evidently the chief beneficiaries, however, for the following year operations ceased and the 40″ engine was sold, some £9,600 having then been called up according to the mining share quotations published weekly in the *West Briton*.[3]

Another more important casualty of 1842 was the Carnon Mine, the seaward continuation of the old stream-works of that name, and a venture which rivalled the Wherry as an example of the perseverance and ingenuity with which Cornishmen pursued their search for tin. All the alluvial ground that could be reached within embankments having been exhausted and a profit of £50,000 having reputedly been cleared on the working, a new company (headed by the Michells of Calenick) decided to erect an engine on the shore[4] and sink a shaft within the protection of a coffer-dam more or less in the centre of the creek. Pumping in this was carried out by means of flat-rods and a series of levels was driven in the tin ground beneath the bed of the creek, the sett comprising the whole of the latter from Carclew Quay downstream to a short distance seaward from Daniell's Point.[5] This was in 1824, shortly after the old streamworks' embankment had been breached by the tides and removed in places to form navigation and tidewater channels for the newly built Redruth & Chasewater Railway's wharves at Devoran. As the mine workings were pushed slowly farther and farther downstream in this working, the virgin shelf of tin ground continued equally rich, the tin produced being of an exceptional purity, and yielding a far higher price than that from any lode. Financially this working was also successful, resulting in a profit said to have been about £28,000, but at length the levels were too far from the shaft and the line of flat-rods was too long for the engine's power. From this cause, the working, officially known as the Carnon-Stream Mine, came to an end and the machinery was sold off in 1830.[6]

[1] RCG 5.5.37. [2] WB 1.3.39.

[3] This was not the end of the Wherry Mine. In 1859 there must have been some prospecting or mining activity there as 19s. 7d. dues were paid to the Duchy of Cornwall (owners of the mineral rights under the sea) in that year for tin ores raised.

[4] In addition to the one on the shore, the adventurers had put up another one on an island which they had formed out in the centre of the creek. Richard Thomas (*History etc. of the Town and Harbour of Falmouth*, 1827) records that a steam engine 'was erected out on an island' adding that 'these operations have been sufficiently successful to induce the adventurers to extend their works half a mile further down: two shafts having recently been sunk through the sand in the middle of the creek and secured by lining them with iron cylinders.'

[5] Lease dated March 1824 (for 21 years, at 1/20 dues). Lemon Records, CRO, Truro.

[6] The equipment of the mine, sold in 1830, (WB 19.2.30, RCG 27.2.30) comprised 24″ and 14″ engines, two horse-whims, and several fathoms of cast-iron shafts, in addition to a 'Horse Ferry Boat' and two others boats 'used for the purposes of the mine'. Almost certainly the 24″ engine stood in the house whose ruins now stand at the seaward edge of Tallack's Creek and the 14″ on the tidal island which stands (and can still be clearly seen) opposite the extreme seaward end of Devoran

Meantime another separate group of adventurers had also been granted a sett by Sir William Lemon in 1827 covering the bed of the creek seaward from that of the Carnon-Stream Mine, extending to where it joined the main estuary of the Fal at Weir (Restronguet) Point.[1] Operations were not commenced here until 1835, when a shaft was sunk close to the tide-line on the spit of land at the end of the seaward side of Penpoll creek.[2] Here a 40″ engine was erected and levels driven out beneath the water as before. In this much wider part of Restronguet it was not possible to carry flat-rods to a second submarine shaft, not only on account of distance but also because it would have obstructed navigation, and it was decided instead to form an island in the centre of the creek as a base for further operations. Sand, shingle and stones were piled up by the barge-load,[3] until there was an island 'about 30 or 40 yards in diameter.' Through this a 12′ diameter shaft, rivetted up from wrought-iron boiler plate, was then driven down to bed-rock by the laborious expedient of covering the cylinder top and shovelling sufficient silt on to it that the weight—as much as a hundred tons—forced it down in successive stages. Thereafter foundation piles were driven into the mud and on this tiny islet more or less in the middle of the creek[4] a small rotary engine was erected. At high tide schooners bound for Devoran passed within a stone's-throw, whilst under the bed of the creek beneath their keels the miners stripped out the shelf of tin gravel. The island, with its steam engine on a wooden-piled staging and the iron shaft, which was 'of such a height as to come above high-water mark,' was described[5] as 'a very conspicuous object'. It was reported that 'an immense quantity of waste was thrown into the water; thousands of tons around the shaft . . . and on one occasion there was a deputation from the Truro Corporation to see if the shaft and works interfered with the navigation'. In the six years after the first sale was recorded in November 1835, 638½ tons of black tin was sold from Carnon Mine[6] with 212 persons employed there at the peak of its prosperity in 1838.

wharves, on the long training wall that contains the channel leading up to Perran Wharf. The foundations of this engine-house can be traced here at low water. From the existence of the boat used for ferrying a horse or horses it is also evident that this island shaft was equipped with a whim for winding.

[1] Lemon Records, CRO, Truro. From later evidence (RCG 11.8.54) one half of the dues were paid to Lemon, the other half to Lord Falmouth whose lands also bordered Restronguet Creek.

[2] This promontory, on a shore then occupied only by the mine buildings, was close to the main tide channel and in the late 1850's a ship building yard was set up there on the site of the old mine. The local name of Yard is derived from this.

[3] Almost certainly from the dredging operations carried out off Devoran wharves.

[4] The remains of this are still visible as a shallow but extensive bank at low water. According to E. S. Tregoning in *A History of Falmouth and its Vicinity* (1865), writing of the various islands from which Restronguet tin had been worked—'the last they formed, and whose workings were finally stopped in 1843, is half a mile above the Ferry-house (i.e. the Pandora Inn) near the entrance of the creek, and 2500 yards below the highest reach of the tide at Carnon Causeway.'

[5] RCG 11.8.54. Considerable detail on the Carnon Mine exists in this source, reporting a famous legal dispute concerning the rights etc. to navigation in the various creeks of the Fal over which Truro Corporation claimed jurisdiction.

[6] Monthly tin ticketings, *West Briton, passim.* (1835 9 tons; 1836 105 tons; 1837 122 tons; 1838 130½ tons; 1839 121 tons; 1840 128½ tons; 1841, to August, 67½ tons). Without exception during these years, every first parcel of Carnon tin realised by far the highest price of any of the tin ticketed.

By 1842, consequent upon the low price of tin, the mine was losing heavily and a loss on the working of about £16,000 had been incurred:[1] ' the seam yielded as plentifully as ever, but owing to the enormous outlay, and vast expenses attendant on working a mine in such a situation, it was at length abandoned . . .'[2] The 40″ engine and two horse-whims were sold in September 1842 and a smaller 16″ engine the following March.[3]

In 1843 as a whole another seven mines of some consequence were ' knacked ' and probably as many more whose demise or sale has gone unrecorded.[4] During the year tin in London reached its nadir of £64 per ton (metal)—a figure at which very few Cornish mines could be continued except at a heavy loss. This was Peel's era of free trade and the threat of the abolition of the import duties on foreign tin ores—a measure which was heralded as virtually the death knell of tin mining in England by those engaged in it—made a major contribution to Cornwall's depression.

Two important mines to be sold up in 1844 were Wheal Ann in Wendron and Wheal Coates in St. Agnes. The former was closed in April or May, primarily on account of the death of Thomas Teague who held 150 of its 320 shares, the engines remaining unsold on the mine four years at least awaiting a buyer.[5] Teague was a considerable adventurer in several mines in the Wendron area and had a majority holding in Old Wheal Trumpet. This was exceptionally rich in the 1820's from one great bunch of tin, producing ore pure enough to be fit for the smelter as it was sent up to grass, and it was said that it was Trumpet's riches that sustained the early days of the great copper mine of Tresavean in Gwennap—Teague being also manager and a prominent shareholder in the re-working of the latter mine at this same time. On one occasion it is recorded that the miners at Wheal Trumpet were ' kept at work breaking the tin and sending it to the surface the day before the monthly pay at Tresavean, in order that Captain Teague might take it to the smelting-house and get his tin bill in time to negotiate it with his bankers, and thus obtain the money necessary for his pay day at Tresavean.'[6] At Wheal Coates, the 60″ pumping engine was for sale in November,[7] although surface and shallow working continued

[1] G. Henwood, *Four Lectures on Geology and Mining*, 1855. Some rather stylised sketches and a plan appear in Henwood but these are of little value.

[2] W. J. Henwood, On the Detrital Tin-Ore of Cornwall. *Journal of RIC*, Truro, 1873.

[3] WB 2.9.42 with 40″ pumping engine, two horse-whims, 10 tram wagons, tram irons, castings for a 12″ steam whim, and 14′ stamps water-wheel etc. Also included were 16″ diameter pumps, a sign that a considerable amount of water was being raised from the workings. The 40″ engine was later shipped to a buyer in Marseilles.

WB 10.3.43 16″ rotary engine and pitwork, plus two boats and about 5 fathoms of a 4-feet 6-inch cast iron shaft. This was almost certainly the island engine and other equipment.

[4] North Towan Tin Mine, St. Agnes, with 60″ pumping engine (WB 24.3.43); West Wheal Lovell, Wendron with horse whim and ' horse-engine ' (WB 19.5.43); Wheal Uny, Redruth, with 70″ pumping engine, two water-wheel whims, two horse-whims, and thirteen sets of water-wheel stamps (WB 11.8.43); Rocks Tin Mine, near Bugle, with 40″ pumping and stamping engine, with 40 heads (WB 1.9.43); Brynn Tye Mine, Roche, with 36″ pumping engine—one of the mines worked by the East Cornwall Mining Association (WB 22.9.43); Bottle Hill Mine, near Plympton, with 45′ and 18′ water-wheels. (WB 27.10.43).

[5] Comprising 48″ pumping engine and 16″ steam whim. (WB 17.5.44, RCG 5.6.48).

[6] MJ 4.2.05. [7] WB 29.11.44.

there until the end of 1847 at least, as 133 persons were then employed.[1]

The five or six years after 1838 had been marked by trade recession throughout Britain as a whole, but nowhere perhaps more acutely than in Cornwall. During this time began the first major wave of emigration from the mining districts, an exodus which was to continue in greater or lesser degree for two generations and more. Earlier, there had been a voluntary and usually temporary exile of men who were tempted abroad by higher wages or by a taste for adventure. English-promoted companies working foreign mining concessions, largely in the New World, had sought their skilled nucleus in Cornwall since the 1820's and before,[2] but the going of these men bore no relation to the emigration caused by the sheer necessity which began in the late 1830's. By 1837 it was reported that 'the expenses of most of our mines are now reduced to the lowest possible scale; and the returns are shaped as just to meet this economical expenditure. The greater part of the tutwork in most of them is suspended, while on the productive ground the hands are now considerably lessened. A great number of our miners are now working on the railway or wherever else they can procure employment; and several of them are selling their little cottages and preparing to emigrate to Mineral Point in North America, in quest of that productive labour so necessary to the support of themselves and families which they cannot now obtain in Cornwall.'[3]

Another report in the *Royal Cornwall Gazette* (30.3.38) concerning the departure of a group of miners to the notorious copper mines in Cuba was evidence of the prevailing state of affairs, for 'the low rate of wages and lack of employment in Cornwall is such as to induce many of our miners to volunteer their services for that land of disease and death.' The stark truth was that even for those miners who remained in work in Cornwall it was not possible to support a family, save on the border-line of starvation, on the average miner's earnings of ten or eleven shillings a week[4]—this moreover only for the desperately short working life of most miners and without taking account of the inherent dangers of his calling. Making things worse was that families were deprived by the mining depression of the extra income brought in by the young children who had been found work at surface. Although pitifully small, this little extra was none the less vital, and large families that had managed to exist on the edge of poverty were now plunged into it. True, in many cases they had a sufficient parcel of land to make subsistence possible, yet even for these, and for those who worked hard and knew no illness, all years were lean ones. Some were

[1] MJ 1.1.48. By comparison, at this date, there were 152 employed at Polberro, 430 at Charlestown United, and 451 at Dolcoath.

[2] Numerous reports exist in the *West Briton* and *Royal Cornwall Gazette* in the 1820's of Cornish men sent to mining ventures in Jamaica, Cuba, Norway, Colombia, Mexico, Brazil, Chile and elsewhere.

[3] RCG 23.6.37.

[4] Wages varied greatly from mine to mine, and from pitch to pitch. A good tributer might make 25s. a week during a lucky 'setting' but a tut-worker could rely on only ten, twelve or at the most fifteen shillings a week as an 'owner's man.'

worse than others and it was during these that emigration was turned to as an escape from the remorseless toil not only of the mine but of the small-holding. In many cases, the migration was of whole families, their passages paid for by a desperate period of hard saving after the bitter decision to quit Cornwall had been made. Alternatively a tributer's windfall from a sudden richness of the lode might enable a provident man to find the money to pull up his roots, whilst in other cases loans came from the many emigration agents who arranged their voyage. Passages in the steerage in the entirely unsupervised small vessels of the early 1840's, packed in ordinary holds with no surgeon and precious little food, were cheap even by the impoverished standards of west Cornwall.[1] There were in addition assisted or even free passages for those prepared to people respectably the convict colony of van Dieman's Land (Tasmania) and also South Australia.

The government of the day, realising that there were greater potentialities for the new continent than as a dumping ground for transported convicts —many of whom were thereby given, ironically enough, a new start in a better land than that they had perforce left—had a survey made of the southern Australian coast, and sponsored by Her Majesty's Colonization Commissioners, large-scale emigration to this area began about 1835. Ten thousand settled there between February 1836 and February 1838,[2] many of them Cornish, enticed by the free passages offered by the Commissioners, who sold land to the more wealthy at £1 per acre to form an emigration fund. Advertisements to would-be emigrants had become a prominent feature of the Cornish newspapers by 1839 at least, with official emigration agents appointed in Truro and elsewhere. At first only selected married tradesmen were accepted, and offered free passage for themselves and their wives but with a charge of £3 for every child. This latter stipulation was subsequently abolished, whilst all classes of general workers were also taken as the demand for labour grew in the rapidly burgeoning colony. Before long, single men, their sisters, and other unmarried females 'accompanied by protectors' were taken and by the early 1840's it was no longer necessary for these fortunates to find the means to reach London or even Plymouth, for many of the emigrant ships were now departing direct from Cornish ports. By 1845 vessels had sailed in some numbers from Fowey, Hayle, Padstow, Penzance and St. Ives, and even from smaller havens such as Boscastle and Gweek, as well as from Restronguet and Malpas on the Truro River. Emigration to North America, primarily to Quebec and New York, followed the same pattern at this same time also, although the assisted passages were in

[1] For example the fare from Fowey to Quebec in returning vessels engaged in the timber trade was 45s. in 1846. At this price, would-be emigrants had to provide their own provisions, as witness one *West Briton* advertisement (27.2.46): 'notwithstanding the fast sailing of this vessel (the *Royal Adelaide*, 800 tons), her last passage being twenty-six days, each passenger must be provided with sufficient provisions for two months.'

[2] RCG 30.8.39.

this case by loan from emigration agents—in some cases acting for shipping companies—and not by any government agency.[1]

Yet from one cause or another, emigration was impossible for many and distress amongst the working population of Cornwall as a whole had reached grim proportions by the early 1840's. One sign of the unrest occasioned by this state of affairs was a major strike by the miners at Consols in Gwennap, a district that was by no means the hardest hit. Some 200 marched to Carn Brea Mines to get the men there to join them and to form 'a working miners' union.'[2] Later, in June 1847, a more widespread 'rising' occurred when there was an immense gathering of tin and copper miners at Penzance. On this occasion the men from St. Just and those from the districts east of Penzance assembled to the west of the town and marched through the streets, 3000 strong, to hold a mass meeting on the Eastern Green. Anticipating trouble from these 'riotous tinners' all premises had been locked, troops had been sent from Pendennis Castle, and two hundred special constables sworn in, although there was not the riot that had been feared.[3] Yet such an occurrence left its mark, and it was not surprising to find the Penzance authorities, and others, anxious to help lift the safety valve that emigration provided for the working classes. In March 1848 there were suggestions that a rate should be imposed to meet the cost of emigrant passages as a solution towards easing the burden of the poor rate once and for all.[4] Fortunately, at this juncture, came the further spur of massive copper finds in south Australia, opening up a whole new world for dissatisfied Cornish miners. This was given yet another fillip by the finding of gold in California in January 1848, an area which drew a large number of the more adventurous Cornishmen who had few, or unwelcome, family ties. The 'rage for emigration' from 1847–49 as a result of Australian copper and Californian gold was indeed phenomenal; in April 1847 it was reported that 'upwards of 700 persons have left Camborne parish within the last ten days for Australia and North America,'[5] whilst two years later one Truro agent alone reported that he had received no less than 631 applications for emigration passages in the space of sixteen days.[6] Nor was it only the young who were anxious to depart, for one case was quoted of a man from St. Agnes aged 78 and his wife nearly 70, who left in March 1848 for a new life in the United States.[7] In three months of 1849, 3690 persons left Plymouth for the Australian colonies and 1058 in one single week for Quebec,[8] whilst the number of direct sailings of emigrant vessels from Cornish ports in 1847 and the two succeeding years told its own story.[9]

[1] Free passages were also granted to the Cape of Good Hope in 1846. (WB 16.1.46).

[2] MJ 26.3.42. Such an occurrence was unlikely to have escaped the observance of the local Cornish press but none of the county newspapers, for one reason or another, recorded it.

[3] RCG 4.6.47. [4] WB 24.3.48. [5] RCG 9.4.47.

[6] WB 16.2.49. [7] WB 18.8.48. [8] WB 27.4.49.

[9] Padstow 8, Falmouth 11, Penzance 15, Hayle 5, Malpas and Restronguet 4. The great majority of these sailings were to Port Philip, Port Adelaide or Quebec, but others went to Natal and Hobart, Miramichi (New Brunswick) and New York, and even to San Francisco direct.

This readiness to quit Cornwall on the part of working miners was understandable enough, for their lot had not been bettered one iota since the beginning of the century. It was still as hard, dirty, and unrewarding a way of earning a bare livelihood as it had been since Waterloo, or indeed since the previous century.[1] Other labourers, such as farm workers and fishermen, had lives perhaps as hard, but their toil at least was healthy and mitigated by the fresh air, the sun and the passing of the seasons. Technically, mining had come a long way in the period from 1825 to 1850—for example, in the case of the steam engine, which had been developed greatly for pumping, for winding and stamping ore—but as regards the conditions under which the men worked, things had gone worse rather than better, if only from the growing depth and heat of the mines. With changing circumstances, the amount of tutwork had increased greatly and compared to tributing, which needed all a miner's intelligence and experience, this was mere underground navvying. The bigger mines too were far more dangerous, for minor falls of rock were all too common in the extensive stopes and deep gunnises where the hanging wall could not be sounded or seen. At Dolcoath in particular, the workings were too wide for normal methods; in 1837 it was reported that 'in working on the sides of these immense chasms the miners commence their operations at some level where timber can be applied, to which they fasten a 'swing stage.' This they contrive to drop against such part of the side of the chasm as they intend to take away; and then letting themselves down by means of 'swing-chain ladders' they blast immense quantities of rock . . . the greater part of it comes down in such large masses as frequently to require blasting again before it can be sent to the surface. . . . In these operations the workmen incur great risk, as many a maimed and mutilated miner, and many a grave-stone in our church-yards, bear witness.'[2]

To offset these increasingly harsh and dangerous conditions, the only significant step forward that had been made for the working miner was the introduction of the safety fuse by William Bickford in 1830.[3] Yet even this was slow to be adopted—partly perhaps because most miners had to buy their own materials and the old hand-made 'fuse' cost so much less—and it was years before its invention ended the toll of blinding, disfigurement and death that had resulted from 'misfires.' Trams were also being gradually introduced underground in place of wheeling the

[1] The harsh truth of the Cornish miner's working life is a story still to be told. Stripped of the romance with which it has been invested now that it is part of the past, it was a miserable, dangerous, and even squalid, existence. Seen through the rose-coloured spectacles of mining historians he is portrayed as a sturdy and independent cottager, sleek with feasting on pasties and seldom without a hymnbook. In reality, if one could exchange the shadow for the substance, he was more frequently the rent-owing occupier of a damp cottage, gaunt on a diet of potatoes and pilchards, and addicted to the local beer-shop.

[2] RCG 26.5.37.

[3] The first advertisement of these new fuses, then termed by Bickford 'Improved Safety Rods,' was dated 7 January 1831 (WB). Their inventor had considerable trouble during their period of introduction and in an advertisement later in the year (WB 14.10.31) he had explained that some of the initial failures of the new fuse in use had been due to the miners 'sending down their tools tied up with the rods, tying their shoes, etc., etc.' before putting it to its intended use.

time-honoured barrow, but this was a development which mattered more to the copper than the tin mines, except perhaps those of the largest size. The man-engine had also just seen the light of day in Cornwall but not in any mine outside Gwennap, and for every tin miner from Drakewalls to Wheal Owles, the inexorable daily burden of the ladder roads remained. Few of the young men sensed or heeded the hidden effects of the ladders upon their constitutions and when they did experience the first signs of chest or heart trouble it was too late. A few more years and their only alternative lay in a job in a shallow mine or in a pitch at a lesser depth. Ultimately, if they were lucky, they might find a job at surface. Nor was it only the ladder-roads that slowly crippled the men: no matter how hard the granite, all boring was done by hand and this in air thick with the lingering smoke and black dust of burnt gunpowder. In small and often crooked levels where artificial ventilation was quite ineffective even where it was tried, where the air was already fouled by human exhalation and vitiated by the burning of innumerable candles, the working conditions can hardly be imagined, let alone described. In some ends, far distant from any shaft or winze that might afford some small natural flow of air, where candles would not burn except on their side—and even then with the faint blue flame that signified a lack of oxygen—it was only the spur of starvation that made men work at all, knowing full well that others even less fortunate were always ready to take their place. Some conception of the worst conditions can be gained from the recorded death of two miners in a level at Wheal Owles in 1842, quite simply, from 'foul air.'[1] But outright death in this way was rare; so commonplace as to go unrecorded was the lingering end of miners in their cottage homes years after the onset of the headaches and dizziness, the chest pains and asthmatic lack of breath that resulted from long hours of toil in hot, close, mines.

It was perhaps no less than justice that the adventurers in tin—in whose service labouring miners' lungs turned black and their faces pale—should suffer financially at this time of depression. Certainly no-one prospered, save perhaps the wily smelters, and even the great Wheal Vor had been hard hit. The ruinously low price of tin and the continued litigation over the disputed shares meant that by the end of 1838, even though £1,240,000 of tin had been sold, a mere £100,000 of this had been available for dividends—a return of $12\frac{1}{2}$ per cent and this on the richest tin mine Cornwall had ever seen.[2] The mine was then 240 fathoms deep and growing more and more costly to work. In 1838 the Carleen section was

[1] MJ 12.3.42. An illuminating sidelight on the state of affairs underground in this respect, humorous had the subject not been tragic, was provided by one John Carberry, a Penzance chemist, who introduced and widely advertised in 1839 a form of lozenge termed 'Vegetable Pectoral Cakes' for miners working in badly ventilated mines. These were devised 'after great labour and long study . . . for preventing those fatal cases produced by foul air, and restoring those who may be labouring under disease from the above cause.' (RCG 5.4.39).

[2] WB 21.12.38. It is interesting to note also that one correspondent of the *West Briton* estimated that the Wheal Vor adventurers lost at least £100,000 by establishing their own smelting works: '. . . they never possessed the secrets of distribution and were for years in the hands of the middle men . . .' (WB 28.8.40).

abandoned as unremunerative at the existing price of tin and the 49″ pumping engine there plus two of the smaller sets of steam stamps (with 16″ and 15″ engines) were sold off. In July 1839 the old 45″ engine and another 27″ stamps engine were also disposed of[1] but, even so, working of the main sett as a whole was continued on a massive scale. In 1845 there were reported to be still 1400 persons employed,[2] although the mine was then barely paying costs and smelting had been discontinued the previous year. Litigation, artfully prolonged, still continued to provide as abundant a harvest as ever for the lawyers, with no less than eight separate chancery suits pending in 1843 for the recovery of the Gundry shares. In July 1844 it was said of the Wheal Vor case that 'the litigation had been protracted, the results doubtful, and it had become important to dispose of the matters in dispute, the lords of the mine refusing to make a new sett to the parties at present engaged in the working'.[3] In October, with the shares then valued at £1,000 apiece despite the existing depression, a compromise settlement was agreed upon by all the parties save one, the daughter of John Gundry. By the time she had been brought into agreement with the others, in November 1845, their value had come down to a fifth of this, to £200, and this state of affairs thereupon caused some of those who had earlier agreed to the compromise terms to withdraw their consent. This was done 'in an attempt to obtain improved value for the property by holding a number of shares in dispute over a period of depression that they might realize better prices on the return of better times.'[4] Perceiving this stratagem, the Commissioner in Bankruptcy directed in December 1845 that matters should be decided upon 'with all convenient speed' to bring the protracted case to a close.[5] Complicating matters was the fact that Trelawny, the principal mineral lord involved—who was reputed to use his share in the proceeds from Wheal Vor in running his celebrated pack of hounds in east Cornwall—had also taken up some of the disputed shares in 1820 and the protracted costs and annoyance of the matter had made him determined not to renew the lease of the sett on its expiry. In 1845, moreover, there was a serious run of ground in the deepest part of the mine where a particularly rich section of the lode had been removed entirely and this 'caused damage to the extent of £3000 to £4000.'[6]

[1] WB 12.7.39. [2] MJ 14.6.45. [3] WB 19.7.44. [4] RCG 7.1.53.

[5] For details of Wheal Vor legal disputes etc. in the 1840's, see RCG 31.12.52 and 7.1.53.

[6] Buried in this was the celebrated underground blacksmiths' shop in the 245 fathom level, set up there in 1841. (WB 19.2.41) Here all the miners' tools were sharpened and repaired to save cost and time in hauling to surface—an idea that had come from the deep Allihies copper mine in Ireland where Captain Mark Reed, manager of Wheal Vor since 1838, had found it successful.

This subterranean smithy has received passing mention by several writers: hitherto unrecorded but in a minor way even more astonishing is the fact that one or more of the Wheal Vor miners was indulging the hobby of gardening deep underground, the *Royal Cornwall Gazette* recording (26.2.41) that 'There are now growing in one of the levels of Wheal Vor mine, at a depth of 214 fathoms from the adit, several remarkably fine kidney and broad beans'. Certainly the warm and moist conditions would be ideal for germination, or for mushroom cultivation, but hardly for vegetable growing almost 1500′ from surface.

For details of this hitherto unrecorded disaster which may well have been partly caused by the setting-up of the underground blacksmiths' shop, see RCG 21.11.45, 28.11.45: 'a shocking accident occurred on Tuesday at Wheal Vor mine, by which three persons lost their lives and two others

In view of this disaster, an accumulated debt due of £17000 and no possibility of lease renewal from Trelawny, the manager was called to a meeting of the adventurers at Helston 'at which he was told to raise all the tin possible in the six months before the end of the lease.'[1] Some time later, unfortunately, the other mineral lords prevailed upon Trelawny to re-grant his part of the sett to the company but nevertheless this agreement came too belatedly, for, in the words of the manager, 'it was then too late, the mine could not then be brought into a proper state without a considerable outlay which the old shareholders as a body, with Gundry's assignees amongst them, could not of course meet. That was the cause of the stopping of Old Wheal Vor.'

The exact sequence of these events in relation to the advertised disposal of Wheal Vor's machinery is not clear but in August 1846 it was reported that on account of the lords' refusing to give up their dues—the company then making considerable losses—the decision had been taken to stop all work below the 115 fathom level, a step that would throw 700 hands out of work.'[2] The next month three pumping engines (80″, 53″ and 49″) plus a 16″ steam whim were advertised for sale.[3] For part of the following year, Wheal Vor continued to be worked on this basis but on 18 August a resolution was passed by the adventurers that 'all the deep workings should be stopped and that the pumps should go to grass.'[4] Black tin was then selling on average at £43 per ton and the mine was simply not worth working; meanwhile part of the sett was to be continued in a small way to shallow depth. Within a fortnight, with the engines stopped, Wheal Vor was full of water to the adits, and the mine, said in its heyday to have resembled a small town, was more or less deserted. The remaining engines,—two big 80″, the old 49″ which had been advertised many times but had never found a buyer, two (20″ and 18″) whims, and a set of 36 heads of stamps—were advertised at the end of the year[5] and Wheal Vor's career as a deep mine came to a halt.

As Wheal Vor declined, still capable of producing tin but not to any profit at the ruling prices of the day, Balleswidden took its place for some time as the leading Cornish tin mine. In terms of dividends Balleswidden was not particularly rich but it was a most important mine as regards size and output. From 1833–41 only £51,960 of tin had been sold[6] but late in 1841 the mine cut rich. Additional shares in the mine were then created, increasing the number from 87 to 145[7], and in connection with these it was stated that the adventurers were 'working on an inexhaustible

sustained severe and extensive injuries . . . a large mass of ground gave way above the [blacksmiths'] shop and falling on it not only crushed in the building, but the weight and force of the falling mass broke away that portion of the ground on which it stood, and the whole was precipitated in one general destruction. From the great depth at which the melancholy occurrence took place, several hours necessarily elapsed before the bodies could be extracted and brought to the surface'.

[1] RCG 7.1.53. [2] MJ 1.8.46. [3] RCG 4.9.46. [4] RCG 7.1.53.

[5] RCG 5.11.47, MJ 11.12.47. In the depressed state of mining only two 80″ and 20″ engines were disposed of, the remainder being subsequently re-advertised RCG 29.6.49.

[6] WB 28.1.42. [7] MJ 14.6.45.

lode, producing upwards of fifty tons of tin per month.' The mine had five engines at work, one for pumping, two for hauling and two sets of steam stamps working 144 heads, as well as two water-wheels working a further 36 heads, plus six horse-whims. No less than 634 persons were employed[1] (including 282 underground), the mine being worked on a different principle to most other tin mines for only tutworkers were to be found underground to remove the lode which was 'large but poor'. In the year 1845 £3478 was paid in dividend, the peak for Balleswidden from 1843 onwards.[2] The mine was interesting for the attempts made to supplement normal Cornish tin dressing practice by chemical means, using sulphuric acid to remove impurities by a leaching process. In March 1844 there was a claim against the adventurers for breach of a patent in using acids in this way and £750 was subsequently paid for a licence for the process.[3]

Other leading mines in the 1840's were St. Ives Consols, Charlestown United and Royal Polberro Consols. The first named had been particularly productive in the 1830's and by 1840 the holders of its 94 shares had received almost £50,000 in dividends[4] and had paid few, if any, calls. Much of this had come from the Great Carbona—graphically described in one contemporary report as 'the house of tin'[5]—but this bonanza came to an abrupt end in April 1844 when fire broke out in the forest of timbering which kept the ground safe to work.[6] This part of the mine was said to be unworkable thereafter and the resources of the main lode eastward were then developed. Charlestown United, which had also been extremely rich in the 1830's, was particularly hard hit by the slump in tin and its 112 shares which had been quoted in the *West Briton* in January 1842 at £650 each, declined steadily in value until by August 1843 they were £250, with few takers. In the eighteen months prior to December 1847 a loss of £7846 was incurred and the group of adventurers, primarily from Tavistock and London, thereupon decided to abandon the working

[1] Then the most in any Land's End district mine. In 1841 Levant employed 507, Wheal Owles 202, Boscaswell Downs 184, Botallack 172, Parknoweth 133, Wheal Caul & Wheal Castle 128, Boscean 100.

[2] From 1844 to 1853 the company paid an average of £3000 per year (total, £25375) except in 1847 and 1848 when £725 and £609 were paid—reductions due to the provision of a new pumping engine etc. Output in the ten years after 1844 was: 1844 499 tons, 1845 452 tons, 1846 611 tons, 1847 549 tons, 1848 515 tons, 1849 533 tons, 1850 499 tons, 1851 469 tons, 1852 418 tons, 1853 277 tons. (Bolitho Records, DD RG 1/26, CRO, Truro).

[3] Chemical purification of tin ores in Cornwall has been rarely used instead of the normal thermo-mechanical methods. For the removal of certain impurities such as the oxides of copper and iron, however, it was more effective, although the process (consisting of treating the ore in wooden tanks of acid) was a relatively expensive one. Wheal Owles also 'had spent a lot of money in endeavouring to improve the tin by the use of acids,' according to a later report (RCG 17.10.67), whilst G. Henwood (*Four Lectures on Geology and Mining*, 1855) records that this method (using nitromuriatic acid) was successfully adopted at Wheal Boys, St. Just, as well as at Wheal Owles.

[4] WB 13.3.40. [5] WB 19.4.44.

[6] RCG 19.4.44, 26.4.44. The fire amid the estimated 500 tons of timber that supported the excavation began on 12 April and was reported to be still not burnt out fourteen days later. On the 19th three men tried to resume work in a pitch not far from the Carbona but 'suffocating gas' soon drove them to the surface, one of them falling from the ladders in their headlong rush and receiving fatal injuries.

of all except the Boscundle section and sell off the materials.[1] The latter portion was kept going until 1850 but in May it was decided to abandon that as well and to sell off the still remaining material on Bucklers.[2] At one time over 800 persons had been employed at Charlestown United and some idea of the scale of operations is evident from there being no less than forty shafts still open to surface on one part of the sett alone in 1858. Polberro Consols, graced at this time by the prefix 'Royal' because it lay on land of which the Duchy had the mineral rights, was similarly hit by the slump, having been worked for centuries and with no immediately accessible reserves on which to draw. Moreover, the connection of several of its chief adventurers—notably de Tastet and Vigurs—with the expensive failure of co-operative tin smelting had been disastrous. The mine was an extensive one and in 1843 there were 24 shafts open to surface, including nine in full working. The sett occupied a full 200 acres and five engines and two large water-wheels were at work when the whole was advertised for sale as a going concern in 1843.[3] Another currently rich mine was Botallack but here, as at Levant, it was primarily copper lodes that were being worked. In the period from 1814 to 1835 the mine had sold only £53230 of tin and copper, and was then 'abandoned on account of a heave in the lode.'[4] In 1836 or 1837 it was re-opened but in the next four years a mere £2055 of copper ore had been raised as against calls of £17500 and the mine was again about to be 'knacked' when a lode of exceptional value was unexpectedly encountered early in 1842.[5] Throughout the remainder of the 1840's Botallack adventurers waxed fat on this find, £23000 in dividends being declared in 1843 alone.[6]

By 1845 the depression brought on by Peel's abolition of the import duty on foreign tin ore was particularly acute and in the light of events to this date it seemed as though the Cornish tinners were right who held the opinion that this signalled the doom of their native industry. Once again the miners perceived in this the domination and self-interest of the smelting fraternity—who did not oppose the measure—and at a county meeting of 'Adventurers in Tin Mines' held in February it was agreed to levy one farthing in the £ on all black tin produced, to form a fund with which to fight the abolition. The relative importance of the tin smelting

[1] Carlyon Records, 5M/EO/8/33, CRO, Truro

[2] RCG 17.5.50. Comprising 32″ pumping engine and two water-wheels on Boscundle: 50″ pumping engine, two (26″ and 22″) whims and several other water-wheels used for stamping, on Bucklers.

The decision to abandon the last sections of Charlestown United was hastened by an accident in October 1849 when old flooded workings at Cuddra were holed into and three miners drowned. It was reported that 'it will take many days to get the mine in fork' and one of the bodies was only recovered fourteen days later. (RCG 19.10.49, 2.11.49).

[3] WB 3.3.43.

[4] MJ 14.6.45. According to another source (Oats Collection, RIC, Truro) Botallack's finances from April 1804 to December 1834 had been—costs £120,203 sales £151,746 leaving a profit of £31543.

[5] 'One of the best, if not the very best, courses of copper ore ever known is this county . . . grey ore, and about four feet solid.' MJ 12.3.42.

[6] MJ 14.6.45.

industry to Cornwall as a whole, compared to tin mining, was pointed out, for about nine thousand persons were employed in tin mining,[1] whereas 'the aggregate number of men employed in smelting the whole tin produce of the county [was] only about sixty . . . one man could convert about 100 tons of ore into pure metal annually.'[2] Certainly it could not be disputed that the tin smelting interest wielded a power, not only in Cornwall but also

Valuable TIN MINES, called the ROYAL POLBEROU CONSOLS, in the Parish of St. Agnes, within eight miles of Truro, Cornwall, with the Engines, Machinery, and Materials thereon.

TO be SOLD, pursuant to a Decree of the High Court of Chancery, made in a cause wherein FIRMIN DE TASTET and ANOTHER are Plaintiffs, and WILLIAM CARNE and OTHERS are Defendants, and wherein the said FIRMIN DE TASTET and ANOTHER are Plaintiffs, and EDWARD BRYANT GAREY is Defendant, with the approbation of Nassau William Senior, Esquire, one of the Masters of the said Court, in ONE LOT, at the PUBLIC SALE-ROOMS of the said Court, at Gray's Inn Coffee House, Holborn, *London*, on FRIDAY, the 31st day of March next, at Twelve o'clock at Noon, all those extensive and valuable TIN MINES, well known as the

Royal Polberou Consols Mines,

Situate in the Parish of Saint Agnes, in the County of Cornwall, which have for many years last past averaged the monthly produce of 30 tons of Black Tin of excellent quality. These Mines extend under a surface of upwards of 200 acres, and contain 24 shafts, 9 of which are now in full working, with about 3,000 fathoms of levels, and are held under grants or setts, of which 13 years and upwards are unexpired, at the very moderate dues of 1-20th,

TOGETHER WITH THE

Steam Engines, Machinery, Erections, Buildings, and other Mining Effects

Thereto belonging, comprising TWO STEAM PUMPING ENGINES of 60 inches cylinder each, five Boilers, Balance Bobs, &c. complete; TWO STEAM WHIMS, respectively 22 and 20 inches cylinder, with 4 Boilers, to one of which is attached a Crusher; a 30 inches cylinder Steam Stamping Engine, with 3 Boilers; 64 Heads Frames, Passes, Pits, &c., complete; 2 sixteen Heads Stamping Mills, with 2 wheels, each 30 feet in diameter, and 3 feet abreast, with Lifters, Passes, &c., complete; every requisite for dressing and calcining Tin, and a Sawing Machine and Water Wheel. The Buildings comprise Sump Men's and Changing Houses, Smiths' and Carpenters' Shops, Counting, Store, Captains, Gig, Engineers, Sampling, Casting, Tin Dressers, and Changing Houses, the latter with 193 Miners' Chests; also, Assay Office, with every requisite; 2 Stables, 1 Powder Magazine, 1 covered Saw-pit, with frames. The Smiths' Shop, which has every convenience for 5 forges, with powerful Punching and Screw Machines, Kibble-mould, and Tools of every description. A large Machine, Turning Lathe, and other requisites in the Engineer's House. Also, all the Pitwork Ladders, Casings, Brass Air Pump, 10 Horse Whims, Sheaves, Kibbles, and Drawing and Tramming Carriages and Apparatus; Weighing House, Beams, Scales, and Weights; Counting House and other Furniture; 3 Dials, 2 Crab Winches, Large Steelyard, 150 Barrows, Capstans and Shears, Capstan and other Ropes, Flat Rod and Stands, Rods, Clack-seat Pieces, Pumps, H and Top Door Pieces, Working Barrels, Plunger Poles, Stuffing Boxes and Glands, Rod Plates, Common Plates, Staples and Glands, Block Pins, Rod Pins, Connexion Rods, Blocks, Cisterns, nearly 2,000 fathoms of Wood and Iron Tram Roads, with all the Leavings, Halvans, Barrows, and other Property and Effects on the Mine (except the Stores and other Articles hereinafter mentioned to be taken at a valuation).

The Store Yards, &c., contain large quantities of Iron, Boiler, and Kibble Plates, Candles, Gunpowder, Hilts, Leather, Tallow, Grease, old Copper and Lead, about 4,000 lbs. of old Brass, 2 Horses, Gig and Harness, 3 Carts and Harnesses, Debenture and other Timber, Coals, Rope, &c., and other Materials of daily consumption; which, with the Tin Stuff and Tin Ores on the Mines at the time of completing the Sale, are to be taken at a fair valuation by the purchaser.

The outlay on this property has been upwards of £100,000, and the Setts, which are very large and extensive, are now in full working, and are well found in every respect.

Particulars may be had, gratis upon application at the Chambers of the said Master, Southampton Buildings, Chancery Lane; of Mr. E. B. GAREY, Solicitor, 24, Southampton Buildings, Chancery Lane; of Messrs. ANNESLEY and READE, Solicitors, 64, Lincoln's Inn Fields; of Messrs. HUGHES, KEARSEY, and MASTERMAN, Solicitors, 17, Bucklersbury, London; of Mr. C. K. VIGERS, Truro, Cornwall; of Mr. J. G. BECKERLEG, 3, Alderman's Walk, London; at the principal Inns in Saint Austell, Truro, Redruth, Camborne, and Penzance, in Cornwall; and at the Mines. The property may be viewed on application to the said Mr. C. K. VIGERS, Truro, the Manager and Receiver of the Mines.

Dated February 24, 1843.

Sale notice of Polberro Consols, St. Agnes, West Briton 3.3.43

in London, that bore no relation to its size as an industry. It was said of the Welsh copper smelters in 1842[3] that ' their works do not derive their importance so much from the number of hands they employ as the aggregate value of the metal they produce ' and the same was equally true of tin smelting.

To the many adventurers who had borne losses and calls patiently for

[1] Compared to about 22000 at this date in copper mining. In the latter, relatively few women and children were employed, the preparation of copper ore for sale to the smelter being far simpler than that of tin.

[2] MJ 1.3.45.

[3] *Royal Commission on the Employment of Children in Mines*, 1842.

some years this throwing open of the market for the duty-free import of foreign tin was seen as the final straw to break the camel's back. As a consequence even more mines were abandoned, including some, like Wheal Vor, of premier importance. In October 1847, Beam Mine, one of the largest in the St. Austell area, was advertised for sale[1]—a mine notable for the very pure ore it produced, which brought a high price from the smelters. The company concerned had taken the sett in 1841 and erected an engine ' on that part of the sett called Good Fortune, on King's Hill '[2] where, according to this same source, ' upwards of £450000 worth of tin was sold by the late proprietors Messrs. Rashleigh Williams & Co.' In 1848, Wheal Grey in Germoe—then a much more extensive mine than is now either realised or recorded—was sold in February, having 53″ and 40″ pumping engines, an 18″ stamps working 36 heads and no less than ten horse-whims.[3] Close to Beam, north of St. Austell, Rocks Consolidated was also for sale, with 70″ and 40″ pumping engines plus numerous water-wheels, whilst in St. Agnes Royal Polberro Consols was again for sale, complete with six engines.[4] In July Wheal Boscean at St. Just was advertised for sale ' in full course of working,' equipped with a 36″ pumping engine and 24″ whim,[5] and in September Wheal Ruby in Wendron, with 58″ engine and 22′ water-wheel working six heads of stamps.[6] The next month Wheal Andrew & Nangiles was advertised, with a newly erected 70″ engine—a mine that had been re-started for tin and copper in February 1844—followed in December by the sale notice of Great Hewas Consols. This latter had been re-set to work on a considerable scale in August 1846—somewhat unaccountably in view of the existing depression in the trade. The sanguine adventurers had high hopes but found on re-opening the mine that extensive damage had been done by tributers who had been allowed to work above adit level in the twenty-five years that the Hewas sett had been idle.[7] The loss when the concern was sold up at the end of 1848 was considerable, for all the machinery, chiefly comprising a 66″ pumping engine, five horse-whims and three water-wheels working 28 heads of stamps, had been ' placed on the mine within the last two years.'[8] In this respect this short working of Hewas provided a typical example of the numerous ill-conceived and short-lived trials that have punctuated Cornish mining history. Even though we now have the benefit of historical hind-sight in assessing the value of such

[1] MJ 30.10.47. [2] WB 15.1.41. [3] MJ 12.2.48.

[4] RCG 4.2.48. Apart from the substitution of an 80″ engine for one of the 60″—this being the largest pumping engine ever to work in the St. Agnes district—these were the same as had been advertised in 1843. At this date Polberro was also variously known as St. Agnes Consolidated.

[5] MJ 15.7.48.

[6] RCG 1.9.48. Wheal Ruby was worked jointly with Garlidna Mine and the two were for sale RCG 23.2.49, with 70″ and 58″ pumping engines, 32″ stamps (with 72 heads), and two (36′ and 22′) water-wheels.

[7] MJ 17.7.47, 29.8.46. This was a danger to which many mines were subject if a mineral lord allowed free-setting tributers, who were largely unsupervised, to fossick through old workings. Ore was ripped out without thought for the future of the mine as a whole, deads dumped into the nearest shaft, no timbering used, and considerable general damage done.

[8] RCG 8.12.48, MJ 6.1.49.

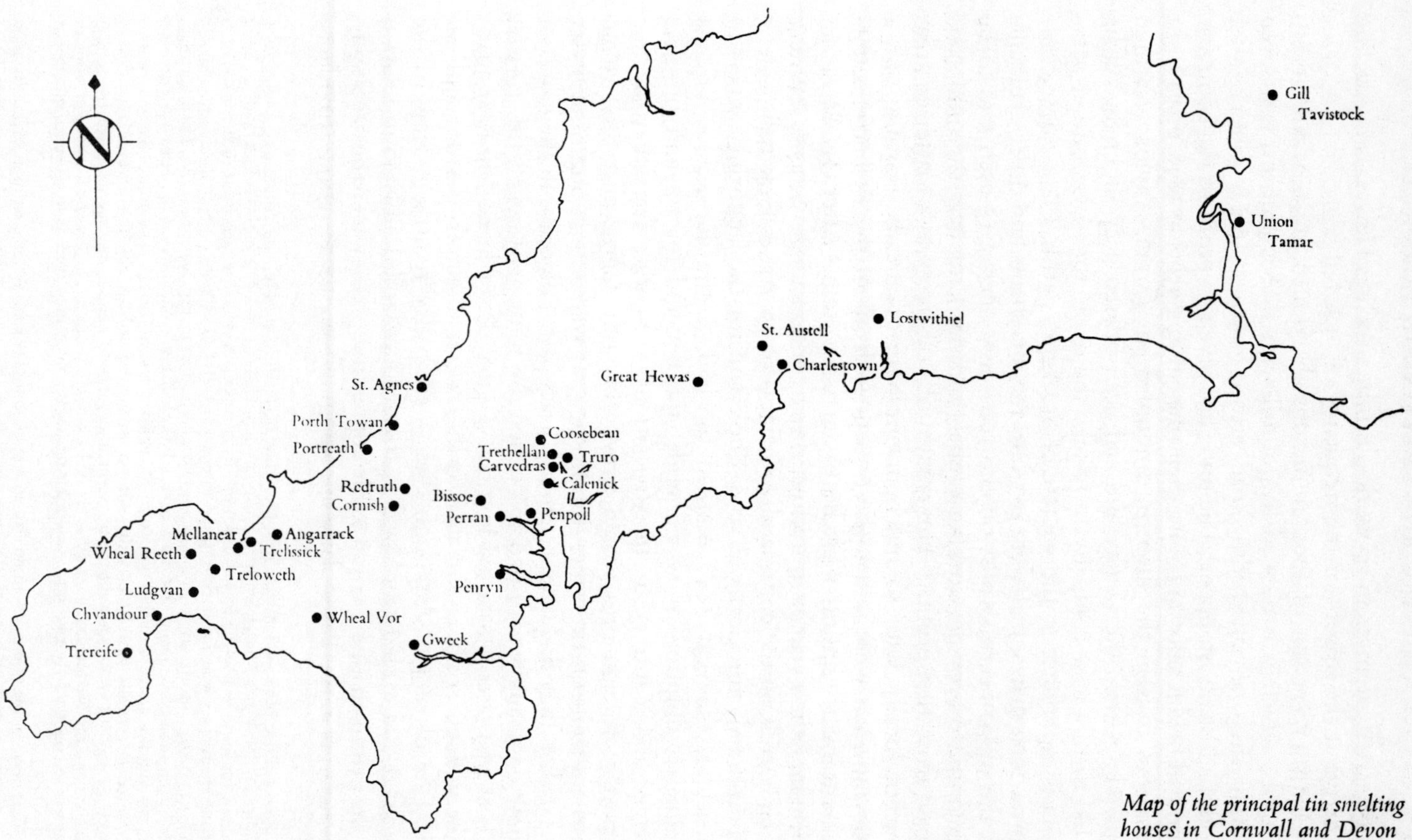

Map of the principal tin smelting houses in Cornwall and Devon

half-hearted attempts to re-work old mines such as Hewas, it is puzzling to arrive at the reason for the frequency of such folly.

One survivor out of these numerous abandoned mines was Polberro Consols, which was rescued by John Taylor in September 1848, seven months after the mine had been advertised for sale, and a few weeks after work had finally stopped.[1] The mine had given no profit to the company who had held it since 1835 and their outstanding debts in 1848, apart from numerous calls in the interim, amounted to £6746. Taylor was then mineral agent to the Duchy and had been offered very slight dues if he could form a new company to re-take the mine, for it was one of paramount importance to the welfare of St. Agnes parish. Of the new company's 2000 shares, half were to go to the merchants and other claimants on the old concern, whilst on the 1000 new ones £15 each was to be called up, to provide working capital, and this latter was the sum to be repaid from first profits.[2] This revival of Polberro was a matter of great concern locally with so much unemployment already existing, and a thousand men were said to be present in front of the Polberro count-house at the first setting of pitches in the re-working.[3] After this, St. Agnes celebrated the event in traditional form, culminating with the burning of tar barrels on the commanding heights of St. Agnes Beacon.[4]

Fortunately for Cornwall as a whole and for the labouring miners in particular, the year 1849 marked the end of the depression. With a gradual resumption of world trade, the demand for tin and tin-plate revived and, contrary to all expectations, the earlier fears that the free import of foreign ores would irrevocably ruin Cornwall were proved untrue. Profits on tin mines during the year were few enough but better by far than they had been for many years past. Polgooth, in the hands of a new company which had re-opened the mine in 1845,[5] paid £5504; Wheal Margaret, re-started in 1842, £4368; Providence Mines £3640; Balleswidden £2922; Great Work £2677; Tincroft £2100, and St. Ives Consols £470. Although these were paltry returns in view of the turnover and capital employed in these mines, and doubly so in relation to the profits made by comparable copper mines, they were heartening to adventurers in tin. Particularly encouraging were the results at Providence

[1] Polberro had been suddenly stopped for a time before this, in May 1847. On this occasion a large number of St. Agnes miners, with starving families, came to the Truro authorities with a pitiful 'request for leave to beg the town for food'. Each was given a loaf, which most of them—though starving themselves—carried back to their homes intact. (RCG 21.5.47).

[2] MJ 13.7.50. Unfortunately Polberro's re-working had an inauspicious start. By December 1849, after fifteen months working, 406 tons of black tin had been sold for £17702 but costs had been £23769, and this added a further £6000 to the existing loss of £6746. The mine, however, was to be worked successfully thereafter by this same company, headed by John Taylor, until 1873.

[3] MJ 23.9.48. 'On Tuesday the agents went underground making arrangements for the setting next day: and on Wednesday, as early as six o'clock in the morning, men came flocking in from all parts of the country.' By two p.m. when setting began, 1,000 men were held to be in attendance.

[4] RCG 29.9.48.

[5] This new company erected an 80″ steam engine (bought from Wheal Darlington) and by 1847 the mine was reported to be worked vigorously down to 76 fathoms below adit (the mine then being 110 fathoms deep) with 250 tributers employed. (MJ 10.7.47, 24.7.47).

Mines in Lelant, which had given no dividend in the fifteen years since re-working began there in or about 1834. At times the mines had been desperately poor and the materials had been advertised for sale on more than one occasion.[1] The lords had foregone any dues for years past and on the occasion of declaring the first dividend in November 1849 it was said that 'but for this liberality of the lords, and the perseverance of the adventurers, the mine would have been long since abandoned.'[2]

Brightening trade conditions were also evidenced by the appearance in 1849 of three new or newly re-started smelting works. The first of these was at Charlestown promoted by H. J. Enthoven who had been a partner in banking and other activities with Ricketts, James, and other members of the ill-fated Copper Miners Co. The first purchases of black tin were made in October 1848[3] whilst at much the same date the Bissoe works was revived—described in November as 'in a forward state of preparation'—by Messrs. Clint & Tregoning. The latter, a family of Gwennap origin, were Liverpool tin-plate merchants and in 1851 established their own tin-plate works at Llanelly, later becoming principal owners of the Bissoe smelter. The third works, commenced in July 1849, was owned by the Union Tin Smelting Co. and occupied part of the existing silver-lead smelting works of the Tamar Smelting Co., at Weir Quay, Bere Ferris. This had been established in 1845, originally by the adventurers in the Tamar Mines to smelt their own ores, but had been expanded thereafter into the purchase of the produce of other mines in Cornwall, Devon and other parts of the country. By 1847 this smelting department of the Tamar Silver-Lead Mining Co. had in operation '. . . 7 calciners, 3 flowing, 3 blast, 4 refining and one reviving furnace, and are building two more calciners and one flowing furnace which . . . are sufficient to smelt about 300–330 tons of ore per month, employing from 80 to 90 men.'[4] The Union Tin Smelting Co. was closely associated with the Tamar Co. and it occupied either an extension of the latter's works or perhaps even utilised a number of its furnaces which were more than were required for the purposes of lead smelting. The confusion is worse confounded by the fact that in 1850–51 the Tamar Co. was also very occasionally listed as a buyer of black tin as well as the Union Co.[5]

By 1849 there were thus eleven tin smelting works in operation in

[1] MJ 25.6.46, with 30″ pumping engine and 8″ water-pressure engine: RCG 29.1.47, with 30″ pumping engine and 18″ steam whim. So far as is known there was no intermission in working at Providence during this time, and it is possible that this was part only of the sett being sold off.

[2] RCG 30.11.49.

[3] MJ 31.10.48. The old works of John Taylor was almost certainly used, although an earlier reference (MJ 21.10.48) stated that "new tin smelting-houses are about to be erected at Charlestown.'

[4] MJ 23.10.47. It is not clear whether, in fact, this Tamar works was newly built in 1845 or whether it occupied the site of an earlier one. The sale of the Beeralston (or Tamar) Mines included a lead smelting-house in 1821 (WB 29.6.21) and although, after deciding to start smelting themselves in February 1845, the Tamar S-L Mining Co. were reported to be building a works in March (MJ 8.3.45), there is a confusing report three months later that their manager, Percival Norton Johnson, was reported to have bought 'the lead smelting works at Beeralston.' (MJ 14.6.45). A biography of this latter person, who was manager of the mining company, unfortunately throws little light on this.

[5] See e.g. MJ 23.7.51.

Cornwall,[1] owned by eight companies—enough, ostensibly, to ensure some measure of competition between them, and thus a fair price for his produce to the long suffering tinner. Yet the bitter experience of the co-operative smelting venture of the 1840's had not been forgotten and the advent of newcomers to the trade was probably only by sufferance of the established ' ring ' and by compliance with their unwritten rules and regulations. As the *Mining Journal* sadly commented after the failure of the Miners Company smelting plans, 'the pure smelters, relieved from competition, proceeded to act upon

> The good old plan,
> That they should take who have the power,
> That they should keep who can'

Thus, unobtrusively but completely dominated by the smelting interest, the Cornish mining industry stood poised at mid-century for great expansion.

[1] Daubuz (Carvedras, Treloweth): Williams Harvey (Trethellan, Mellanear): Bolitho (Chyandour, Angarrack): Batten (Trereife): Tregoning (Bissoe): Enthoven (Charlestown): Michell & Bolitho (Calenick): and the Union Company (Tamar). In addition there was also Higher Blowing House still in active operation on a small scale, owned by Daubuz, at St. Austell.

Missing from this list—although not strictly a Cornish smelter—was the name of Gill & Co. (Devon Smelting Co.) of Tavistock, and perhaps Calstock, which appears to have fallen victim to the depression of the early 1840's. In 1834 it was recorded as having coined 686 blocks, in 1835 545, in 1836 297, and in 1837 1149 blocks, but no record is known of its output in the years thereafter.

11

DEEP TIN AND THE RECESSION IN COPPER

WHEAL VOR AND THE RISE OF DOLCOATH: 1850–1869

'For a long period tin mining has been very unprofitable to its numerous adventurers, and until within a very recent period tin mines were looked on with distaste, especially by out-adventurers.'

Mining Journal, 17 September 1859

AT MID-CENTURY, something like normality had returned to the money and metal markets after the depression of the 1840's—the hungry forties of the labouring poor. In Cornwall the number of dividend mines in 1850 increased by a half, from thirty in 1849 to forty-five, and the amount they paid in an even greater proportion. This increase, however, was confined almost entirely to copper producers and tin continued to be virtually a side-line compared to copper output, both in value and tonnage. As had been the case before, in the 1820's, tin remained very subsidiary to copper in the great upsurge that took place in metal mining in the early 1850's. Several reasons may be advanced for this: first and foremost that the prizes in copper mining in the south-west were immeasurably greater, as witness the fabulous career of Devon Great Consols which had paid £187,528 in dividends since its discovery only six years before—more than all of Cornwall's tin mines put together in that same time—or the equally extraordinary revival of fortune of Wheal Buller, where the shares rose within months from a value of £5 to £1500 apiece when a massive new lode of copper was unexpectedly struck in the Grey Bird part of the sett.[1] Copper lodes, too, were more dependable than tin, the mining of the ore, its dressing and subsequent sale more easily understood by the outsider. Thus the greater attraction of a twelve-foot lode of copper ore over even a rich two-foot lode of cassiterite was obvious, and even by 1853 the value of the black tin produced in the two south-western counties was only one quarter that of the output of copper ore.[2]

Nevertheless, in the expansion in mining in and after 1850[3] a number of

[1] MJ 4.3.65.

[2] 8866 tons of black tin value £277,186: 181,944 tons copper ore, value £1,155,167 (G. Henwood, *Four Lectures on Geology and Mining*, 1855).

[3] Some conception of the number of mines opened or re-opened at this time in Cornwall and Devon can be formed from the colossal outlay on them in 1851–2. The calls in the former year were £302171, the dividends £214319. In the following year, the calls were £468749, the dividends £255029. (WB 31.12.52.)

tin mines were re-started. As before, the smelters played an important part in this, expanding their black tin purchases in order to meet the increasing demand from the rapidly growing Welsh tin-plate works as well as to maintain their control over the industry. One of the first big mines to be re-started was Pednandrea, in Redruth, in October 1850.[1] This was followed in 1851 by the resumption of Wheal Uny, which had been worked for copper in the 1840's by the Foxes of Falmouth as part of Redruth Consols; Mill Pool; Penzance Consols; Wheal Jane; East Wheal Margaret; North Levant; Chypraze Consols (St. Enoder); Trannack & Bosence; Great Wheal Fortune; Wendron Consols; Wheal St. Agnes; Gaverigan Mine (St. Columb); Nancegollan Mine; and Bottle Hill near Plympton. The last named had earlier been one of the most important of the tin mines in Devon—although unimportant by Cornish standards—and at the re-opening had eight shafts open, five to the 40 fathom level and three to the 110.[2] In 1851 also a new company took over Polgooth where it was reported that the sale of the first parcel of tin 'raised during the first twenty-four days of the new company's operations has just taken place, and realized £1190 11s. 11d., giving a profit over the working costs of £136 19s. 1d.'[3]

In 1852, with the price of tin still rising, others followed: Leedstown Consols, with a nominal capital of £30,000; West Ding Dong; Lelant Consols, under the revived name of Wheal Kitty; South Wheal Reeth; Great Hewas; East Polgooth; Carnyorth; Wheal Kitty, St. Agnes.[4] In Gwennap, St. Day United Mines was formed to re-work Poldice, Wheal Unity and Wheal Gorland for copper and tin, whilst south of Tavistock, tin mines such as Rix Hill, Wheal Anderton, East Crowndale, Plymouth Wheal Yeoland and others were also started. In the following year several other mines were floated but in terms of tin the major venture was the re-working of Great Wheal Vor. This was to be on a scale that had been seen before in Cornish copper but not in tin, and her career was to dominate the pattern of tin mining for the next decade. At a meeting held in December 1852[5] the scheme for the company's formation, as Great Wheal Vor United, was outlined: all of the old shares had been bought in, including the former holding of John Gundry over which the lawyers had wrangled for a quarter of a century, and thus the mine was free of any threat of further litigation.[6] The capital proposed was

[1] RCG 4.10.50. In the prospectus it was stated that great quantities of tin had been left unwrought and that when the materials and pitwork were being drawn up 'four men who went down and never returned during three days, realized a profit on their proportion of the Tin they brought to the surface, of £18 per man.'

[2] RCG 13.6.51. Two large water-wheels and a 50″ engine for pumping were to be installed.

[3] WB 20.6.51.

[4] The re-starting of Wheal Kitty was an important local event and was duly celebrated in traditional style; 'the church bells rang merrily and the St. Agnes amateur brass band enlivened the church town by their masterly performance.' (RCG 3.12.52.)

[5] RCG 31.12.52.

[6] This, at long last, enabled the matter of the Gundrys' bankruptcy to be settled. This had outlasted their lifetimes, for they had still not been discharged when they died in the late 1840's. Their final dividends were declared by the court in November 1853, thirty-three years after bankruptcy pro-

£200,000, of which £50,000 (£7000 in cash and £43,000 in shares) was 'to be paid to persons buying up and for propitiating adverse interests'[1] and the re-working was to be managed by the two brothers, Henry and Edward Crease, sons of the late lessee of the Duchy tin rights.[2] At a meeting held in Helston there was jubilation at the prospect of the re-working, for the mine had contributed greatly to the business of the town during the years of prosperity. The future of Wheal Vor it was said was not only rosy, it was certain: 'with the present high price of tin, the resumption of the mine . . . is not a speculation but an investment.[3]

The major problem of re-starting a mine of this size was draining the old workings, a gigantic task on which hinged the whole success of the venture. It was decided to install two big engines for pumping, an 85″ on the old Trelawny's Shaft and a 100″ on Crease's (formerly Borlase's) Shaft. The latter, in keeping with the whole scale of the re-working of Wheal Vor, was larger than any engine ever used on a Cornish mine.[4] These engines were ordered from Harvey & Co. of Hayle—major shareholders in the new company—in August. By December an immense amount of work had been done in bringing the mine into an efficient state. When taken up by the new company, the sett presented 'an appearance of great desolation, ruined buildings, deserted cottages and large mounds of refuse ores.'[5] A start was made on clearing up and re-timbering the two adits throughout—the main one about 800 fathoms long—on re-opening their air-shafts, putting ladder-roads in the main footway, hauling and engine shafts, making roads and surface drains, levelling burrows, putting in launders to carry water for dressing purposes, building houses for the pumping and whim engines, re-building the count-house, store rooms and engineers' shop, re-thatching the timber houses and erecting a saw-mill, digging out three ponds for storing water, as well as renovating and extending the dressing floors at the main South Stamps and at Mount Pascoe.[6] To enable immediate returns of tin to be made, the adjoining

ceedings had been commenced. Some of their creditors received settlement in full, and most of the others 17s. in the £., Thomas Gundry's debts having been £17,138 and John Gundry's £16,720 in 1819. (RCG 8.7.53, 4.11.53). What occupation they turned to during this time it would be interesting to know. A Thomas Gundry is recorded as being master at Wheal Vor school (*Collectanea Cornubiensis*, 1890) and if this was one of the brothers, as seems probable, his feelings under such circumstances can be imagined—teaching miners' children on the very mine which he had once managed and which had bankrupted him before it became the richest for tin that Cornwall had ever known.

[1] MJ 1.1.1853.

[2] The first named was in fact the chief promoter of this working of Wheal Vor. His naval rank of Captain gave him a somewhat misleading advantage in mining affairs, except to Cornishmen. He had also been one of the directors of the proposed Cornwall Tin Smelting Co. of 1836. (See RCG 23.9.36).

[3] RCG 7.1.53.

[4] A few 90″ engines had been installed, for re-working copper mines, chiefly in Gwennap. An engine with 100″ cylinder was almost half as powerful again as these. See D. B. Barton, *The Cornish Beam Engine: its History and Development*, 1965.

[5] Although most of Wheal Vor was abandoned and its appearance evidently ruinous, part at least was re-working in 1850, when the *Royal Cornwall Gazette* reported (5.4.50) that ' in consequence of a re-action in the [tin] market, she [Wheal Vor] has been re-commenced and within eighteen months a standing debt of about £12,000 has been paid off and a dividend of £30 per share was paid, the mine being we are informed at present in a position to pay dividends for years to come.'

[6] MJ 24.12.53.

sett of the Flow was acquired for £8500[1] whilst the sett of Wheal Metal was also added during the year. At this juncture it was decided to restrict the working capital to £100,000,[2] although money continued to be spent like water on the property during 1854. By June tin sales, chiefly from the Flow, were £12233 and operations at Wheal Metal (then 50 fathoms deep) were giving promising results. In August the first of the two big engines (the 100″) was put to work, months past its contract date, and the task of draining old Wheal Vor began. Work was started alongside it erecting the 32″ whim and man-engine to work in the same shaft. Elsewhere in the sett renovation was continued; three more large condensing and dressing water reservoirs constructed, a steam capstan installed to drop the pitwork in Crease's Shaft, the building of two calciners as well as a dry and changing-house for the miners, the overhauling of the 56 heads of steam stamps and the 16 heads worked by water. By November the Chairman, H. W. Schneider,[3] could report that ' what they had done had been done with great energy and crowned with signal success.'[4]

After another six months, by June 1855, old Wheal Vor had been drained to the 70, and mining operations were in progress to this depth. Metal Engine Shaft had also been sunk to a similar depth below adit. 465 tons of tin had then been sold, value £26414, as against an outlay of £43922. The second big (85″) pumping engine was in process of erection, whilst about this time negotiations were started for a lease of Angarrack works for smelting Wheal Vor tin.[5] Eight horse-whims were in use and to handle the ore at surface more expeditiously a railroad 400 fathoms in length from the main shafts to the stamps was laid in. The latter had been increased to 88 heads, with 32 more to be coupled in as output increased, worked by a 36″ engine.

Six more months and the debit balance against the old mine stood at £62274, with the Flow showing a profit of £6060 and the Metal section just paying costs. Sales to this date were 648 tons of tin, value £38332. The main work accomplished in the half year was the provision of a 26″ whim at Wheal Metal and a double 9″ portable ' puffer ' engine for drawing from Trueman's Lode, and the excavation of yet a seventh reservoir, bringing the water storage capacity on the mine to no less than 3,505,756 gallons. In addition to the main tin dressing floors, with their 314 frames etc., and where 5000 tons of tinstuff could be treated a month, copper ore floors were put in, with the requisite cobbing houses, bucking mill, etc. This was on the strength of copper ore which it was hoped to raise from lodes in the northern part of the sett close to the old Godolphin

[1] That ' it is expected that with the expenditure of a moderate sum of money, it may of itself, within a short period of time, become a good paying proposition.' (MJ 24.12.53.)

[2] RCG 30.12.53.

[3] A Londoner who had made great sums in the previous few years in copper mining in Australia and who was later a major shareholder in the Furness iron mines.

[4] MJ 23.11.54.

[5] MJ 20.3.58. This lease was drawn up but never signed by the adventurers. Angarrack was advertised for sale RCG 11.11.53. (See on, page 129).

Bridge Mine which had formerly been fabulously rich for copper.[1] The heaviest expenditure was now almost complete, for the 85″ had been started in September and the rate of forking was expected to be double thereafter. The mine was drained to the 144, although this had cleared only the least extensive part of the old workings and an immense pool of water still lay below to be pumped. Some trouble with the pitwork was encountered, for the shaft was found to be smaller than had been thought and, with 4′ longer stroke in the pumps, bob-plats had to be cut back considerably and the whole shaft enlarged in places in what was extremely hard and costly ground.[2]

By the summer of 1856 Great Wheal Vor had been in commission again for a full three years and although progress was steady and the prospects encouraging, the financial position of the company was not very satisfactory. Main Lode had been drained only to the 164 and this was delaying the returns that had originally been calculated. Even though Metal Lode was proving 'in itself a first-rate mine,' being particularly rich at the 80 fathom level,[3] it was necessary to issue more shares, amounting to £53334 in value, to repay outstanding loans. Somewhat belatedly, it was realised that there was the need to 'advise the utmost economy in the application of any further capital.' Output to this time had been 845 tons of black tin, sold for £53990.[4]

The proposed new issue of shares proved to be a failure, for there were very few subscribers, and in October the shareholders had the unpleasant experience, not of the rich dividends which they had so confidently expected long before this, but of a call to clear off the outstanding debts of £46,000.[5] Attention was still being devoted almost entirely to the problem of unwatering the main lode and even though 1154 persons were employed in all (700 men, 218 boys and 236 females) only 68 men were working underground in Wheal Metal whence the chief returns were coming.[6] The company's aim and object was still the great lode 'which all authorities represent as having been left in the bottom, of undiminished richness,'[7] a lode elsewhere described, and disputed by no-one to this date, as 'the finest tin lode in the world.'[8] Nor did the company neglect the numerous side lodes, as later writers have said; this had been one of the avowed objects of the re-working as declared in the prospectus, and in 1854 it was stated that 'by short cross-cuts the deposits of tin in the adjacent side lodes' were also being tried.[9]

Even though the returns from the Wheal Metal section 'already far

[1] MJ 16.9.54. [2] MJ 22.11.55.

[3] The produce from here was so rich 'that a great part of it had to be brought to the surface in sacks.' RCG 28.11.56.

[4] MJ 19.7.56.

[5] The working capital of the company and the prospect of almost immediate ore sales had been considered so ample for the re-working at the outset that the company was not a true cost-book concern, and it had to be re-constituted at this date to enable calls to be made legally on the shareholders by Stannary jurisdiction.

[6] MJ 25.10.56. [7] MJ 1.1.53. [8] RCG 4.8.54. [9] RCG 31.10.56.

exceeded the highest estimate made,' with reserves laid open worth over £40,000[1] the painfully slow progress in draining the main workings was still the company's chief concern. By December 1856 the water level was only down to the 188 and six months later to 227 fathoms below adit. True, this was near to the old 236 fathom (bottom) level wherein lay the chief object of the re-working, but it would still be necessary to fork beyond this for permanent working and meantime, during the interruptions from breakages in the pitwork or during the period when pumping had to be stopped to send down and fix another lift of pumps, the water level rose each time and had again to be re-forked.[2]

The progress of Wheal Vor had inevitably been the chief mining topic in Cornwall since 1853 but the scale of the re-working and the enormous outlay brought almost a nation-wide interest in the project. It was seen as the phoenix-like resuscitation of England's tin Eldorado, achieving even the distinction of being featured in the *Illustrated London News*, and by November 1856, when the first signs of danger became apparent, the *Mining Journal* commented, 'It would strike a blow at our national vanity not to succeed in this undertaking, which is not only great but sure to be prosperous.'[3]

By the summer of 1857, however, a blow to national vanity or not, the first tentative proposal to stop the old mine was made at one Wheal Vor meeting. This would save an enormously heavy monthly cost, sometimes approaching £10,000, and would enable the company to concentrate its efforts on the Wheal Metal section where a probable profit of £20,000 a year could be made. Yet, with only a few fathoms left to drain, it seemed folly to forego the main prize after such an outlay and the work of draining was continued throughout the remainder of 1857. By December the gigantic task was complete, with Crease's and Trelawny's Shafts dry to their sumps and the work of clearing up the bottom of the mine—found to have been left in a very disordered state from the previous working—commenced. Meanwhile another 48 fathoms had also to be drained of water in the deepest section of all, at Bounder's Shaft, an estimated further seven months' work.[4] The fall in the price of black tin, from an average of £83 to £63 per ton, was now adding to the company's heavy losses and at the end of the year there came a reckoning of what had, and had not, been achieved at Wheal Vor in the past five years. Several of the principal shareholders[5] had expressed criticism of the mine's management and they appointed a Committee of Investigation whose findings were

[1] MJ 20.12.56, 17.1.57. [2] MJ 20.6.57. [3] MJ 8.11.56.

[4] MJ 19.12.57. This was not an engine shaft and was pumped by a system of flat-rods carried through one of the levels from Trelawny's Engine Shaft.

[5] Prominent amongst these were Messrs. Harvey & Co., who had been major holders since the outset, taking shares in part payment for the engines and materials supplied. They were the only Cornishmen financially involved in the venture and as merchants supplied two-thirds of all the mine's purchases. This was a quantity so great that it was the mainspring behind the leasing of Porthleven harbour by Harvey & Co. in September 1855 (perhaps partly also to forestall Crease, who would certainly have acquired a harbour at that date if he thought that Wheal Vor needed one).

made known in January 1858. These could be summed up in the condemnation that 'the Committee of Management have wholly failed in exercising that salutary supervision over the company's affairs which their duty to the shareholders imperatively required . . . from the very commencement there seems to have been extraordinary, and almost uncontrolled, powers allowed to the managers.'[1] There were also allegations of irregularities in the mine's accounts, criticism of the excessive expenses that had been incurred and of the cost of London offices, whilst more serious charges were made about the way that the Flow section had been sold to the company by Crease for £8500 even though it was adjudged to be part of the original grant of setts to the company when it was formed. There was also the unsolved question of the mysterious acquisition of the adjoining Sithney Wheal Buller sett by Crease for the Wheal Vor company for £8685 from persons unknown, acting through nominees.[2] Henry Crease had an answer to most if not all these charges and defended himself stoutly in a pamphlet published later that year,[3] although he could not explain away the cold figures of the income and outgoings of Wheal Vor as a whole. To 31 December 1857 the expenditure on the mine had been £294769 against which there had been the following receipts:

Capital in shares	104716
Call 31 Oct 1856	44714
16 Oct 1857	11335
16 Dec 1857	675
Sale of ore	120852
Sundry receipts	1024
Total	£283316

leaving a current deficit of £11453 at this time.

The outcome of the investigation was that Schneider was replaced as Chairman, Henry Crease was compulsorily retired as Manager, and the whole direction of the mine was changed, in a clean sweep that even extended as far as the replacement of the clerks in the count-house. The new committee was determined that there would be no more extravagances such as the enormous wines and spirits bills, nor any false entries in the cost sheets, and an inspecting auditor was appointed to visit the mine weekly.[4] The first major change in the working of the mine itself was the overhauling of the pitwork[5] and by September the whole of the old mine down to the lowest section at Bounder's Shaft was free of water. 'A good course of tin' was then reported in the bottom stopes 'and all may be deemed very satisfactory.'[6] These heartening reports encouraged the company greatly, particularly in view of the continuing richness at Wheal

[1] MJ 16.1.58. [2] MJ 16.1.58, 23.1.58.
[3] *Great Wheal Vor United Mines: A Letter to the Adventurers*, H.P.P. Crease, 1858 (CRO, Truro).
[4] MJ 20.3.58.
[5] Including the substitution of wooden pump rods for those of cast-iron, which had given endless trouble in the working of the 100″ engine.
[6] MJ 18.9.58.

Metal where 'some extraordinarily rich work' had been encountered at the 110, and they bought an 80″ engine from Gwallon Mine (near Marazion) for the Carleen section in order to develop the mine eastward.[1]

Unfortunately, by December 1858, it was evident that the appearance of the lode in the bottom of the mine had flattered only to deceive. In the words of the Chairman, 'It was evident that the alleged existence of a great extent of rich and profitably productive ground in the bottom of the mine, forming the principal if not the sole object of the present adventurers in resuming its working, was an exaggeration, and that the state of the lode in the deeper workings did not justify the expectations entertained of its great value . . . it ought never to have been re-opened as a commercial speculation.'[2] Yet despite this outright condemnation of the mine's prospects, it was still decided to persevere and to give the bottom of the old Main Lode workings a fair trial after all the expense to date, and in June 1859 it was stated that there were 'still hopes of eventual success.' The mineral lords gave up their dues (of 1/36) on the old Wheal Vor section and meanwhile a 60″ was put up on Wheal Metal (then 122 fathoms deep) as well as the 80″ on Carleen. This latter was deemed necessary if the mine was to be kept dry throughout a bad winter, for the 100″ and 85″ even working at their nominal maximum of 8 strokes per minute were insufficient. The Metal 60″ and Carleen 80″ were started in October and December 1859 respectively but within a matter of days the 80″ was stopped when the eastern ends fell off abruptly in value.[3]

From January to October 1859 costs had been £47407 as against tin sales of only £25302 and a heavy call was required to make up the deficit. The amount of money laid out in Wheal Vor as a whole was now £390,415, as against receipts of £182,617 and in January 1860 a meeting was held between the adventurers and the mineral lords as to its future working. The average produce from the old mine was only 28 lbs. of black tin per ton and the obvious course was to abandon this section where the heavy loss was being made, and concentrate entirely on Wheal Metal.[4] Early in February the decision was taken and the 100″ engine stopped pumping. The 85″ was kept at work for a short time whilst the pitwork was drawn up—'an anxious operation,' for it was a case of stripping the shafts before the rapidly rising water.[5] At the sale of materials held in June the sum of £10776 was realised, although both big engines remained unsold and the company retained the 60″ and 26″ engines at Wheal Metal, the 80″ at Carleen and the 36″ steam stamps.[6]

Thus old Wheal Vor reverted again to desolation, with the two big engines left in their houses awaiting some stout-hearted buyer. The greatest single loss ever made to this date in the history of Cornish mining

[1] MJ 18.9.58.

[2] MJ 11.12.58. Expenditure to this time had been £343,008 and the sales of tin £157,315. (The Flow £35283, Wheal Vor £41889, Wheal Metal £80143). (MJ 18.12.58.)

[3] MJ 17.12.59. [4] MJ 4.2.60.

[5] MJ 25.2.60. That in Bounder's Shaft was abandoned *in situ*. [6] MJ 16.6.60.

had been incurred—over £207000—a financial disaster that 'put a stop for a long time to the clearing up of old deep mines.'[1] Even though the company had, as a result, the very promising Wheal Metal as a by-product of the main re-working, the scheme had been a gigantic failure. It was of course easy to be wise after the event and say, as many did, that the re-opening of old Wheal Vor was doomed from the outset. In fact there had been nothing except congratulations voiced to the company when the

GREAT WHEAL VOR, CORNWALL.

THIRTY THOUSAND POUNDS WORTH OF MINING MACHINERY AND MATERIALS FOR SALE.

The LARGEST and BEST ASSORTED lot of MINING MACHINERY and MATERIALS ever offered for sale in the county of Cornwall.

MR. GREENWOOD has been favoured with instructions to SELL, BY PUBLIC AUCTION, in the early part of the month of April, at GREAT WHEAL VOR MINE, near Helston, Cornwall, the following superior MINING MACHINERY and MATERIALS, viz.:—

ONE 100 in. cylinder PUMPING ENGINE, 11 ft. stroke, equal beam, with seven boilers 11 tons each, the wonder of the world.

TWO 85 in. cylinder PUMPING ENGINES, 11 ft. stroke, equal beam, with five boilers 11 tons each.

ONE 32 in. DRAWING WHIM, with capstan attached, 10 ft. stroke, with two boilers 11 tons each.

ONE 22 in. DRAWING WHIM, with capstan attached, 4 ft. stroke, one boiler 11 tons.

ONE 38 in. DRAWING WHIM, without capstan, 9 ft. stroke, one boiler 11 tons.

Also, several smaller engines and boilers, together with about 1000 fms. of superior pitwork, varying in size from 8 to 19 in. pumps, with every requisite to match, and equal to new. Also, every other article required for mining purposes.

The engines and machinery require no comment but to say they were all built by Harvey and Co., of Hayle, whose capabilities for engine building are known in every quarter of the globe.

Full particulars will be given in hand-bills and future advertisements.

For cards to view, apply to G. NOAKES, Esq., Great Wheal Vor Mining Office, Gresham House, London; Capt. GILL, the manager, on the mine; or to Mr. GREENWOOD, the auctioneer, at his mining offices, Pydar-street, Truro. R. GREENWOOD, Auctioneer.

Truro, March 9, 1860.

sett was secured in 1852. Moreover the circumstances of the earlier working, hampered by litigation and in far from united hands, made the re-working a speculation of more than usual interest. Added to this there was the testimony, amongst many others, of Captain Mark Reed (manager of the mine under the old company) of the rich lode of tin left in the bottom of the mine—and Reed was himself one of the captains in the new working until about 1858.

Setting aside the question of extravagance and mis-management,[2] the history of Wheal Vor in the 1850's showed clearly that the re-opening of an old deep mine was an undertaking fraught with great financial risk. No tin mine of comparable size and depth had been re-started before, or was to be re-started thereafter, and in this respect it was true to say that 'the working of the sett has done great injury to Cornish mining.'[3] The whole venture might be seen perhaps as an attempt to emulate in tin what had been done in copper as a result of re-working old mines, although even

[1] MJ 24.9.59.

[2] It is doubtful in fact whether the working was mismanaged. Certainly Crease could not be said to have re-worked Wheal Vor in an un-miner-like way. Quite the reverse; 'some of the best mining authorities of the day, in thoroughly investigating the present state and progress of the works, all of whom had arrived at the unanimous conclusion that the money laid out had been exceedingly well expended . . .' (MJ 1.11.56.) Crease, better than any man, realised that the reverse of the old saying that a good bal makes a good captain was no less true.

[3] MJ 24.9.59.

here there were some examples of financial failures as well as spectacular successes. But the net result was that it taught the whole mining world a lesson never entirely forgotten and the name of Wheal Vor became one that was associated with enormous losses as much as with enormous gains. In fact the loss on the 1853-60 re-working more or less equalled the profit that had been made in the earlier 1820–45 period, and the only persons to gain substantially were again the mineral lords, the merchants and the smelters.

The story of Wheal Vor's disastrous rise and fall is, in bare outline, well known today. Less familiar are the other similar occurrences, albeit on a smaller scale, which also marked the history of Cornish tin in the 1850's. There was Porkellis United, re-started at mid-century, where sales of tin to the end of 1855 were only £39606 as against costs of £62689.[1] This was a big but shallow group of old mines where 460 were employed in 1856. A 60″ pumping engine, 18″ whim, numerous horse-whims, a 32″ stamps (80 heads) and two sets of water-wheel stamps (20 heads) were then in use. By 1858 the loss had swollen to approximately £40,000 and when, in September, a disaster hit the mine—caused by the collapse of an extensive slimes pit into one of the shafts, killing seven men underground[2] —the adventurers lost what little remaining heart they still had. Most of the sett was then abandoned, although a new company re-started the southern part of it under the new name of Basset & Grylls.[3]

The re-working of Pednandrea fared little better, for here the adventurers' outlay had been £27750 by the end of 1857. The mine was then re-opened to the 90 fathom level and had been equipped with a 70″ pumping engine in 1852, a steam whim in 1855, and a set of steam stamps (48 heads) in 1856. 460 were employed by 1859 and the mine by the end of that year was just paying costs.[4] The adventurers had what was considered a 'permanent' mine but any return by way of dividend for their outlay was still a long way ahead.[5] A briefer career, typical of many smaller mines

[1] RCG 27.6.56.

[2] RCG 10.9.58. The exact cause of this collapse is not clear but it was generally considered that a lode had been worked up too close to surface beneath the dressing floors in the notoriously soft ground of Porkellis Moor. The whole area had been burrowed through at shallow depth and after the accident 'a great pool of dirty water' almost half an acre in extent remained to mark the graves of the seven unfortunate men.

Nor was this the only collapse of ground that Porkellis United had suffered. Five years earlier three men had been killed and two seriously injured when a run took place from the 24 fathom level to grass. (RCG 23.12.53).

[3] An additional factor in the decision to re-name this part of the mine was the prejudice which was reputed to prevail against the entire Wendron area amongst up-country 'capitalists': 'the old name of Basset & Grylls—Porkellis United—acted most seriously against the getting up of the new company.' (MJ 10.1.63.)

[4] MJ 31.12.59.

[5] In fact Pednandrea was to lose over £100,000 in the twenty-five years after its re-opening in 1850. (see page 169).

It is essential to remember that dividends under the cost-book system were not dividends as we know them today. Until a cost-book company had received back its outlay these might have been more accurately termed 'return of capital.' A mine might pay at one period 'dividends' amounting to tens of thousands of pounds yet still be a losing concern overall, and one which—far from paying true dividends—had been a source of financial loss to the adventurers in it. It is only when the 'calls' (i.e. the money put into the running of the mine by the adventurers) are set against the 'dividends' (i.e. the money received from the mine by the adventurers) that the true dividends in the modern

up and down west Cornwall was Tregonebris and Carnebone Fatwork Tin Mine, in Wendron—a venture that was to become better known and was to be attended with far more profitable results some years later as East Wheal Lovell. This was started by a London company in 1853 and three years later, after an outlay of £7000 and the erection of a 36″ pumping engine, had sold only £56 of mineral[1]—about as much perhaps as would have paid for the cost of the candlelight. Another example was East Wheal Vor, worked from July 1852 to October 1856, and here all that the adventurers had to show for an outlay of several thousand pounds was £800, realised at auction for the 40″ pumping engine and other materials.[2] At Wheal Margery (St. Ives)—re-started in 1853—£27400 had been required in calls by the end of 1857[3] and even at Wheal Kitty (St. Agnes), one of the most favoured re-workings of the period, the outlay by September 1856 had been £48621 and of this the adventurers' calls had been £22475.[4] Other failures of the 50's were Wheal Lewis—a mine where 341 were employed in 1853—Gilmar, Wheal Wellington, and Trevenen, whilst in the St. Austell area, the re-working of Great Polgooth, East Polgooth, Great Hewas United, Great Beam[5] and Wheal Tristrem—a sett just east of Charlestown United that had been worked from 1853–56—all incurred heavy losses. At Polgooth in 1853, despite an already considerable loss, it was decided to sink a further and final sum of £10,000 in developing a new part of the sett. After three years, the mine was still poor and working ended in 1857 when the materials and machinery were advertised for sale.[6] Although this was not to be the last of Polgooth, it was the end of working there on a very large scale. The East Polgooth company had a shorter career, in which the loss to the closing in 1856 was only about £15,000. Great Hewas United, re-started in 1853, made sales of tin totalling £18086 by the end of 1856, whilst the adventurers' outlay over and above this had been about £40,000.[7] Part of this loss admittedly was due to the defalcation of the company's secretary in or about 1854 with some of the company funds but when the mine closed in 1860[8] the loss was much greater.

It took, of course, a great deal of time and money to re-open and re-equip a tin mine, more so than one of copper. A heavy outlay was certain, doubly so if the mine was to be given a fair trial: what was not certain, and never could be, was when to call halt to the expenditure, for no-one ever knew whether another couple of fathoms sinking or driving might

sense can be arrived at. This is a matter of paramount importance in any consideration of the finances of Cornish mining in the old days and this confusion between what may be termed nominal and actual dividends has given rise to several misconceptions as to the true profits made in Cornish mines.

[1] MJ 6.6.56, RCG 8.8.56. [2] MJ 18.10.56. [3] RCG 29.1.58.

[4] MJ 27.8.56. The first dividend to offset this deficit was paid in 1857 and in subsequent years up to about 1873 Wheal Kitty paid good dividends.

[5] For sale RCG 21.11.56, with 50″ pumping engine, 30″ stamps (24 heads) and 36′ water-wheel.

[6] RCG 20.3.57. Comprising 80″ and 67″ pumping engines, 35″ (120 heads) and 24″ (48 heads) stamps, three 22″ whims. According to a subsequent report (MJ 30.7.64) the accumulated deficit on the suspension of working was £3829.

[7] R. Tredinnick, *Cornish and Devon Mining Enterprise*, 1857. [8] RCG 6.7.60.

make the mine cut rich. Lured on by hope compounded with avarice, and reluctant to lose entirely what had already been ventured, the adventurers would thus continue to pay calls until such time as either the death of a major shareholder, a decline in tin prices, trouble underground or the like, knocked the heart out of them.

Almost all of these failures of the 1850's were London-owned ventures. Although the original promoters were often Cornish, there was little Cornish money lost in them. It was of course quite logical that Cornishmen

ANNUAL OUTPUT OF TIN IN CORNWALL, AND AVERAGE ANNUAL PRICE, 1850–72

	Black Tin tons	Metallic Tin tons	Price per ton of black tin	Price per ton of metallic tin
1850	10383	—	—	£ 81
1851	9455	—	—	£ 87
1852	9672	—	—	£ 92
1853	8866	5763	£68	£115
1854	8747	5947	£64	£114
1855	8947	6000	£68	£120
1856	9350	6177	£71	£133
1857	9783	6582	£76	£136
1858	10618	6920	£63	£119
1859	10670	7100	£75	£131
1860	10462	6695	£72	£136
1861	11640	7450	£62	£122
1862	14127	8476	£60	£116
1863	15157	10006	£64	£117
1864	15211	10108	£61	£107
1865	15686	10039	£55	£ 97
1866	15080	9990	£48	£ 89
1867	13649	8700	£51	£ 92
1868	13953	9300	£55	£ 98
1869	14725	9760	£70	£123
1870	15234	10200	£75	£127
1871	16272	10900	£79	£138
1872	14266	9560	£87	£153

would not neglect any likely prizes lying idle on their own doorstep—least of all the smelters who had the capital, the enterprise and the waiting outlet for black tin. The names of the Cornish tin smelters such as Daubuz, Williams, Bolitho, Michell and others, were conspicuous by their absence in each and every one of these failures. Better than any man they had a nose for tin, and, equally important, a nose for knowing where there was no tin. It is probably true to say that there was not a tin mine of any consequence, before or after this date, in which smelters did not hold an important, and in many cases, a majority share. In part this explained the low rate of dividend on Cornish-held tin mines, the profits coming at the smelting not the mining stage. In 1856 it was estimated that tin smelting profits exceeded £30 per ton of metal[1] and it is possible that even this was well below the true figure.

[1] MJ 10.1.57.

Certainly the dividends from tin mining in the 1850's were paltry in comparison to those from copper mines in the south west, and it was from the latter that the big fortunes were made. How subsidiary tin dividends were to those of copper may be seen from the dividends of any single year, as for example 1850. The ten best tin mines then paid a total of £20527, the ten best copper mines £120,620. Nor was this exceptional, for in 1855 the comparable figures were £30105 and £222,178. Three copper mines (Devon Great Consols, Wheal Basset and Wheal Buller) each individually surpassed the whole of Cornwall's tin dividends in this and other years. Even by 1860, on the same comparative basis of the ten richest producing mines for each of the metals,[1] tin profits were £51150 as against copper profits of £195,720. This relationship in size and in richness between tin mining and copper mining in Cornwall cannot be overstressed: one of the big copper mines in its heyday gave as much profit in a year as a comparable tin mine in ten: Wheal Buller in 1853 gave £45440—not its maximum, for its profit in the following year was £53760—more than St. Ives Consols which yielded £32900 in the eleven years from 1850–60 or than Polberro which only gave £31608 in the thirteen years after 1850.[1]

Perhaps the most heartening feature of the period for Cornwall—if only to offset the debacle of Wheal Vor—was the return into the dividend list of Dolcoath. This great mine, it should be remembered, had been first and foremost a copper producer since it was re-started in 1799 after a period of suspension (from 1787) during the great Anglesey-Cornwall copper war in the last decade of the century. It was on copper and copper alone that Dolcoath was re-established at that time, becoming the county's richest mine within a matter of three or four years from the re-commencement.[2] These great riches lasted into the 1820's but gradually declined thereafter and the last dividend on copper was paid in 1836, £157,541 having then been paid out. Dolcoath at this time was 210 fathoms deep and as in many

[1] It is interesting to note that the numbers employed in mining the two metals in the west of England did not reflect the much greater importance of copper at this date. In 1854 the following were the numbers employed—

		Males	*Females*	*Total*
Copper mining	Cornwall	15608	3684	19292
	Devon	1185	162	1347
Tin mining	Cornwall	12255	1876	14131
	Devon	583	6	589

These figures also serve to illustrate the relative unimportance of Devon as a tin producer, a reversal of the mediaeval position when Dartmoor had been regarded as the most valuable tin bearing area. In 1855 there were fourteen tin mines (mostly of very small size) in Devon as against 129 in Cornwall, and only four as against 134 in 1858. In the former year Devon output of black tin was only 320 tons out of a total of 8947, and only 54 tons out of 10618 in the latter year. (*Mineral Statistics U.K.*, 1854, 1858). In the nineteenth century mining on Dartmoor was never of any very great importance and in 1862 H. C. Salmon (*Mining and Smelting Magazine*, vol. 2, p. 108) wrote of it—'During eighty years of working, the Dartmoor district has only made one profitable mine—Wheal Friendship—against an expenditure of probably a million of money.'

[2] In 1803 the first profits of the re-working (£2000) were divided. Two years later a profit of £35764.5.2 had been divided from sales of ore value £114121.5.1, whilst in one month of 1808 alone a profit of £7040 was made. (W. Herbert Thomas, The History of Dolcoath, *The Cornishman*, July—October 1893.) These were figures compiled from the Dolcoath account books by another member of the Thomas family who were concerned in the management of Dolcoath during most of its life.

another big copper mine the tin that had evidenced itself had been almost ignored. Captain Charles Thomas, the chief agent, saw that the mine could be kept open for tin, using tributers—with hopes at the same time that further copper lodes might be opened up—and he convinced the adventurers that some pumping should be continued.[1] On this basis, helped very greatly by some fortunate discoveries of copper in the western part of the sett early in the 1840's,[2] Dolcoath was kept going, although desperately poor. The dues were entirely given up by the mineral owner, Lady Basset, whose share holding was increased as less stout-hearted adventurers dropped out.[3] Gradually the results from tin became promising enough to encourage a resumption, in stages, of re-draining the entirely abandoned lower levels of the mine. In this poor period during the change from copper to tin, the mine was held together and put back on its feet by careful use of tributers—a classic example of the value of this mode of working in times of adversity.[4] By 1849 Dolcoath was re-drained and sinking resumed under the 210 fathom level, and in April 1853, helped by rising prices for metal, the first of many dividends was paid on tin.

The transition of Dolcoath from a copper to a tin mine was a gradual process and no sudden momentous decision to continue working the mine purely for tin was taken. Thirty years later it was said that ' not one man in a thousand for one moment thought there was the slightest chance of any profit being made '[5] but in fact there had been newspaper comment which ran counter to this as early as 1836, when the *West Briton* commented ' we believe several of the copper lodes in our deepest mines, as Wheal Abraham, Dolcoath, etc., *went out in tin* and the workings were subsequently discontinued. Tin was then worth much less than copper, now it is worth more. We doubt not . . . these workings may be profitably resumed.'[6]

Dolcoath's dividends in later decades were to be a testimonial to the accuracy of this comment but initially they were slight: 1853 £2595,

[1] ' In consequence of its poverty Dolcoath adventurers have resolved on drawing up about 40 fathoms of pumps . . . foreboding that it is a preliminary to stopping the mine.' WB 6.1.37.

[2] Copper ore sales of Dolcoath during this time were—1842 3305 tons, 1843 3970 tons, 1844 3858 tons. After 1847 this output dropped swiftly, e.g. 1847 2057 tons, 1848 1254 tons, 1851 801 tons.

[3] In 1846 the market value of the mine's 186 shares was a mere £4000. It was at this time that the Fox family of Falmouth, who had been the mainspring behind the re-starting of Dolcoath in 1799, sold their interest (partly to the Williams of Scorrier, who were to gain greatly thereby). The amount of dues given up in the period from 1836–52 was £17000 (see MJ 1.6.61). Even in 1852 Dolcoath shares were quoted at only £26–£28, or about £5000 for the whole mine. (WB 10.9.52.)

[4] Tribute work, unless properly supervised, was not necessarily to the benefit of a mine, for tributers left to themselves could destroy the long term prospects of a mine by leaving deads and other waste rock haphazardly behind them in levels and other workings in order to secure a good, if temporary, return. This was frequently the case in the last few months of a mine's life when once it was known that it would have to be abandoned, with all payable ore ripped out and chaos left behind—a state of affairs that many companies formed to re-work old mines discovered to their cost in subsequent years.

[5] MJ 15.4.71.

[6] WB 8.1.36. It is also possible that Carn Brea Mines provided a stimulus to the Dolcoath adventurers to persevere for tin. They had been re-started with great success for copper in 1834 and in 1838 it was reported that '. . . towards the bottom and going west under the old Wheal Fanny mine they have a fine tin lode which provides constant employment for the new steam stamping mill of 72 heads.' (RCG 15.6.38.)

1854 (a year when there was considerable outlay on new machinery) £537, 1855 £1074, 1856 £5101, 1857 £6444, 1858 £7518. In the latter year the sale of 635 tons of black tin brought £43487, 593 tons of copper ore £2942. Henceforward, for over half a century, Dolcoath was to sell more black tin than any other Cornish mine. Even by 1859 it was said that it had ' more tin in reserve than any three mines in the county ' and with a dividend in that year of £17542, Dolcoath's triumphant revival may be said to have been some compensation to Cornwall for the contemporary disaster of Great Wheal Vor.

Dividends of the major tin mines in the period 1850–60 were:[1]

	1850	1851	1852	1853	1854	1855	1856	1857	1858	1859	1860
	£	£	£	£	£	£	£	£	£	£	£
Balleswidden	3451	3247	1948	2307	406	—	—	—	—	—	—
Botallack*	1000	1500	2550	6500	11100	8000	5800	5600	3000	3000	500
Carn Brea*	13000	9000	6000	12000	4000	4000	—	6000	8000	10000	14000
Dolcoath	—	—	—	2595	537	1074	5101	6444	7518	17542	21489
Great Work	4165	2975	3451	2082	1785	—	3867	892	—	—	—
Levant*	4000	3040	320	320	960	960	1280	1280	2240	1600	800
W. Margaret	3136	1568	392	952	2016	1904	3920	6720	4256	10976	5376
W. Owles	—	—	996	4040	3400	920	1720	1840	800	800	1800
Providence	2576	1680	—	840	1540	6300	10080	10080	10080	5040	11634
St. Ives C.	282	1222	1974	1410	752	—	1880	1880	2585	11750	9165
Tincroft*	3150	—	3150	3150	—	—	4500	3000	4500	6000	1500
Polberro	—	2680	—	3283	2525	2416	3295	4480	1506	2436	5000
East Pool	—	—	—	—	640	1280	3200	2800	960	—	—

Amongst several of the producers of both copper and tin, the same pattern of change in depth from one ore to the other took place as at Dolcoath. This was particularly marked in the Illogan district, in Tincroft, East Pool, Cook's Kitchen, Carn Brea and others, although the transition varied in its mode of occurrence from mine to mine. In Carn Brea, for example, 2780 tons of copper ore were sold in 1858 for £13989, plus 542 tons of black tin for £33586; East Pool in the same year sold £5817 of copper and £2593 of tin; Tincroft £10667 of copper and £11140 of tin. In almost every case the proportion and value of the tin produced by these mines increased at the expense of copper. One exception was Levant where the output of the two metals varied in their proportion from time to time throughout the remaining history of the mine. Another major tin and copper producer was Par Consols where sales in 1860, for example, were 3804 tons of copper ore value £33341 and 291 tons of black tin value £23160.

The improved price of black tin in the late 1850's, largely as a result of the greatly increased tin-plate exports from Britain,[2] brought increased

[1] MJ 30.7.64. * Copper and tin.

[2] Tin-plate exports through Liverpool alone rose from 460,194 boxes in 1850 to 811,051 boxes in 1859. Much of this was due to the great expansion in the use of cans for the nascent petroleum industry at this time, whilst another important early use of tin-plates prior to the introduction of galvanised sheets was in place of slates for the roofs of wooden buildings in the pioneer regions of America.

dividends in the established mines and another rash of interest in new ones. Operations were resumed at Rosewall Hill & Ransom United in April 1857, setts which had not been worked in depth since 1817. £12,000 was expended in bringing them into commission,[1] with 40″ pumping, 22″ whim and 26″ (32 heads) stamping engines erected. The same year the re-named East Wheal Lovell was started by a new Cornish company, whilst in Lelant East Providence[2] was re-opened and in St. Agnes Penhalls sett[3] was taken up as a separate entity by the Wheal Kitty adventurers. But the chief upsurge in tin activity came in 1859, sparked off by the greatly increased dividends at Dolcoath, St. Ives Consols and Wheal Margaret. The last named had paid £21952 in three years and the richness of the sett was responsible for the starting of both East and West Wheal Margaret. The latter, comprising the old Tincroft (Lelant) sett, was re-commenced in 1859 with a 32″ engine for pumping, whilst the nearby Trencrom Mine was also put to work at this same time. In Breage, North Great Work was started in April, although this was a small affair compared to Great Work itself. The latter was still working on a big scale, with 424 persons employed in 1860, although it was a venture of value more to the mineral lords and the merchants than to the adventurers as a source of dividends. In St. Just the old Boscaswell was re-started, and also Wheal Hearle, east of Trewellard. At Botallack, currently rich for both copper and tin, a new incline shaft and tram-road was being put in to work the 180 fathom deep Crowns submarine section more economically. In the Camborne-Redruth area—described at this time as an almost new district for tin—Dolcoath was proving a great stimulus. At Pednandrea, where 460 were employed in the late '50's, the adventurers cheerfully paid their quarterly calls, with the mine sinking below the 100 fathom level. At Wheal Uny, still not paying costs, tin was also assuming greater importance, with sales in 1859 of 100 tons of black tin for £7763 as against 258 tons of copper ore worth only £1093. The same was true at Wheal Buller which was described as ' very low for copper and the lode changing to tin . . . they are preparing for a tin mine and have a steam stamps working which is not yet fully supplied.'[4] In the next year Wheal Buller's sales were 1716 tons of copper ore, value £10852, as against 105 tons of black tin, value £83365, and this once fabulously rich copper mine looked assured as regards its future for tin. At South Condurrow the main shaft was down 20 fathoms and an engine was erected in 1859 for pumping. Tin and copper were being

[1] This was the estimate according to the prospectus and it was only £84 under the actual figure when the first sales of tin were made in June 1859. This moreover even though the depth of the main shafts from the previous working was not known with any accuracy.

[2] In this re-working East Providence became notorious for being extraordinarily badly ventilated, being often so full of powder smoke that the men could not remain underground ' for more than four or five hours at a time.' (MJ 26.12.63.)

[3] Penhalls, a mine with its origins in the eighteenth century if not before, had last been worked prior to this in the depression days of the 1840's—but not primarily for tin as its recommencement then was entirely consequent upon the discovery of ' a large and valuable copper lode ' in Trevaunance Consols, adjoining on the west. (WB 8.7.42.)

[4] MJ 24.9.59.

Trewellard Zawn
B.M. 116·5
Stones
F.P.
Trewellard Bottoms
Zawn Brinny
Levant Mine
(Tin & Copper)
Old Shaft
Old Shafts
Wheal Geevor Mine
(Tin)
Levant Zawn
Boscregan Zawn
Carn Du
Cockle Zawn
Trewellard South Cliff
Stone
Shaft
B.M. 271·6
Quarry
278
Geevor Mine
(Tin)
Carn Vellan
284
275
Stamps & Jowl Zawn
Whealcock Zawn
Roscommon Cliff
Higher Levant Mine
(Tin)
East Levant Mine
(Tin, Disused)
Levant Road
Wheal Cock
(Tin & Copper, Disused)
Spearn Moor Mine
(Tin & Copper, Disused)
Botallack Head
308
Manor House
Spearn Consols Mine
(Tin, Disused)
Trewellard House
Smithy
Chapel
Chapel Terrace
449·6
B.M. 830 3
Roscommon
Nineveh
The Crowns
Crowns Mine
(Tin & Copper, Disused)
Carnyorth Mine
(Tin, Disused)
L.B.
Meth. Chap.
B.M. 442·8
De Narrow Zawn
313
Zawn a Bal
W
School
Carnyorth Moor
Carnyorth
327
Botallack Mine
(Tin & Copper, Disused)
Botallack House
(A.D. 1665)
B.M. 424
Falmouth Place
362·4
Stone Circles
G.P. (Site of)
Creswell Terrace
348
365
B.M. 380·0
Chapel
Old Quarry
429
Loe Warren
L.W.M.O.T.
West Wheal Owles
(Tin & Copper, Disused)
Botallack
Queen's Arms
(P.H.)
Wheal Edward Zawn
Wheal Edward
(Tin & Copper, Disused)
Parknoweth Mine
(Tin, Disused)
B.M. 435·2
Range 200
300
500
Ranges 400
600 Yards
Wheal Owles
(Tin, Disused)
Truthwall
442
Carn Praunter
Ford
Stamps
Kenidjack
Lower Boscean Mine
(Tin, Disused)
Wheal Drea
(Tin, Disused)
Higher Tregeseal
413
426
Boscean
Wheal Boys
(Tin, Disused)

Polberro Mine
(Tin, Disused)
Trevaunance Cove
Harbour
Pier
Quay
Landing Rock
Star Rock
Pier Ho.
Penhall's Mine
(Tin, Disused)
Trevaunance Coombe
Wheal Friendly
(Tin)
Convalescent Home
on Site of
Manor House
Trevaunance Road
Wheal Kitty
(Tin)
Engine House
Holgates Shaft
Sunny Corner Shaft
Bluehills Mine
(Tin, Disused)
Trevellas Coombe
Trevellas
on Site of
Mansion
Wheal Betsy
(Tin, Disused)
Stream Works
(Tin)
Jericho Stamps
Quarry
West Kitty Mine
(Tin & Copper)
Foundry
Disused
Church
Peterville
Coastguard Station
School
Meth. Chapel
St. Agnes
Goonlaze
Water Lane
Gooninnis Mine
(Tin & Copper)
Goonvrea
Rosemundy
Rosemundy House
New Polbreen Mine
(Tin & Copper)
North Shaft
Vicarage
Polbreen Mine
(Tin, Disused)
Western Shaft
Footway Shaft
Old Clay Pit
Sand Pit
Trevaunance Mine
(Tin, Disused)
Old Quarries
Higher Bal
Coastguard Lookout
Tumulus
Tumuli
Bawden
Ropewalk
St. Agnes Beacon
St. Agnes
Truro

raised in quantities roughly equal in value, although the chief hope of the adventurers was to cut the eastward extension of the copper lodes that had proved so rich in West Basset. West Condurrow was a similar new mine, down to 24 fathoms depth only, whilst Wheal Grenville—at this date still almost entirely a copper producer—was 90 fathoms deep.

Mine share-dealing by this time had grown enormously, many of the London brokers having private agents in Cornwall as well as issuing printed leaflets and advertisements which set out particularly recommended mines. Aided by the new electric telegraph, by which means the real or supposed changes in a mine's prospects could be given within hours instead of days, mining share speculation began to boom. In December 1863 the new Redruth Mining Exchange was opened—in itself an inauspicious augury perhaps for the future of worthwhile Cornish mining—to take advantage of the *Mining Journal's* prediction that 'large as the present mining business in shares undoubtedly is, it is as nothing in comparison to the proportions it may hereafter assume.'[1] Tin mining was, unfortunately, a field *par excellence* for promoting and share-pushing, for 'puffing' and jobbing, as the price of shares could be made 'to run up-hill like water' and, for no very apparent reason, like the price of tin itself—could be made to 'rise like a rocket and come down like its stick.' In no other financial field in fact were there to be such possibilities for dubious and semi-fraudulent manipulation.

Statistics reveal the relative importance in 1859 of the two chief metals raised in Cornwall and Devon, with 177,864 tons of copper ore sold for £1,018,195 and 9961 tons of black tin sold for £633,501.[2] Thus although tin was a growing industry it was still the junior partner in West of England mining. This was doubly the case in terms of financial return, even though Cornwall's copper mines as a whole were in a state of slow decline. An analysis of the dividends in the two decades prior to 1863 showed how few were the tin 'prizes' compared to those of copper. No less than half-a-dozen copper mines had each paid over £200,000 in dividends whereas the most profitable two tin mines had only paid £117,359 (St. Ives Consols) and £112,062 (Dolcoath, from 1853).[3] The dividends in the sixteen years from 1847–63 of five of the biggest tin mines—

Wheal Margaret	dividends	£69328	outlay	£ 8848
Providence Mines	,,	£79511	,,	£ 5600
Wheal Owles	,,	£26732	,,	£11569
Balleswidden	,,	£26148	,,	£23433
Great Work	,,	£21894	,,	£11900

[1] MJ 31.12.59. The Mining Stock Exchange in London had been established in 1853.

[2] Of this total, 9906 tons came from Cornwall and only 55 tons from Devon, where virtually the only tin mines of any consequence were Wheal Sidney and Bottle Hill, near Plympton. (MJ 31.12.59.) The value of the output given in the Memoirs of the Geological Survey (the annual Mineral Statistics) do not agree, being £1,506,835 for copper ore and £807,502 for black tin, due to these figures being taken from the averages of prices realised by the ore.

[3] MJ 30.7.64. Mines producing both copper and tin, such as Levant, Carn Brea, and Par Consols, are not included in this comparison. To this date these were copper producers first and foremost.

could be contrasted with the dividends for a similar period of smaller and far less well known copper producers[1]—

North Pool	dividends	£61450	outlay	£4500
Trethellan	,,	£48441	,,	£1861
Bedford United	,,	£54700	,,	£9333
South Tolgus	,,	£39168	,,	£4318

In view of this disparity between the dividends of tin and copper mines, it is interesting to compare the financial career during part of this same period of one of the county's leading tin smelters. This was Williams Harvey, operating the Mellanear and Trethellan works, whose profits far exceeded those of any Cornish tin mine.[2] In ten years the company divided £233,956 and it was understandable that figures such as these were closely guarded secrets in mining circles. If these figures are quadrupled, to include the smelting activities of the Bolithos, Daubuz and the other lesser firms, a valuable comparison can be made of the relative rewards of Cornish tin mining and tin smelting. Only a hundred men or so were employed in the latter, as against thousands in the former; whilst not one-tenth part of the capital outlay or financial risk was involved, yet the returns were incomparably greater.

To the outside world, however, this was not apparent, or was only guessed at, and Dolcoath's dividend of £21489 in 1860 was hailed as an

[1] Comparison with the five richest copper mines during this same period is even more illuminating—

Devon Great Consols	dividends	£781960	outlay	£1024
Wheal Basset	,,	£281560	,,	£2624
Wheal Buller	,,	£244672	,,	£1280
South Frances	,,	£183173	,,	£9393
West Seton	,,	£154600	,,	£19000

[2] Profit and Loss Account of Williams Harvey & Co., 1850–61. (Bolitho Records DDX. 104, CRO, Truro.)

	Profit
Balance at June 1849	£44343 . 0 . 2
twelve months ended	
June 1850	£20384 . 14 . 9
June 1851	£18080 . 8 . 0
June 1852	£16615 . 12 . 6
June 1853	£16617 . 8 . 0
June 1854	£13552 . 10 . 1
June 1855	£18434 . 9 . 5
June 1856	£30602 . 5 . 3
June 1857	£22288 . 16 . 4
June 1858	£26750 . 10 . 5
June 1859	£29877 . 0 . 2
June 1860	£19620 . 9 . 10
June 1861	£10298 . 1 . 7
six months ended Dec. 1861	£ 6520 . 7 . 8

Out of this profit of approximately £243,120 made in the twelve years from June 1849 to June 1861, dividends of £233,956 were paid. The annual average dividend was thus about £20,000 (varying between £16,000 and £30,000) far more than that of any contemporary Cornish tin mine. Very few records survive of actual smelting profits in Cornwall and these figures are of considerable significance in any study of the industry.

It is interesting to speculate how Williams Harvey's chief competitor in the trade secured these detailed figures. It is probable that they were compiled in connection with the legal dispute in 1862 between various members of the Williams family, when their interests in copper smelting (Williams Foster & Co.) and in tin smelting (Williams Harvey & Co.) were re-apportioned between them (see RCG 15.8.62). After this date the principal partners in Williams Harvey (until the 1880's) were W. F. and G. Williams. (RCG 14.10.64).

example of what could be done by an old copper mine on the strength of deep tin. In that year the Camborne-Illogan mining district as a whole was said to have 'prospects before it greater than it had a century ago.'[1] The year was a good one for tin producers generally and in addition to the dividends paid by the mines listed on page 113, Great Wheal Fortune paid £899, Polberro £5000, Trumpet Consols £1600, Wendron Consols £3584, Wheal Kitty (St. Agnes) £1500, Wheal Kitty (Lelant) £1024, Wheal Mary £2000 and Boscean £1200. In addition there was Penhalls, which sold £4380 of tin in 1860 and had divided about £1200 profit—'one of the most successful speculations in St. Agnes, having been brought into a paying state with an outlay of only £6000.'[2]

Another new mine rich almost from grass was Wheal Grylls in Perranuthnoe, close by the old Falmouth Packet Inn on the Helston—Penzance turnpike road. Operations had been re-started here in 1857, chiefly by R. R. Michell of Marazion, and in 1859 a profit of £1536 was cleared on sales of only £3993.[3] Two years later a 41″ engine had been put in, although most of the working was still confined to depths at or a little below the adit, 40 fathoms deep. Until early 1862, when a 32″ engine working 24 heads of stamps was added, Wheal Grylls had no stamps and the tinstuff was sold in the stone. Some of this was rich enough, however, to be sold unstamped, 2 tons 8 cwt. of ore being sent to the smelting-house in 1861 in the same condition that it was hoisted to surface,[4] whilst by 1863 226 were employed and even with only 24 heads of stamps at work 66 tons of tin were being sold each quarter. Wheal Grylls was, in fact, a rare nineteenth century example of a tin mine brought to a profitable state without the provision of any working capital, an echo of the very different and immeasurably richer days in previous centuries when this had been the norm and not the exception in Cornish tin.

The promise of great riches from Wheal Grylls, allied to the booming state of the Wheal Metal section of the nearby Wheal Vor, was responsible for the starting or re-starting of many other setts nearby. East Grylls, floated in 1862; St. Aubyn & Grylls, later named Great Wheal Grylls; West Grylls, in 1860; South Grylls in 1863; and Wheal Florence.[5] This may be taken as typical of the way in which a rash of new setts appeared around any mine that cut rich, or bid fair to cut rich. Every great mine had its imitators and satellites, named after every compass point around the original, sometimes in directions almost contrary to what might be expected. There was a West, South, North and New Dolcoath as well as a Dolcoath United; a South, West and East Wheal Margaret in Lelant; a South, West, North, East, New and the original Great Work, to instance but a few, whilst there were to be even more numerous variations

[1] MJ 29.12.60. [2] MJ 2.2.61. [3] MJ 31.12.59.

[4] MJ 16.11.61. For a time, during the year, it was thought that Wheal Grylls had the makings of the richest tin mine ever discovered in Cornwall.

[5] Commenced in June 1861, adjoining Wheal Grylls on the west, in old workings sunk about 30 fathoms from surface. The tinstuff was stamped and returned at the nearby Leeds & St. Aubyn Mine.

a few years later on the name Lovell and Wheal Metal. As in the history of all mining fields, the richness of one mine always caused a rise in the shares of all the others around it, as well as in the market value of the other setts in the same district. This was true irrespective of whether they were on the same run of lodes or not.

Meanwhile, 1860 and 1861 saw the re-starting of a further number of mines. These included Nangiles, in the Carnon valley, in January 1861; St. Ives Wheal Allen, in 1860, immediately west of St. Ives Consols and earlier worked as part of the latter by flat-rods from one of its engines; Tyringham Consols, where a 40″ engine was erected in 1860 to re-try the old Bussow Mine; Prosper United, at Marazion, started on a large scale with two 70″ and other engines, for both copper and tin; St. Just United, floated as a company at the end of 1861 to re-work what had formerly been Bounds and Bellan Mine; Carn Camborne, started in 1860 and where an ornamental stack and engine-house were erected to grace the summit of Camborne Beacon. Other mines re-started in 1860 were Bosworthen (Sancreed), Prospidnick, South Ding Dong, Worvas Downs (Lelant), Wheal Charlotte (St. Agnes) and South Wheal Kitty (Lelant), whilst in the Wendron area four names were added to the already long list of active mines—New Trevenen, Trumpet United, West Wendron, and Garlidna United. The latter was the largest of these, a union of the old Wheal Ruby and Garlidna setts which had been worked on a considerable scale in earlier years, Wheal Ruby having had a 70″ pumping engine installed in the 1840's, the largest in the whole area.

This increase in the number of mines being started, of which the foregoing is but a small selection, was given fresh impetus by the passing of the Companies Act in 1862 which introduced the principle of limited liability into the Stannaries. This removed one of the main obstacles to investing in Cornish mining in many eyes and opened the door to widespread small-scale shareholding in mining companies. This was a development that was to prove both good and bad for Cornwall. It meant more numerous minor entrepreneurs who would ' venture ' in the working or re-working of a sett but amongst them there were few, if any, who could distinguish between good and bad mining. It paved the way for even more widespread ' bal-selling ' and share-pushing, for purely speculative and in some cases—judged by modern standards—fraudulent mining promotion. By an issue of shares under limited liability the raising of capital was made much more easy but the strict control of its use was made more difficult. In a small cost-book company every partner saw to the good use of the monies—setting aside the in-adventurers' interests—but in a limited company, with numerous, scattered shareholders whose only contact with the mine was an annual report and balance sheet received through the post, much more depended upon the mine manager and a committee.

Thus there grew up in the '60's a whole new crop of small mines, hopefully or artfully started to rival Dolcoath or Wheal Metal. It became the era of carefully phrased and invariably optimistic reports from Cornish mine captains; of larger and larger prospectuses in the *Mining Journal* for the re-working of old setts, with bolder and yet bolder statements therein of the tin riches foolishly left unwrought by an earlier generation of miners; of specimens of rich tin exhibited in the windows of Cornish mine company offices in London. It was rapidly becoming the case that there was more money to be made—and later, to be lost—in mining shares than in mining itself. Capital appreciation, by an increase in share value, became the yardstick by which a mine was judged, not its dividends, and to keep this market not only alive but fermenting as well, the *Mining Journal* grew swollen with lengthy mine reports and with less innocuous recommendations on ' what to select and what to avoid ' from a host of mining share-dealers in London, Redruth and elsewhere. Some of the latter even acquired setts themselves to be worked as market ventures, and given the right name—such as West Dolcoath, for example—up-country investors were all too easily gulled. Most of this parasitic activity was within the law, although a few cases of downright fraud are recorded. In one celebrated share-pushing case, two Perranporth mines were involved; according to the evidence at their promoter's trial, shareholders had eventually discovered that one of them, Bolingey Hill Mine, was ' simply a hole twenty or thirty feet deep, fitted with a windlass and bucket . . .'—more in fact a well than a mine. The second, South Wheal Leisure, was a rather more elaborate deception, for ' when visitors [shareholders] were expected, the engine fire was lit and men hired, so as to make it appear that the mine was in work. When the visitors went away, the fire was put out and the men discharged . . .'[1]

The opportunities for trade in all this mine speculation, as well as the stray chance of the occasional prize being turned up, brought an inevitable proportion of Cornish participation in the less outrageously speculative of these ventures. No firm could stand apart from it: ' Merchants and tradesmen of every class—from the leviathan proprietors of a monopoly port down to the pettiest village huckster—take shares with the object, avowed and unconcealed, of pushing their trade.'[2] The interest of merchants and other ' in-adventurers ' had always been a strong one in Cornish mining and a considerable proportion of the mines at mid-century were run for their prime benefit, unknown to the world at large. In-adventurers, of course, made their profits even when a mine was still in the initial stages of re-opening and even when it was continued unprofitably thereafter, whilst in more than a few rich mines they took good care to see that the lion's share of the profits—as opposed to the dividends—came to them and

[1] RCG 13.8.68.

[2] MJ 12.9.63. This was a reference to Harvey & Co. and the Copperhouse company of Hayle. The former had been involved in a libel action against Richard Tredinnick, editor of the *Mining Review*, who had said the same thing but quoted chapter and verse by instancing Harvey's financial interest in the contemporary re-workings of Wheal Vor, Great Wheal Alfred and Wheal Busy, where a total of over £450,000 had been lost. The company were successful in their action but damages of only 40s. against Tredinnick were awarded. (RCG 21.3.62)

not to the outside shareholders who had also contributed capital. The very depth and extensive nature of these 'merchants' mines' made them all the more valuable to those who supplied coal, timber, candles, castings, or rope. Nor was this all, for in such a mine banker-adventurers would ensure either a handsome overdraft or a handsome balance, smelter-adventurers would see to the sale of tin by a quietly renewed private contract, engineer-adventurers would ensure more than adequate machinery, and all would see that agents and captains received the occasional gift of one hundred guineas from the company for their devotion to duty. In such mines, exploration might even be carried on merely to employ more labour and materials to increase the profitable turnover and not the net profit. Such mines, with big output and small dividends, were also favoured by mineral lords, whose main concern inevitably was the amount of dues they could obtain. Not a few mines were re-started with the primary aim of in-adventuring in view; at the worst, if the venture failed abysmally, there were also the pickings when the mine was abandoned, whilst a mine brought to the verge of paying its way could also be bought at times very cheaply when at length London shareholders tired of the repeated calls. The provision of lavish surface buildings and machinery long before the ground had been tried was almost a hall-mark of the more speculative ventures,[1] again solely to the profit of the supplying merchants and foundries, and even further to their profit when the whole lot was knocked down ultimately for next to nothing. This is perhaps painting a black picture but it should be remembered that this was business, and even though not all mines of tin or copper were run on this basis, there is ample evidence that it was a widespread feature. Distant out-adventurers engaged in Cornish mining were fishing in deeper and more troubled waters than they realised.

Such were the financial possibilities of mine promotion, as opposed to true mining, that a great number of existing setts of solid worth were sub-divided in order that ground might be sold to newly formed companies. In this way a great deal of 'hiving-off' of unworked parts of large setts took place. In 1863, for example, West Drakewalls was formed from a portion of Drakewalls; Polgine from Wheal Grenville, under the name of South Grenville; Wheal Sparnon, as a separate entity from Pednandrea;[2] Wheal Crofty from South Crofty.[3] In every mining district

[1] 'After an extravagant outlay of thousands in management and mis-management—in razing old buildings and erecting new ones, such as counting-houses, materials-houses, and even houses and sheds for dressing and returning their tin and copper; and that when they had not a 'hat-cap-full' of either to dress . . .' RCG 20.10.37.

[2] The Pednandrea adventurers saw this as a possible way of cutting down their calls, which still continued with depressing monotony. In February 1863 the manager promised, in desperation, that 'if they made a call upon the present occasion it would be the last Pednandrea required.' MJ 7.2.63.

[3] This was the old Long Close Mine section, sold in 1863 by South Crofty for £1100 with 36″ pumping engine and 22″ whim. (MJ 21.3.63). This separate entity did not survive long and a few years later it was again bought back by the South Crofty adventurers.

of any size, every square mile and, in some favoured areas, every acre of ground was taken up for mining purposes—all of them being described in glowing terms in mine prospectuses, usually as ' the finest piece of unexplored ground in the parish.' New or newly re-worked setts in the Breage area were particularly numerous, largely on account of the current richness of the Wheal Vor company at Wheal Metal. In the mid-60's there were ' Metals ' and ' Vors ' in profusion; Sithney Wheal Vor, Penhale Wheal Vor, West Wheal Vor—re-started for the second or third time in 1864—East Wheal Vor, New Vor & Metal United, Pellor Wheal Metal, New Wheal Vor and East Wheal Metal, and no less than two North Wheal Metals, one each in Breage and Sithney parishes. As if this crop was not sufficient to confuse investors, there was also the artfully named Great Wheal Metal Limited, floated in October 1864, whose only claim to greatness rested in its title.

By the end of that year those with some knowledge of the true state of affairs were beginning to realise that all this mine promotion, and the share jobbing and gambling that went with it, was doing great harm. The year was described by one of the more upright and long established London sharedealers, Joseph Yelloly Watson, as ' made up of disappointment, great losses, disasters and disgust. Large sums of money it is true, have been made by a few promoters of new setts around ' stars '—or what were supposed to be stars—by taking advantage of ' got-up ' excitements to divide and sub-divide untried grants, each part being made to represent an enormous premium for the projectors; but without exception these . . . have proved heavy losses to those who embarked in them, have added to the disgust of the general public, and to the injury of those connected with mining on the market.'[1] Two examples may be cited of this bal-selling, or mine promotion, both in the far west—Cape Cornwall (St. Just) Consolidated Tin & Copper Mining Company in 1863, and East St. Just United Mining Company in 1864. For the first-named, the promoter—a local man—received £8750 purchase price (including £2500 in cash) for the lease and the work done; for the second (which had formed the old Bosorne and Bollowal setts) the vendors received £13500 premium, including £6000 in cash.[2]

During the course of 1864, moreover, the price of black tin fell heavily and it was questioned ' whether Cornish holders will have the stamina to carry on, in any great measure, the vast number of poor mines now working in the county.'[3] Certainly an analysis of the financial results of a typical cross-section of tin mines in the ten years to 1863 made depressing reading: North Great Work £200 dividends against calls of £14100; St. Aubyn & Grylls £896 dividends against calls of £7493; East Wheal Margaret £256 dividends against calls of £18688; Wheal Lewis £2500 dividends

[1] MJ 31.12.64.

[2] H. Salmon, *Mining and Smelting Magazine*, vols. V, VI. Both these promotions resulted largely from the successful flotation of St. Just United—the old Bounds and Bellan—in 1862.

[3] RCG 27.1.65.

against calls of £32500; Carnyorth £1739 dividends against calls of £8704; West Treasury (Crowan) £1024 dividends against calls of £10487; North Dolcoath £1250 dividends against calls of £14625; Boscaswell £936 dividends against calls of £8424; West Wheal Darlington £256 dividends against calls of £13952; Rosewall Hill & Ransom £3000 dividends against calls of £16800; Budnick Consols £208 dividends against calls of £8190; Great Wheal Fortune £10338 dividends against calls of £32903; Polbreen £512 dividends against calls of £4096; Wheal Friendly £500 dividends against calls of £7000; Wheal Hearle £256 dividends against calls of £12288; Drakewalls dividends £8960 against calls of £26240.[1] Well might the *Royal Cornwall Gazette* comment —although it is difficult to decide whether in despair or in derision—that ' hope on, hope ever, is a capital motto for mine adventurers. We cannot suggest a better.'[2]

The fall in the price of tin was due in part to this increase in the number of working mines, but even more to the depressed state of the tin-plate trade—by far the largest consumers of the metal in Britain or elsewhere—as a result of the American Civil War. As this conflict spread throughout the American states in the years after 1861 this closed, in increasing measure, the prime market for English tin and by 1865 mining was in a state of acute depression in Cornwall. Apart from a drastic thinning of the ranks of the speculative new mines, quite a few young and promising ventures in Cornish hands were stopped. Casualties that could be directly traced to this war were numerous, amounting to several dozen tin mines suspended in the three years 1862–5. Most of them were only small and many of them had been opened only months before, in the brief 1859–60 boom. In 1862 the following mines closed: Praed Consols and Carrack Dews (St. Ives), Wheal Lewis, Mill Pool (near Relubbus), West Wheal Margaret, Downderry Consols (St. Stephen), St. Austell Consols, Tretoil (near Bodmin), and St. Dennis Consols: in 1863, Tyringham Consols and Brea Consols (St. Ives),[3] South Ding Dong, South Wheal Kitty, North Wheal Vor, Prospidnick, South Gernick (Crowan), West Lovell, Great Dowgas, East Beam, Cornubia (Roche), West Polmear, Boscundle, and Wheal Messer (Lanivet): in 1864, East Releath and Calvadnack (Wendron), South Providence (Sithney), Goonbarrow & Molinnis, and Gooninnis (St. Agnes). Nor was this the end, for another dozen mines closed down in 1865.[4] Even established mines such as Trumpet Consols, Great Work,

[1] MJ 30.7.64. In these sixteen mines the dividends had thus been £33,343 and the calls £256,180.

[2] RCG 1.5.63.

[3] The prevalence of the title Consols amongst these mines was significant. Even though a mine was far removed from any ' consolidation ' of a group of old setts, and had perhaps only one main shaft, the soubriquet was a favourite one. To unthinking investors it had, by mere association, the same ring of solid worth as the true ' consols ' (Government stock) whilst to speculators it held out the hopes of a repetition of what had been done in the two richest mines in the south-west, Consols in Gwennap and Devon Great Consols near Tavistock.

[4] Wheal Hearle, East Trevelyan, West Tolvaddon, East Busy, North Hallenbeagle, Polhigey Moor, Rosewarne & Herland, Roskearnoweth, Trumpet United, and Wendron United. (*Min. Stats. U.K.*, 1862–5). To this list could be added Carnyorth at St. Just, where since working had

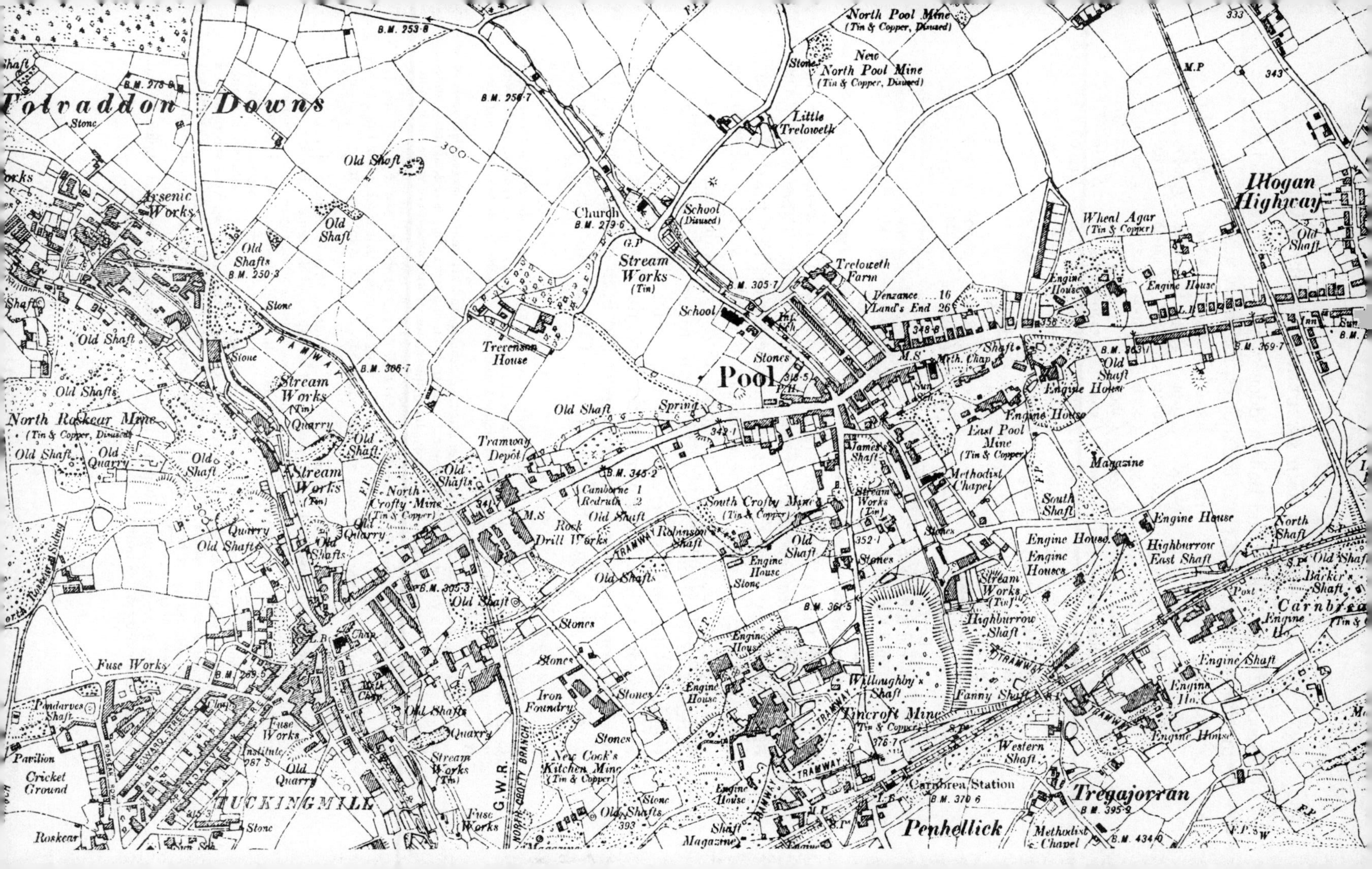
Tolvaddon Downs
Illogan Highway
Pool
Tuckingmill
Penhellick
Tregajorran
Carnbrea
North Pool Mine (Tin & Copper, Disused)
New North Pool Mine (Tin & Copper, Disused)
Little Treloweth
Treloweth Farm
Wheal Agar (Tin & Copper)
School (Disused)
Church
School
Stream Works (Tin)
Trevenson House
Arsenic Works
North Roskear Mine (Tin & Copper, Disused)
Tramway Depot
North Crofty Mine (Tin & Copper)
South Crofty Mine (Tin & Copper)
Rock Drill Works
East Pool Mine (Tin & Copper)
Methodist Chapel
Magazine
Engine House
Highburrow East Shaft
Highburrow Shaft
Tincroft Mine (Tin & Copper)
Willoughby's Shaft
Fanny Shaft
Western Shaft
Carnbrea Station
Robinson's Shaft
James's Shaft
Iron Foundry
New Cook's Kitchen Mine (Tin & Copper)
G.W.R.
North Crofty Branch
Fuse Works
Institute
Pendarves Shaft
Pavilion
Cricket Ground
Roskear
Edward Street
Tramway
Old Shaft
Old Shafts
Quarry
Stones
Penzance 16
Land's End 26
Camborne 1
Redruth 2

Boscean, Ding Dong,[1] and Balleswidden[2] were making serious losses and the dividend list of 1865 was the shortest for many years:

	1861	1862	1863	1864	1865
Wheal Vor	£9600	£2954	£4431	£14722	£20678
Dolcoath	£16826	£16110	£16826	£14678	£7308
Tincroft	£4500	£6000	£13500	£12750	£6000
Carn Brea	£2000	£4000	—	£2000	—
Providence	£4480	£5040	£5320	£4760	£4620
Botallack	—	£2500	£2800	£1600	£1200
Gt. Whl. Fortune	£899	£3596	£4945	—	—
St. Ives Consols	£2115	£1410	£2350	£470	—
Wheal Grylls	—	£1536	£4096	—	—
Wheal Margaret	—	£4704	£896	—	—
Wheal Owles	—	£2000	£1800	£1000	—
Drakewalls	—	£960	£1920	—	—
East Lovell	—	—	£715	£1668	—
Wheal Kitty (S.A.)	—	—	£1250	£4295	—
Wheal Kitty (L.)	—	—	£1024	£768	—
Basset & Grylls	—	—	£3000	—	—
Great Work	—	—	—	£1785	—

Although the root cause of this depression lay in America, it was the smelters who received much, if not all, of the blame. There was a great outcry that they were the ones responsible for the low price of tin, an obvious fallacy in view of the domination of the world market in the metal by the Banca sales. Up to about 1860 Cornwall's tin had been a sufficient proportion of the world output to be a factor in regulating supply and demand but thereafter it was the price of foreign tin that ruled the price of English. Moreover, by this date, the theory that English tin was superior, grade for grade, to foreign had been exploded, despite its careful nurturing over the years by the Cornish smelters.[3] The latter,

re-commenced in 1852 (by the same set of adventurers as in Spearn Consols) the calls had amounted to £7680 and the dividends to £2536 on the sales of 1187 tons of black tin. In 1866 the mine, with its 30″ pumping engine and 24″ stamps/whim, was sold to the Botallack adventurers for £1000. (Oats Collection, RIC, Truro.)

[1] Ding Dong, after a century and more of intensive mining, and before that of countless years of sporadic 'old men's workings,' was nearing exhaustion. Even in a sett as big as this, about 1½ miles long and above three quarters of a mile wide, and working numerous lodes, there was a limit to the mine's economic life. By 1861, Ding Dong was reported to have been worked throughout the length and breadth of the sett, and to 115 fathoms below adit at the deepest part in the east 'but the water has not for many years been kept below the 70 and the workings below that level are said not to be very extensive.' (MJ 5.10.61). The mine was then equipped with a 40″ pumping engine, 24″ stamps and 17″ whim.

[2] Despite its size, Balleswidden had done little more than meet cost since mid-century. As a result of relinquishments of numerous shares, it was decided to dispose of the mine in January 1858 but carrying it on one month more 'to afford parties the opportunity of seeing the mine in full working.' This appears to have prevented completion of the installation of gas lighting here, the adventurers having resolved eighteen months earlier 'to accept Mr. Alexander Wright's proposal to light the mine above and below ground by GAS at £450 per annum' and the plant was shortly thereafter sold to the town of St. Just. Although no buyer for the mine was found, working was continued on as large a scale as ever, with a 80″ pumping engine (bought from Wheal Vor for £1867) erected in 1861. By 1865 losses were heavy, relinquishments occurring, and an appeal was made to the lords to abate or forego their dues. (Oats Collection, RIC Truro.)

[3] MJ 30.9.65. Foreign tin, up to about 1850, was very indifferently smelted and English tin had once an acknowledged superiority.

to meet competition, lowered their prices readily enough, cutting back in proportion the price they would pay for black tin. Moreover by the 1860's they were acting in a tight and well-formed 'ring.' It was well known to the mining world at large that the price of black tin was fixed at the smelters' monthly meeting held at Truro[1] but what was not publicly known was that even the annual tonnage of black tin to be smelted was apportioned rigidly amongst them. Competition was, in fact, entirely non-existent and the mining interest lay completely in the grip of the smelters' combination.

This harmonious state of affairs amongst the smelters had not arisen overnight, nor without some troubles in Cornish smelting in the period from 1850 to this date. The Truro smelting works, which had been successively held by Daubuz, Vigurs and others, had been for sale since 1849[2] and an attempt to re-start it was made by a London company in 1851.[3] The Union Tin Smelting Co. was also not entirely successful and in 1854 the company was wound up,[4] although tin smelting was continued there by one J. G. Tyrie[5] until his death in 1863.

The Trelissick works at Hayle had also passed from the scene in the early 1850's, having been held by a variety of owners from 1847 onwards. In November of that year Williams, Daubuz and Bolitho jointly took over the lease from Charles Pascoe Grenfell and it is probable that smelting there was temporarily stopped. In September 1852 these three sold the lease for £1,560 to one F. A. Aylwin, from Kent, and Trevenen James Junior, whilst a few months later (in August 1853) it was re-assigned to a Peter Cameron. Finally, the latter disposed of the lease to Harvey & Co. for £1,050 in December 1855, who subsequently demolished the furnaces and used the premises for foundry use and for storage etc.[6]

From 1860 to 1863 the output of Cornish black tin rose by almost fifty per cent (see page 110) and to take advantage of this, two new concerns were started during this time. These were the Redruth Tin Smelting Co.,

[1] Held, according to the minutes of one of the 1865 meetings 'with a view of preventing excessive competition and making the Tin Trade fairly remunerative.' (Bolitho Records, DD BL(4)17, CRO, Truro.)

[2] RCG 29.6.49.

[3] Prospectus MJ 18.10.51, RCG 24.10.51. The works was described as having its own dock and 'three furnaces, capable of smelting upwards of three tons of black tin every six hours . . . they are built on granite foundations, the superstructure being secured with iron bracings and posts in the best manner.' How long this company survived or whether in fact the works were put into commission is not known, for it was again advertised to be let 'lately in the occupation of the Governor and Company of Copper Miners in England . . . capable of smelting 10 to 12 tons of black tin per day . . . recently constructed at great expense' only two years later. (RCG 2.8.53). In 1859 the premises were again for sale, still capable of smelting 10–12 tons of black tin daily, and described as 'formerly in the occupation of Messrs. Revel Vigors and Co., and now unoccupied . . .' (RCG 18.3.59.)

[4] RCG 24.2.54. The bankruptcy of Peter Stainsby, the principal partner, precipitated this. (see RCG 18.4.56).

[5] Chairman of Tincroft. Much of the black tin from here, as well as from Drakewalls, was smelted at the works.

[6] Leases, Harvey Records, DDH 587, CRO, Truro. The works had been advertised for sale (RCG 3.8.55), with residue of the lease of 'all those extensive and desirable Tin Smelting works situated in the town of Hayle and called Trelissick smelting works . . . capable of 1200 blocks a quarter.'

set up in 1862[1] and the Truro Smelting Co., re-started in the above-mentioned old Truro works in July 1863 by a merchant and shipowner in the town, John Gatley. At the same period, Trereife passed to the ownership of R. R. Michell of Marazion[2] who had been a shareholder from 1849 onwards in the Union (Tamar) Company.[3] Michell rebuilt and

BLACK TIN.

Mines.	Tons	c.	q.	lbs.	Price per ton.	Amount.	Purchasers.
Sold on the 24th March.							
Trevenen, &c.	2	17	0	14	.... £82 0 0	£ 235 4 3	—Enthoven & Sons.
ditto	0	11	3	27	.... 51 0 0	30 11 7	— ditto
Sold on the 31st March and 5th April.							
E. Wh. Lovell....	—				.... —	1099 9 8	— —
Treworlis	—				.... —	137 3 6	— —
Sold on the 12th April.							
Mill Pool	6	0	2	27	.... 76 10 0	461 16 6	—Chyandour.
Sold on the 13th April.							
Great Treveddoe..	2	14	0	8	.... 61 15 0	—	—Higher Blowing.
St. Dennis Cons...	3	5	2	27	.... 76 0 0	249 16 4	— —
ditto	0	12	1	3	.... 67 0 0	41 2 6	— —
Sold on the 18th April.							
Tincroft	10	0	0	0	.... 70 7 6	—	—Enthoven & Sons.
ditto	5	0	0	0	.... 70 5 0	—	—Bissoe & Co.
ditto	5	0	0	0	.... 70 5 0	—	—Calenick Co.
Sold on the 19th April.							
Bottle Hill	3	15	0	0	.... £77 12 6	—	—Williams & Co.
ditto	1	15	0	0	.... 70 12 0	—	—Daubuz.

Tin Ticketings Returns, March—April 1860 (Mining Journal). *At this date the Cornish smelters' 'ring' maintained a fixed tin standard and also apportioned smelting tonnages.*

considerably enlarged the works[4] after his acquisition of it early in 1861. From the outset he was not prepared to join the 'ring' and not only did he outrage its members by buying a greater tonnage of black tin than was considered a fair proportion for Trereife but he actually bid at prices in excess of the standards laid down by Williams Harvey and Bolitho. These two controlled the trade and it was tacitly accepted that the smaller fry would follow their line. In 1862 Trereife smelted 1842 tons of black tin, more than Daubuz at his Truro and St. Erth works, and Michell refused to join in an agreement signed at Truro in November 1863 whereby the four smaller smelters in the county restricted themselves to the following monthly tonnages—Calenick 140 tons, Bissoe 110 tons, Charlestown 100 tons, Redruth 100 tons. However, the depression in the tin market in the succeeding two years, allied perhaps to commercial pressure from

[1] The principal partners at the outset were Sampson Waters, John Little and William Elford, all connected with mining in the Redruth area at this date. The works was situated immediately below the junction of the Falmouth and Helston roads where they came into the outskirts of Redruth (the site at this date being in almost open country). The works was only a small one—probably with two furnaces at the outset. At the bankruptcy proceedings of its builder, Richard Wellington of Truro, in the following year it was stated to have cost £735 to build, taking four months. (RCG 24.10.62).

[2] Not to be confused with R. & J. Michell of Truro who owned the Calenick works jointly with Bolithos. Although possibly they were distantly related, there was no business connection between the two houses.

[3] RCG 27.4.72.

[4] Trereife had been advertised for sale in 1858 (RCG 25.6.58), having been used 'for the last four years as a white lead manufactory but still fitted up with machinery etc. . . . smelting operations were successfully carried on there for upwards of 50 consecutive years, and only abandoned when the then proprietor gave up that branch of their business.'

Trereife's larger rivals, brought unanimous agreement on the benefits of apportioning rather than competing for Cornwall's produce of black tin. In November 1865 it was recorded that 'the smelters of the county met at Truro . . . and after having shook hands over past differences, advanced the price of tin £2 in the standard.'[1] Thereafter, although the world at large was not informed, the following monthly tonnages were agreed upon and were adhered to by all the Cornish smelters—Williams Harvey 290 tons, Bolitho 290 tons, Daubuz 165 tons, Trereife 130 tons, Bissoe 104 tons, Enthoven 104 tons, Redruth 103 tons, Truro (Gatley & Co.) 60 tons, Calenick 54 tons; with any annual increase or decrease to be shared in like ratio.[2] Although the first named of these firms was commercially the most powerful at this date, the Bolithos were regarded as the patriarchs of the trade. Apart from outright ownership of the Chyandour and Angarrack works[3] they had a major share in Calenick—by this date a much smaller works than formerly,[4] whilst in matters of policy which affected the Cornish tin trade they were empowered to speak on behalf of Enthovens. Although the two firms had formerly had mutual banking interests in west Cornwall and were on friendly terms, Bolithos had no financial share or interest in Enthovens. The latter smelted lead ores as well as tin at Charlestown, buying cargoes of ore from mines in the Isle of Man, Northern Ireland and elsewhere in the early 1850's. The firm was also one of the first in Europe to buy Bolivian ores, from 1859 onwards, giving them a measure of independence from the rigidly controlled division of the Cornish supplies.[5]

[1] RCG 30.11.65.

[2] All this information relating to tin-smelting in the 1860's is extracted from the Bolitho Records (DD BL/17) now at CRO, Truro.

Another source (RCG 14.10.64) lists the quantity of black tin smelted annually as—

Bolitho Sons & Co	Chyandour, Calenick & Angarrack	3900 tons
W. F. & G. Williams	Trethellan & Mellanear	3800 tons
L. C. & W. Daubuz	Carvedras & Treloweth	1750 tons
R. R. Michell & Co	Trereife	1650 tons
Enthoven & Sons	Charlestown	1500 tons
Tregoning Sampson & Co	Bissoe	950 tons
Gatley & Bawden	Truro	800 tons
Waters, Lanyon & Co	Redruth	650 tons

In addition to these, but with outputs too small to be of consequence, were two of the blowing-houses at St. Austell (New and Higher), the latter still owned by Daubuz. Both were still buying tin in 1863 (RCG 20.3.63, 17.4.63) but appear to have been closed down sometime thereafter during the slump of the mid-1860's.

[3] Angarrack, which then had three furnaces, had been advertised for sale (RCG 11.11.53), when the Bolithos took the lease. In 1856 1916 cwt. of black tin was smelted (530 blocks) and in 1857, 2125 cwt. (585 blocks). The works was then re-built and additional furnaces put in (together with a new stamps water-wheel acquired from Ding Dong). Typical outputs thereafter were—

1858 5621 cwt. black tin (1550 blocks)	1871 14220 cwt. ,, ,, (3950 ,,)
1863 9603 cwt. ,, ,, (1835 ,,)	1875 7378 cwt. ,, ,, (2050 ,,)
1868 11875 cwt. ,, ,, (3300 ,,)	(Bolitho Records, (DD BL.21), CRO, Truro.)

[4] The scale of business of the Michells, as merchants in Truro, as tin smelters at Calenick, and as lead smelters at Penpoll had been greatly reduced after the death of Robert Michell in or about 1861. The Penpoll Lead Smelting Works of R. Michell & Son was acquired by a newly formed limited company in 1865 and in this the Michell family had only a part share.

[5] By Cornish standards, Bolivian black tin was foul (i.e. of a complex nature). Various other smelters in the county had tried to emulate Enthoven's success with South American ores but lost money by it during this period. At this date in the development of Bolivian tin production, ore could be obtained there incredibly cheaply and Enthoven sent Cornishmen out to superintend dressing

By 1865 the Cornish smelters were thus a tightly knit band, for which the freely used term 'ring' was appropriate enough. Yet whilst they assuredly did nothing to raise the price of black tin except when they were forced, they were not the prime cause of the depression which was afflicting Cornwall. World production was in excess of demand and this position was to continue until American industry was restored to normal after the Civil War. This, moreover, was a long time coming after the end of hostilities and even though the exports of tin-plate from South Wales soared in 1866 there were still large stocks of tin metal on hand—in Banca, in Holland, in Cornwall and in South Wales itself. The position was summed up in the *Mining Journal*: 'The accumulated stocks of tin during the stagnant period of the American Civil War, the increased shipments of it from the Indian Archipelago, and the enlarged development of the English tin mines would appear to offer supplies beyond the immediate demand of the market, notwithstanding that the requirements for consumption have largely increased'.[1]

In every Cornish tin mine operations were curtailed in 1865, to cut the losses being made on every ton raised, and in the summer of 1866 it was said that 'great gloom still pervades the county in all our mining districts,'[2] whilst the official view of the situation, summed up in the 1866 *Mineral Statistics*, was that 'the distress occasioned by this condition of things has been, and continues to be, very great.' In the year more than a score of tin mines were abandoned; Wendron Consols, Tolvaddon, South Wheal Lovell, East Treskerby, Retanna Hill, West Wheal Vor, Woodley Lane Tin Works (Lanivet), Great Wheal Grylls, St. Just United, Trencrom, North Hallenbeagle, Garlidna United, Mulberry Hill (Lanivet), Unity Consols (Gwinear), Clowance Wood, Sithney & Carnmeal United, West Wheal Jane and Wheal Harriet. The seriously reduced price of copper as a result of foreign competition was also bringing grave depression to that side of the Cornish mining industry, further endangering those mines being worked for both metals. Great Wheal Busy, which had been re-started ten years earlier, closed down in July and the major group of Gwennap mines known as St. Day United the following May, in each case with heavy losses.[3] The latter had sold as much tin as copper since about 1860 and over 600 persons were employed. In June it was acquired by Sir William Williams for £5,700[4] and re-set to work. Wheal Buller was in similar straits, the adventurers being buoyed up here by a tin

floors which they set up near Oruro where the barilla (tinstone concentrate, the equivalent of Cornish black tin) was prepared. (MJ 8.8.63). (See also *The Tin Deposits of Bolivia*, MJ 27.6.63 *et seq.*)

By 1862 Enthovens were importing entire cargoes of 'Peruvian' tin ore (so-called at this date as they were shipped from there, although originating in Bolivia), including, for example, one of 126 tons into Liverpool in July 1862. This was far larger than any previous consignment into Britain, all of which had been of only a few tons each. (Bolitho Foreign Ores Ticketings Book, RIC, Truro).

[1] 24.3.66.

[2] RCG 5.7.66.

[3] The St. Day United engines comprised 85″, 80″ and 65″ pumping, five steam whims (26″, 26″, 20″, 18″, 18″), a 15″ rotary and 36″ stamps engine working 64 heads. (RCG 23.5.67).

[4] RCG 20.6.67.

discovery in 1865 which, in the nick of time, had saved the mine from being completely abandoned.[1] Further east, another big tin and copper mine to go under at this same time was Par Consols, sold up in November 1867, whilst in Crowan, Crenver & Wheal Abraham United had earlier changed hands in May.[2]

It was providential indeed for Cornish mining that tin could to some extent take the place of copper and that many mines, like Dolcoath, had completed, or were in sight of the completion of, the transition from one to the other. And it was on this basis of copper mines turned to tin that much, and eventually all, of future Cornish mining history was going to turn. Dolcoath, and the richness of Tincroft during the late '60's, provided the example which all the Camborne, Illogan and Redruth mines followed. In 1865 it was said 'shafts that were stopped sinking years ago have been resumed . . . mines which a few years ago were reckoned as about the poorest in the county have gradually developed themselves.'[3] More than that, it had kept some of them going: 'Camborne Vean, Stray Park and Cook's Kitchen would have been abandoned long ago but for Dolcoath.'[4]

These results in depth had a greatly heartening effect upon adventurers whose hopes, and pockets, were wearing thin and the theory that it was only necessary to sink below the exhausted copper zone into the granite and so into the tin anywhere in this district received further encouragement by the results at Cook's Kitchen. Work was resumed sinking the engine shaft there in 1854 after four years in which the mine had been carried no deeper. By 1857 costs were being met by the returns and in 1861, *mirabile dictum*, after 51 years (since November 1809) another dividend was paid by Cook's Kitchen. True, it was only £980 as against calls of £41742 made in this same period (under seven different managements) from 1809–61 to keep the mine afloat but it was construed as an augury for the future.[5] The transition at East Pool was also similarly rewarding, and what was more important, was destined to be rewarding for very much longer than the brief resumption of dividends at Cook's Kitchen. East Pool had been re-started for copper in February 1835 and a good lode was found almost at once in the old 16-fathom level of the former working. A 36″ rotary engine was erected and by June 1837

[1] In January 1865 the adventurers had tried to sell Wheal Buller as a going concern, the mine being in only 256 shares which made the calls particularly heavy. These had been more or less constant since the last dividend was paid on copper in May 1861. It was then decided to abandon the sett (RCG 17.2.65) but only a day or two later a fine lode of tin was struck in the 70 cross-cut near Kistler's Shaft. It was then reported (RCG 17.3.65) that the lucky tributers who had made this completely unexpected discovery would share £200 the next pay-day. Another week and Wheal Buller shares had risen from £5 to £25 and by May (RCG 12.5.65) to £70. For a time there were hopes of a second Dolcoath but this was something of a flash in the pan, although it heartened the adventurers to persevere in the sett for another ten long years.

[2] RCG 23.5.67. With 90″ 80″ and 70″ pumping engines, two 30″ and one 26″ whim, and a 14″ horizontal engine.

[3] MJ 14.3.65. [4] MJ 24.9.59.

[5] MJ 11.5.61, 1.6.61. In 1861 £1592 was paid, and £1715 in 1862, although by 1863 the mine was again not paying costs.

£7256 had been paid in dividends.[1] A 60″ pumping engine (bought from Angarrack Consols) was installed in 1838 in order that the mine might be sunk deeper and, on copper, considerable profits were made.[2] For a time in the 1840's the mine was poor and in 1846 it came near to closing, with the water let in to the 70 and all work restricted.[3] On account of this endangering the adjoining East Crofty, Lady Basset (the mineral owner) insisted that the sett be fully worked or else it would be revoked.[4] Thereafter copper returns increased again, only to diminish once more late in the 1850's. It was then decided in or about 1862 to emulate Dolcoath and Cook's Kitchen and find the necessary money to recommence sinking the engine shaft in search of tin. The shaft came into granite at about 135 fathoms below adit and almost at once a rich copper-tin lode was cut.[5] This was a discovery of some importance also to Wheal Agar where activity had hitherto been confined to the eastern (the old Wheal Fortune) part of the sett and where, since the mine had been re-started in 1834 with a 50″ engine, success had been conspicuous by its absence.[6]

In a more gradual way, the district south of Carn Brea was similarly becoming one of tin rather than copper. The long string of mines extending from Wheal Grenville[7] east to Wheal Uny near Redruth had been immensely rich copper producers in the late 1840's and 1850's, and they experienced the same slow transformation into tin mines. The case of West Basset may be taken as typical, the first tin sales being made in 1857, with the amount sold and its value rising slowly but steadily throughout the ensuing years. As the sales of copper fell off, so did the sales of tin increase and by November 1866, with the mine 154 fathoms deep, tin was of equal financial importance to the company as copper. In May 1867, on the strength of the tin ground being opened up, part of the abandoned

[1] For this information on the early history of East Pool see MW 22.4.22. In the original working up to about 1800 the mine was sunk to only 16 fathoms below the adit which was itself 16 fathoms from surface at its deepest point. Pumping was carried out by means of a long train of flat-rods from a water-wheel 'to the south of Pool village.'

[2] Only nine months after the mine was re-started, it entered the dividend list, paying £12.10.0 per 1/128 share in December 1835. (Cost Book 1835–43, AD 184, CRO, Truro). In copper mines, such results were not unusual, in contrast to the majority of tin mines which required a great deal of capital outlay in shaft sinking and dressing machinery, as well as considerable time, to be brought to a dividend-paying state.

[3] MJ 29.8.46. [4] MJ 12.9.46. [5] MJ 21.3.63.

[6] One of the last mines in this district to turn to tin was South Crofty which, as late as 1868, sold only 4½ tons of black tin as against 3170 tons of copper ore. In 1869 it was said that 'the transition is gradual and decided, and the same character in our neighbouring mines has in every instance been borne out' (RCG 2.10.69), whilst as late as July 1871 the mine was still only 'fast merging from copper into tin.' (RCG 29.7.71). At this date tin was still being sold 'in the stone' (i.e. unstamped) to owners of stamps and other buyers, as no dressing equipment had been installed for returning black tin.

Carn Brea Mines, to this date, provided almost the only exception to the promising results. In the 1860's, after great riches in the previous decades, the ends were all poor and the losses so heavy that she came near to being stopped in 1865. Dividends from 1834, largely on copper, had then been £273,500 on initial calls of only £1500. (MJ 3.6.65).

[7] Wheal Grenville made its first regular sales of black tin in 1863 and in 1864 119 tons realised £7561 compared to 259 tons of copper ore sold for £1988.

North Basset sett was acquired (for £250) and sixteen heads of steam stamps, worked by one of the whims, were erected to supplement the existing sixteen worked by a water-wheel. By November 1869 it was stated that West Basset 'is destined to become a deep and lasting tin mine.'[1] Nor was it only in the central mining district of Cornwall that this transition to tin took place. Phoenix United, north of Caradon Hill, followed the same pattern of rich copper gradually giving way to rich tin in the 1860's. Here again black tin sales rose as rapidly as copper sales declined, with the tonnage increasing steadily year by year from 15 tons in 1862 to 421 tons by 1869.[2]

This turning from one metal to the other in many mines received considerable impetus during the copper slump of 1866–67, a period which marked the first major downward slide for Cornwall from her position of pre-eminence in the metalliferous mining world. The value of the copper ore produced in the West of England had been almost halved in this brief two year period and in 1867 tin regained its old premier position as the most important mineral raised in the south-west. In terms of Cornwall alone, where the decline in copper had been felt sooner and more markedly than in the less aged mines of Devon, the total output of tin had in fact become more valuable than that of copper five years before, as the table overleaf shows.[3]

During this period of slump in copper, the tin mines were themselves far from prosperous and unable to absorb many of the displaced miners and surface hands. As a result another great wave of emigration swept through western Cornwall, the majority of those who departed from the county leaving behind them families who faced something approaching destitution. Emigration had been a feature of Cornish mining life for half

[1] Book of Annual Reports and Balance Sheets, West Basset Mine, 1852–1890, CRO, Truro. West Basset had been commenced for copper in 1835 and was notorious for its lawsuit over the boundary dispute with the adjoining South Frances. The costs of this, from the first placing of the matter in the solicitors' hands in September 1857 to its settlement in July 1871—after a hearing in the House of Lords in 1866—amounted to £13321.4.6. West Basset claimed £42115 from South Frances for ore wrongfully taken (the working by the latter of the disputed lode continued for several years after West Basset claimed the encroachment) as well as for accrued interest and for the law costs. The dispute was made more complex because Basset and his steward were major shareholders in South Frances, and the former was the mineral lord of both setts. West Basset eventually won its case but it was a technical victory only, for by 1872 the original South Frances lease had expired and the old company who were responsible disbanded, and settlement consisted only of an agreement to let West Basset have a part of the South Frances sett plus the right to make use of one of its shafts for drawing ore.

What is not generally known is that in 1872 South Frances was again 'up to her old tricks' (as one RCG correspondent expressed it) when dialling revealed that encroachment had again taken place on the West Basset sett. It was then realised why Captain Pascoe had departed hurriedly for a post in Utah a few weeks before, and after the discovery 'as a consequence, there was quite a panic. Some of the knowing ones telegraphed urgent messages to certain shareholders to sell out immediately . . . the shares dropped from £15 down almost to as many shillings.' (RCG 7.9.72). Despite this initial panic, however, this second encroachment was subsequently found to be only a minor affair.

[2] 1862 15 tons, 1863 28 tons, 1864 132 tons, 1865 178 tons, 1866 274 tons, 1867 358 tons, 1869 421 tons, 1870 405 tons.

Phoenix United did not conform to the normal pattern of deep tin below copper, being an interesting case of double deposition, with the stanniferous veinstone occurring alongside and in contact with the copper lodes originally worked.

[3] *Min. Stats. U.K.*, annually.

a century but men had once gone from temptation, lured by better money, a spirit of adventure or the tales of gold. Now, as in the 1840's, bitter necessity was the spur: 'Eight thousand miners have recently left Cornwall in search of bread, having 20,000 persons dependent on their energies elsewhere . . .'[1] As a result, worse distress than had been known before was common throughout the western mining districts. In one parish

OUTPUT AND VALUE OF TIN AND COPPER SOLD IN CORNWALL AND DEVON 1861–71

	Black Tin	Copper Ore	Copper Ore from Cornwall alone
1861	£ 725560	£1364728	£816582
1862	£ 843382	£1216775	£728299
1863	£ 963985	£1100554	£642944
1864	£ 925969	£1155471	£659919
1865	£ 867435	£ 927938	£572618
1866	£ 731946	£ 759118	£431083
1867	£ 694734	£ 557431	£413533
1868	£ 770205	£ 501753	£373005
1869	£1027805	£ 402420	£316364
1870	£1002357	£ 326323	£242227
1871	£1030834	£ 284434	£205025

alone there were 1413 individuals existing 'on an income of 17½d. each per week.' Many of them, moreover, had less than one shilling and the distress was said to be 'in some cases, too appalling for publication.'[2] A County Distress Fund had been set up in 1867, with central and district committees appointed to administer the funds for the relief of the worst cases. Hardest hit were the large families left by the miners who had gone abroad, and food, clothes, bedding and coal were given to these.[3] Even amongst miners still in work, times were cruelly hard and it was recorded as nothing unusual for a man to go underground entirely without food, or at best to have had nothing for breakfast except a slice of fried turnip, and with only a piece of barley bread in his pocket for mid-day croust.[4] The great curtailment of mining operations affected all the labouring classes save perhaps the agricultural, for a wide variety of ancillary trades—smiths, masons, carpenters, day labourers, clerks, carriers, bricklayers, foundry-hands, and the like—depended upon the continuation of the parent industry of mining for their livelihood. In March 1867 there were said to be ninety mines of either copper or tin that had been floated under limited liability being wound up in the Stannary Court. The flood of copper from Chile and from Michigan was now being felt in real earnest, although it was still not yet apparent to Cornishmen that, for the red metal, their race was run.[5] The tin dividends of 1867 were also

[1] RCG 2.1.68. [2] MJ 1.6.67.

[3] At Botallack, a Visitors' Fund was started, to which contributions were made by those who visited this celebrated mine. This yielded '120 pairs of sheets' for the needy of St. Just in 1867.

[4] RCG 2.1.68.

[5] One defiant *Mining Journal* correspondent (Captain Thomas Treglown of Marazion) wrote 'Although we are but a little country, we will show we are not to be swamped, with the world against us.' (22.6.67).

abysmally low, totalling little more than £30,000,[1] and a number of old and once rich tin mines were stopped at this time. These included Wheal Reeth, which had been re-started in 1859, then 240 fathoms deep below adit—one of the deepest mines in west Cornwall—and was sold up in May 1867.[2] The adjoining Billia Consols was stopped in March 1867 after a particularly chequered history,[3] whilst in March 1868 East Great Work—commenced five years before—was sold up,[4] followed in April by Trevenen Mine in Wendron. This latter was another deep old mine re-started in the late 1850's where almost £40,000 had been vainly expended by 1863.[5] St. Ives Wheal Allen stopped the same month[6] and Great Wheal Fortune four months later. The latter, close to Great Wheal Vor, had been re-worked on a big scale, with 568 persons employed in 1863 plus five engines.[7] Even Wheal Vor itself, from being amply rich in the mid-1860's, had declined greatly and by 1869 was in low water, described as 'decaying, and a few months, should no discovery take place, will witness the suspension of the greater part of her.'[8] Another smaller mine stopped in 1868 had a brief career which may be taken as characteristic of many at this time. This was Great East Lovell which had been started in April 1864 with an initial call of £1 on each of its 3000 shares. The sett lay immediately south-west of East Lovell[9] on what it was hoped were parallel lodes, and a 50″ engine was erected on a shaft in 1865. Operations were confined to

[1] Great Wheal Vor £8862, Dolcoath £6802, Tincroft £4500, East Pool £3200, Providence Mines £2520, Trumpet Consols £2000, Wheal Kitty (St. A.) £859, Creegbrawse & Penkevil £509, Ding Dong £328.

[2] With 40″ pumping engine, 20″ whim, 30″ stamps (28 heads), and 30″ man-engine. RCG 23.5.67.

[3] This had originally been Giew Consols formed in March 1838 which had sold 2,612 tons of black tin for £191,449 by February 1859. The mine had then to be abandoned as the main (Robinson) shaft had collapsed where it passed through a cross-course. The company then transferred their engines to the Durlo Lode where, as the Durlo Mine, they sold 513 tons of black tin for £36312 in the period March 1859—January 1865. Attention was then transferred to yet another section of the sett under the new name of Billia Consols but from January 1865 to March 1867 only a further 135 tons of black tin were raised and sold. (MJ 30.10.69). After another attempt to re-start the mine under the name of Reeth Consols in 1869, the machinery was sold off early in 1870, comprising 36″ pumping engine, 22″ whim and 27″ stamps. (RCG 19.2.70).

[4] With 40″ pumping engine etc. (RCG 19.3.68).

[5] MJ 1.8.63. A serious run of ground in the mine almost closed it in August of that year.

[6] With 30″ pumping engine, 20″ whim and stamps (16 heads). (RCG 23.4.68).

[7] Comprising 70″ and 40″ pumping, 24″ and 20″ whims, 36″ stamps (48 heads), for sale RCG 8.10.68.

[8] RCG 11.2.69. In 1863 the Wheal Metal section of Wheal Vor had been particularly rich '—exceeding their most sanguine expectation . . . should this fine course of ore hold out in length and depth as it now promises, Wheal Metal may become second to none in the county.' (MJ 19.9.63). Three months later Ivey's Shaft had been carried down another 14 fathoms through ore ground valued at £300—£500 per fathom, with £70,000 worth of ore laid open in under twelve months. (MJ 19.3.64). In the succeeding two years 1623 tons of black tin were sold and by September 1866 the mine was 204 fathoms below adit. In June 1867, a typical month, 733 tons of tinstuff were raised, producing 56½ tons of black tin, viz 1 cwt. 2 qtrs. per ton of stuff. A few winzes were then valued at no less than £800 per fathom. (MJ 22.6.67). During the year skip-roads were put in at both the main shafts to help keep the levels clear and in 1868 232 were employed underground—entirely on tutwork—out of a total of 606 then employed on the mine. Since 1853 £330,227 of tin had been sold, with dividends of over £67,000 in the past ten years. (MJ 21.3.68). Later that year, a man-engine was put in (to 120 fathoms depth) but during the latter part of 1868 there came a marked falling off in the value of the stopes, although the mine was still some years off closing.

[9] Plan of The Lovells in Wendron, dated November 1869, now at RIC, Truro. The other mines shown on this are East Wheal Lovell, New Wheal Lovell, East New Lovell, Lovell Consols and Wheal Lovell.

sinking this engine shaft, with eighteen men and three boys employed on the mine, and in 1866 cross-cuts were driven north and south from the 25 fathom sump of the shaft. A maximum of thirty-six were employed at one time but the adventurers steadily lost heart throughout 1867 as the total of their calls mounted. By January 1868, still with no tin sales whatever, £7296 17s. 0d. had been vainly expended on the sett and orders were then given to pull the pitwork to surface and stop the mine.[1]

The close of 1868, however, saw an end to the depression in tin, although the inexorable decline in copper continued unabated. Tin had re-assumed its ancient position as the West of England's most valuable mineral after having been ousted from that position by copper for more than a century and a half. This change in the relative value of the two staples of the Cornish mines was to continue thereafter in even greater measure as a result of the withering away of the once great copper mining industry of the south-west.

THE GREAT TIN BOOM: 1870–1872

'Tin mining is going ahead at a tremendous pace . . . New mines, and mines which were abandoned during the long and severe depression, when only very large returns of ore would meet working costs, are being set in motion every week in West Cornwall . . .'

The Times, 2 December 1871

IN THE LATTER part of 1869, after a period of depression engendered by the American Civil War but continuing long afterwards, there came a swift up-turn in Cornish tin mining fortunes. Several successive rises took place in the price of the metal and for the cause of these it is necessary to turn away from Cornwall and consider the state of tin production in the Malay states, for the abnormal conditions there were to be the main factor that determined the pattern of Cornish tin in the next few years.

From about 1860 onwards, sporadic but serious fighting had occurred in Penang and various other states between the two Chinese secret societies (the Hai San and the Ghee Hin) which strove to rule the tin-workers there. In the previous two decades labourers in their thousands had come from China to the Malayan workings and these coolies became organized into two rival groups. Added to this there were quarrels between the various Malay rajahs, with one or other of whom the enormously powerful secret societies were allied. The skirmishes and local outbreaks of fighting grew in extent and seriousness until by 1867 almost the whole of the west coast, whence nearly all Malayan tin was then obtained, was bordering on civil war. This was at its worst in the two chief mining states of Perak and Selangor and the continuation of this unrest,

[1] The company's major asset, the pumping engine, was subsequently sold for £415 to East Lovell sometime before Great East Lovell was finally wound up in May 1872. Many of the adventurers were common to the two companies and joint count-house meetings and dinners were held by the two. (Great East Lovell Cost Book, RIC, Truro).

allied to outbreaks of rioting and widespread piracy, closed many mines and prevented full output from almost all the others. Malayan production was to suffer in this way until a state of anarchy existed, and it was finally necessary in 1874 that Britain should take direct military measures to restore order.

The effects of this Malayan chaos on the European tin market were apparent at the bi-annual Banca sale held in Holland in April 1869, for only 1,600 tons were then on offer, said to be the smallest quantity on record.[1] This, moreover, was at a time when the demand for tin was rapidly increasing, chiefly from the United States where the canning industry had made enormous strides. In South Wales, tin-plate orders were flowing in faster than they could be handled and in the five years 1868–72 production rose by almost one half in Britain, from 107,000 to 143,000 tons.[2] As a consequence of the steadily increasing demand, the price of tin in London rose from £98 to £123 in the course of 1869 and in the next two years it continued to rise yet further to a peak which had only once before been exceeded, in 1810.

One of the first Cornish mines to benefit was Wheal Owles, where a large quantity of tin was sold that had been stocked throughout 1868 in anticipation of this rise in price, enabling a dividend of fifty guineas to be paid on each of the mine's shares.[3] Elsewhere the effect was less immediate but by the end of the year the benefits of the much higher price for their produce was reflected in the accounts of all the tin mines in the county. At West Frances, in April, the first dividend was paid after twenty-two years' working. Sustained and buoyed up by the earlier copper riches of the adjoining South Frances and West Basset, the adventurers here had sunk over £50,000 into the mine, but it was on tin sales that this initial West Frances dividend was paid in 1869.[4] At North Levant the first dividend was similarly paid in September after £57,752 had been expended.[5] The

[1] RCG 4.3.69.

[2] W. E. Minchinton, *The British Tinplate Industry*, 1957.

The tin-plate trade, then as now, was singly the greatest consumer of tin from Cornwall and it is of interest to note the increase in the number of works in Britain in the period from 1860 to 1885—1860 40 works, 1865 47 works, 1870 59 works, 1875 75 works, 1885 96 works (see P. W. Flower, *Origin and Progress of the Manufacture of Tinplates*, 1886).

[3] RCG 25.2.69. Wheal Owles was a larger mine than is now commonly realised and was reported to have twenty-nine miles of levels etc. opened up by 1868, plus three miles of adits, with about an extra mile a year then being driven. From April 1834 6425 tons of black tin had been sold for £401,643 and the dividends (on an initial outlay of £5767) had been £28487. In relation to the profits, the dues paid, amounting to £16015, showed how heavy an item of costs this was. (RCG 27.2.68).

Three pumping engines, four whims, and a set of steam stamps were then in use, together with five sets of water-wheel stamps in the valley below Tregaseal.

[4] RCG 21.1.69, 22.4.69. The total expenditure at West Frances to this date had been £99487, met by ore sales (copper and tin) of £43033 and calls of £56454 (£110.5.0 per 1/512 share). The dividend paid in April was £768 (30s. od. per share), the mine being then 129 fathoms deep.

[5] RCG 4.9.69. The sales of black tin (752 tons) had realised £47345, over and above which there had been calls of £5.1.0 per 1/2000 share. The initial dividend was 10s. per share.

The following year a 45″ pumping engine was put in at North Levant, paid for out of profits, and the mine was then described as ' one of the most kindly bals in the parish '—a fitting enough description of a sett that was to survive in active form longer than any other west of Camborne. (RCG 12.2.70). Two years earlier the mine had been the scene of a disaster when the old workings

scale of operations everywhere was expanded, for the high price of tin made it worth working a great deal of ground that had been laid open but left unworked during the previous few years. New shafts were commenced and surface machinery renovated or replaced: for example, the new Millett's Shaft was started from surface at St. Ives Consols, intended to open up the mine more economically eastwards; a man-engine was put in at Providence Mines, in a shaft that—unusually in Cornwall—had earlier been partly lined with masonry;[1] the first of the new rotary calciners[2], patented by Oxland and Hocking, was put in at Wheal Basset; and at Carn Brea the dressing floors were entirely re-modelled, their original state having been picturesquely described as 'a disgrace to civilisation';[3] whilst in a hitherto unprecedented number of mines new sets of steam stamps were put in. These latter had been regarded as the prerogative only of the biggest and most permanent tin mines and small sets of a few heads of stamps worked by water power had been regarded as sufficient.[4] Now, however, with tin approaching £150 a ton and a seemingly boundless vista of profit opening before adventurers' eyes, the more stamps at work, the more tin could be returned. With expanding output and the turn from copper (where stamps were seldom used) to tin, mines such as West Basset, West Frances and South Condurrow erected their first stamps batteries during the year. South Condurrow was typical of the embryo state of many mines in this respect in the late 1860's, having a mere eight heads of stamps available in 1862 which enabled only three tons of tin a month to be returned.[5] This number was increased to 44 by

of Wheal Maitland, which formed part of North Levant, were holed into and five miners killed by the resulting inrush of water. It had been thought that Wheal Maitland had been entirely filled with waste rock and slime, for the main (Long Tom) shaft was solidly filled to surface but in fact the levels and winzes were still open and full of water. The inrush was described by survivors as like the bursting of an engine boiler but it lasted only fifteen or twenty minutes and its effect upon the subsequent working of North Levant—normally a very dry mine—was negligible. (RCG 4.4.67).

[1] Dunstan's Shaft, walled with granite to 90 fathoms depth at a cost of £600. This work was completed in 1868 to clear a section of weak ground which had given trouble in the past. Higgs' was the main engine shaft, Dunstan's being sunk primarily for ventilation purposes.

Providence Mines were regarded as particularly well-run and equipped. A big new dry had recently been built for the 250 or so men working underground, heated by steam pipes connected to the boiler of a winding engine. The workings had been carried down to 115 fathoms below adit (50 fathoms) but few men were employed in the bottom levels and most of the ore came from the 60, 70 and 80. Thus it was probably the shallowest mine in Cornwall equipped with a man-engine. A 40″ pumping engine was sufficient to keep the mine dry, with another of 30″ cylinder working 48 heads of stamps. All the dressing floors (on which 150 were employed) were roofed-in, a far from universal feature at this date. An interesting feature of the mine was the various 'carbonas,' at least seven in number, that had been worked and that had yielded much of Providence's wealth. Each was given a number, e.g., No. 7 Carbona. (See RCG 12.3.68).

[2] For roasting the black tin, to remove its arsenic content. This new type, basically a sloping cylindrical furnace which slowly rotated, did not entirely replace the older 'Brunton' calciner.

[3] RCG 15.6.72.

[4] The two great disadvantages of most water-wheel stamps were their distance from the mine, which in some cases occasioned heavy cartage costs, and the intermission in their use caused by drought or, more rarely, by severe winter frosts. Dry summers in the St. Just area at this time saw one temporary solution: 'Messrs Holman & Son of St. Just have come to the rescue of the idle 'heads' and unoccupied boys and girls. They have a travelling steam engine which rolls alongside the disused stamps, buckles itself to the axles . . . and very quickly sets the heads to rattle and crush.' This was at Spearn Consols but was doubtless used elsewhere. (RCG 13.8.70).

[5] At this same date, moreover, South Condurrow was also drawing all ore by horse-whim, but with a steam whim erected the following year.

the acquisition of a 36 head set (worked by a 36″ engine) in 1869, acquired from the recently closed Great Wheal Fortune. At this time too there were the first Cornish trials of a rock-boring machine,[1] Tincroft installing one patented by F. B. Doering on a twelve-month trial in 1867.[2] Machine boring was three or four times faster than similar work done by hand, although much more costly, and Dolcoath followed Tincroft's lead by installing another of similar make on trial in 1869.[3] Other technical innovations aimed at increased output were the installation of gas lighting on Dolcoath dressing floors in May 1872, to enable work there to proceed by night, and the building at Boscaswell of a one-mile long wire tramway ' on Mr. Hodgson's patent principle ' to connect the various shafts to the stamps.[4]

Financially the prize of this era was East Lovell, a comparatively new mine in Wendron. This had been ineffectually tried as Carnebone & Fatwork United from 1853–6 (see page 109) and its career subsequent to being taken up by a new set of Cornish adventurers in 1857 provided an admirable example of the fluctuating fortunes of all metal mining ventures. Initially only 43 men and boys were employed in 1857 in re-opening and sinking the mine but after tin sales commenced in March 1858 this number rose rapidly. By August there were 101 employed, 152 by January 1859 and 186 (116 men, 47 boys, and 23 females) by April. The Wendron district was proverbially a bunchy one in terms of mining and East Lovell was no exception, as the great variation in annual output shows:[5]

EAST LOVELL BLACK TIN SALES 1859–76

To March	1859	50 tons	sold	for	£2999	To March	1869	79 tons	sold	for	£ 4755
,,	1860	57 tons	,,	,,	£5547	,,	1870	238 tons	,,	,,	£17992
,,	1861	57 tons	,,	,,	£4619	,,	1871	227 tons	,,	,,	£18219
,,	1862	26 tons	,,	,,	£1095	,,	1872	97 tons	,,	,,	£ 7849
,,	1863	6 tons	,,	,,	£ 394	,,	1873	49 tons	,,	,,	£ 4060
,,	1864	66 tons	,,	,,	£4250	,,	1874	63 tons	,,	,,	£ 4675
,,	1865	43 tons	,,	,,	£2512	,,	1875	140 tons	,,	,,	£ 8193
,,	1866	92 tons	,,	,,	£5156	,,	1876	85 tons	,,	,,	£ 4258
,,	1867	113 tons	,,	,,	£5587	,,	1877	14 tons			
,,	1868	134 tons	,,	,,	£7376						

[1] This somewhat cumbersome name applied more fittingly to the pioneer efforts at machine mining than the smaller, handier, rock-drills of a decade or so later. Early rock-boring machines were massive and mounted on a carriage moving on rails. With stretcher bars to steady them, and complicated mountings, they took hours to set up in position for drilling.

[2] RCG 30.1.68, 13.2.68. This was in the 184 fathom level and apart from the expense its use was considered highly successful. The machine itself cost £560—the contemporary price of a new steam whim—apart from requiring the provision of an engine at surface to drive an air-compressor, and an engine-man and machine-men to work it.

[3] RCG 15.4.69. Just over a year later, three of Doering's machines were in use there, with a fourth due to be added. (RCG 9.9.70).

Despite this initial success, the high installation cost, minor teething troubles, and above all, the inherently conservative nature of almost all Cornish mine agents, the introduction of the boring machine made very slow progress for some years thereafter, until changed economic circumstances brought them back from sheer necessity. (See page 173).

[4] RCG 7.5.70.

[5] East Lovell Cost Books, R.I.C., Truro.

In 1863 the mine was so poor as to be all but abandoned but by 1868 its fortunes had revived and the following year ' the finest and most valuable lode ever discovered in the parish of Wendron '[1] was opened up 70 fathoms from grass in a new shaft which had been commenced in or about 1864. In October ' 14 tons of tin, one fortnight's returns, were risen by one pare of shaftsmen ' and in a nine months' period in 1870 about £17000 worth of tin was raised and sold at a working cost of only £4286—a state of affairs 'never exceeded in any tin mine in Cornwall'.[2] This was true enough, for only about 80 were then employed at East Lovell[3] and for its size it was undoubtedly the richest tin mine in nineteenth century Cornwall. Results such as these were probably not uncommon in the eighteenth century but after 1860 they were rare indeed. Well might the jubilant adventurers decide, on the discovery of the lode in September 1869, to share some of their good fortune with the miners by giving them all one week's wages as a donation.[4]

By July 1871 the main shaft was 80 fathoms deep and was being sunk squarely on the back of ' a lode six feet wide and worth £400 per fathom.'[5] A 36″ pumping engine was installed but the mine was equipped for raising ore solely by a water-wheel and horse-whims. It was reported that ' want of drawing power and the excessive dryness of the season prevented us from stamping out tin stuff'[6] and a steam whim was put in to remedy part of this deficiency early in 1871. East Lovell used three sets of water stamps in the valley below the mine, Glebe Stamps, Wood Stamps and Treloquithack Stamps, working sixteen heads altogether. The last were used jointly with Trumpet Consols but later were bought outright by the East Lovell adventurers.[7] In 1870 the latter divided £15248 profit and East Lovell became a great favourite with share speculators. The shares had risen from £8–£9 in April 1869 to £24 by September and to £37–£38 by the end of the year. The prices fluctuated widely, due to rumours as to the sudden falling off or the increased richness of the ' lode.' In one minor panic in November the price dropped within hours to £16—a fall in the value of the whole mine of something like £10,000.[8] The turnover in shares was enormous and inspection by agents, acting for 'bears,' 'bulls,' brokers, honest adventurers, would-be speculators, and others, became so frequent as almost to hinder the working of the mine,[9] whilst throughout this period

[1] RCG 10.7.69. [2] RCG 16.10.69, 16.4.70, 26.11.70.

[3] 38 men, 31 boys, and 10 girls, in October 1868. East Lovell Cost and Pay Book No. 3, RIC, Truro.

[4] East Lovell Cost and Pay Book No. 3, RIC, Truro. This is the only known instance of this sort of *ex gratia* payment in Cornish mining.

[5] RCG 9.7.70. It was in fact not so much a lode as a deep, almost vertical pipe of ore roughly shaped like a cylinder about 12′ x 7′ in section, not dissimilar to the carbonas of the St. Ives area. This was not realised at the time and only later was it appreciated that it was not a lode of normal lateral extent. The average yield from this deposit, from the 40 where it was first encountered to the 110 where it ended abruptly, was three tons of tin ore per cubic fathom.

[6] RCG 26.11.70.

[7] East Lovell Cost Books, RIC, Truro.

[8] RCG 6.11.69.

[9] On one occasion no less than nine well known mine captains from other parts of Cornwall

there was also a weekly full-scale inspection of the mine with published reports. At £27 per share purchase price, the annual yield was reckoned at 25 per cent in March 1871, and from October 1869 to May 1871—the peak of East Lovell's prosperity, with sales of 494 tons of black tin in that time—the profits were £26684.

East Lovell's inspiring progress in 1869 played no small part in the starting of the boom in Cornish tin and its dividend of £15248 was the talk of all mining men in 1870. The other major dividends of the year—Tincroft £32250, Dolcoath £25776, Providence £5600, Penhalls £4500, Wheal Kitty (St. Agnes) £4294—also served to inspire the adventurers in dozens of other mines. Tincroft's profit, increased from £12000 the previous year, also added weight to the theory now widely held that most, if not all, old copper mines would similarly prove rich again for tin merely by sinking. On the strength of this assumption a new company took over Crenver & Wheal Abraham United late in 1869[1] and an extract from the prospectus then published illustrates how firmly the belief was held that deep tin existed beneath virtually any proven copper ground: 'the conclusion, strongly entertained in Cornwall, that most of the Cornish mines have only to be deepened to enable rich strata of that metal (tin) to be reached . . .'[2] At the great Clifford Amalgamated Mines in Gwennap, in the greatest run of copper ground in all Cornwall, there were also hopes of turning to tin as an alternative to the frightening losses that had been made here in the previous few months on copper.[3] In February 1870 it was decided, with the lord's permission, to draw up all pitwork to the 104 fathom level and to work above there for tin. Clifford whim was to be altered to a stamps engine with 84 heads and an estimated three months would cover the re-modelling of the dressing operations on this, the largest of all Cornish copper mines.[4] Another of the big Gwennap concerns, Poldice, was also turning more and more to tin, for the production of the metal for which the parish was world famous was no longer economic: 'Gwennap miners, who used to have a profound contempt for tin and steadily ignored its existence wherever it showed itself, are now compelled to have recourse entirely to this metal.'[5] The same was also true at Devon Great Consols beside the Tamar, where a major attempt was made in 1872 to locate tin beneath the immense copper lode that had made this one of the fabulous mining prizes of the century.[6] Here, on the

jointly inspected and reported upon the mine. (RCG 7.5.70.)

With the benefit of hind-sight we can now dismiss this furore at East Lovell in 1870–71 as slightly humorous. At the time, however, with the price of tin high and going higher, and with the true extent of the East Lovell 'lode' not yet discovered, it seemed that a second Dolcoath had been found.

[1] RCG 16.10.69. These great mines had been re-started for copper in 1864 and a loss approaching £90,000 had been made when they were stopped three years later during the financial crisis.

[2] MJ 9.10.69.

[3] This was no new thing, for in 1867 there had been one tin pitch at least worked at Clifford Amalgamated, 'producing from £80 to £100 of tin per fathom.' (RCG 24.10.67.)

[4] RCG 26.2.70. A few weeks later, however, it was decided to abandon the sett entirely.

[5] RCG 25.2.71.

[6] Devon Great Consols, from 1845–65, had paid dividends exceeding £975,000. This probably exceeded the sum total of all Cornish tin mine dividends in this same period.

occasion of the lease renewal in 1872, the Duke of Bedford stipulated that a shaft be sunk a further 70 fathoms below the existing sump (230 fathoms below adit) and the general consensus of opinion was that Devon Great Consols' future as a tin mine was merely a matter of time and outlay.[1]

Although 1870 was only the onset of this brief boom, it represented to some extent the apogee of Cornish tin. Output was to rise slightly higher in 1871, to a peak of 16272 tons, but Williams' *Mining Directory* of 1870, although by no means complete, is nevertheless valuable as a survey of the mines then active and, judged by the good comparative yardstick of the numbers employed, of their size. In the St. Just area: Botallack & Carnyorth United 530; Balleswidden 470; Wheal Owles 450; Boswedden & Wheal Castle 150; Boscean 120; Spearn Consols 80; Spearn Moor 120; Levant 202; St. Just Amalgamated 230;[2] North Levant 160; Pendeen Consols 100;[3] Boscaswell 150; Ding Dong 222; Cape Cornwall Mine 12.[4] In the St. Ives—Levant area: St. Ives Consols 360; Rosewall Hill & Ransom United 130; Trelyon Consols 100; Providence Mines 450; East Providence 70; Wheal Kitty 190; Wheal Mary 200; Wheal Margaret 100. In the Marazion—Breage area: Great Wheal Vor 606; Great Work 460; North Great Work—re-named West Godolphin at this date—100; West Great Work 175; South Great Work 25; Penhale Wheal Vor 90; Leeds & St. Aubyn 70; Great Western Mines—a group formed from Wheal Grylls, East Wheal Grylls and other adjoining setts in 1869 by R. R. Michell—270. In the Wendron area[5]: East Wheal Lovell 50; New Wheal Lovell 54; Lovell Consols (formerly South Wheal Lovell) 20; Balmynhear 25; Trumpet Consols 320; East Trumpet 85; Basset & Grylls 260.[6] In the Camborne area: Dolcoath 950; South Crofty 210; Cook's Kitchen 330; Tincroft 400; Carn Brea 720; East Pool 300; North Crofty 200; Wheal

[1] See RCG 1.6.72 for the terms of this lease and the search for tin on the east bank of the Tamar. The comment of the managing director of Devon Great Consols at this time, that 'they had tin all over the mine, and had always had it' was proved utterly untrue.

[2] St. Just Amalgamated had formerly been East St. Just United, re-named in 1867.

[3] Pendeen Consols was on the point of being abandoned and the 40″ and 24″ pumping engines, a 19″ whim and 32″ (32 heads) stamps were advertised for sale in October 1870. About £40,000 had been expended here 'well and perseveringly in the previous twenty years or so but with results insufficient to warrant further effort and expenditure.' (RCG 1.10.70.)

[4] Cape Cornwall (recommenced in 1863) had been more or less abandoned and was advertised for sale in December 1869 (RCG 18.12.69) after an outlay said to be over £20,000. Two engines, a 30″ for pumping and a 26″ for stamping and winding, had been put in, the stack to the former being at first erected on the summit of the Cape (where it still stands today) until it was replaced by one beside the boiler house in the normal way when it was discovered that the draught through the long flue up the slope was too great. Early in 1870 the mine was again re-started.

[5] Williams' list of active mines in the Wendron area was by no means complete. There was Great Wheal Lovell, where a 66″ engine was put in during October 1869, and North Lovell, described as 'adjoining the rich East Lovell on the south,' where a 50″ was erected in December. Trevenen was also re-started on a small scale, together with the adjoining Tremenheere early in 1869. (RCG 14.1.69).

[6] A detailed list of the employees at Basset & Grylls exists and is useful for the light it throws on the various trades and occupations on a Cornish mine at this date. Apart from 56 men underground on tutwork and 52 men and 8 boys on tribute, there were the following at surface—4 men and one boy, carpenters and sawyers; three men and one boy in smith's shop; ten engine-men; eleven men and two boys as 'surface-men, spallers, etc'; five men, landers and fillers; dressing tin, nine men, 40 boys and 44 girls; total 150 men, 52 boys, 44 girls. In addition there were the agents, captains and count-house clerks, totalling a further 14.

Agar 50; Wheal Seton 509; West Seton 600; Stray Park 120; Pendarves United 521; South Condurrow 100; Wheal Grenville 150; West Basset 350; West Frances 260; Wheal Basset 728; East Grenville 100; South Frances 160. In the Redruth area: Wheal Buller 170; Pednandrea 380; Wheal Uny 250; East Carn Brea 170; and eastward of these, Poldice 400; Wheal Jane[1] 300; Nangiles 90. In the St. Agnes area: Penhalls 238; Polberro 250; Wheal Kitty 206; Wheal Friendly 17; Trevaunance 30; Perran Consols 60. In the central mining district of the county, centred on St. Austell, the chief tin mines were: Charlestown United 400; Cuddra 124; Great Polgooth 70; Wheal Eliza Consols 200; Mulberry 100; New Pembroke 130, whilst east of these there was Phoenix United 500; Drakewalls 340; and Prince of Wales 120. Apart from the other small mines east of the Tamar, the three largest mines active on Dartmoor itself were King's Oven & Water Hill, in the parish of North Bovey, employing fourteen; New Birch Tor & Vitifer Consols (formerly the old Birch Tor Mine) employing 102; and New Vitifer Consols, employing thirty. All were worked solely by water power and were frozen up in many winters for weeks at a time. In a previous working of Birch Tor, which lay several miles from any village, it was said that 'on account of the desolation of the place and the severity of the climate, the mine is chiefly worked by labourers who have absconded from other districts for petty offences' and that in the few wretched hovels available nearby the pressure on accommodation was such that 'the beds never cool.'[2]

In addition to these there were in Cornwall an enormous number of smaller mines, many of them newly started or re-started for tin in this boom of 1869–70. Amongst these were New Wheal Fortune, promoted in March 1869 to take up the sett of Great Wheal Fortune in Breage abandoned little more than a year before; Blue Hills, which had earlier been worked under the name of East Polberro, with a 60″ engine;[3] Garlidna, recommenced for the fourth or fifth time in the century in September 1869; Trevarrack Mine on the eastern extension of the Trencrom lode, where a 36″ engine was erected in 1870; Gooninnis, Polbreen and North Wheal Charlotte, in St. Agnes. In 1871 the price of black tin continued to advance, attaining the hitherto undreamed of figure of £80 and more per ton, a price almost double that of the lean years of the 1840's and the highest since the end of the Napoleonic wars. Through-

[1] Wheal Jane, re-started in 1850, was by this date almost entirely a tin producer. Originally the output had been chiefly blende (black jack or zinc ore) and the mine was profitable enough to pay for the erection of a 50″ engine on the old Wheal Durant section in July 1853 without making any call. The ores raised yielded a variety of minerals and in that same year the shares rose to £30 from a matter of shillings on the strength of an analysis which showed there was 2 oz. of gold per ton of gozzan. Considerable excitement prevailed and trials were also made for gold at the adjoining West Jane and Kea Tremayne Mines. (RCG 16.12.53). By 1860 the less valuable but less elusive mundic (arsenical pyrite), then in demand for the manufacture of sulphuric acid and other chemicals, was the chief produce and in that year for example 7178 tons of mundic were raised and sold for £6675. A variety of other minerals was also still being sold, including lead, copper, gozzan, blende, black tin, and tin in the stone (i.e. unstamped and not dressed). RCG 22.6.60.

[2] W. J. Henwood, *Trans. Royal Geol. Soc. Cornwall*, 1845.

[3] MJ 24.10.63.

out the county an unprecedented number of mines were promoted in what was to become the greatest tin boom Cornwall had known: Wheal Daniell and Creegbrawse & Penkevil United,[1] near Chacewater; Restronguet Creek Tin Mine, to try again the alluvial deposits beneath the tidal waters near where the old Carnon mine had worked earlier[2]; Wheal Osborne and Tucker's Downs, both near Townshend; North Pool in Illogan; North Rosewarne, near Camborne; West Polbreen, New Wheal Charlotte, West Wheal Coit and West Wheal Trevaunance in St. Agnes; West Wheal Margaret in Lelant; Trumpet & Trenethick, South Wendron and West Trumpet, in the area north of Helston; Furze Hill and East Bottle Hill in Devon; Wheal Peevor near Redruth; Carnelloe[3] and Morvah Consols in Zennor; Gobbet Tin Mining Co. on Dartmoor[4]; New Treveddoe and Hobb's Hill in St. Neot; New Hingston, near Gunnislake; Florence & Tonkin Mining Co. (later to become known as Callington Consols) on Kit Hill; New Levant, to work the 'knacked' Pendeen Consols, and East Boscaswell in St. Just; Wheal Florence United in Crowan; Georgia Consols, Towednack; Wheal Lucy, West Wheal Lucy and East Wheal Lucy[5] at Hayle; South Dolcoath; Castle-an-Dinas, Brynn Mine, Brynn Royalton Consols and Roche Consols, in the Goss Moor area north of St. Austell.

The flood of new ventures continued unabated in 1872: Wheal Ruby, in Ludgvan; East Mulberry in Lanivet; Penhale Moor in St. Enoder; Wheal St. Just and East Balleswidden; Carzise and South Carzise & Maria, near Camborne; Polcrebo, near Crowan; Parbola, in Gwinear; Polhigey Moor; Unity Wood, West Gwennap Consols and South Tolcarne in Gwennap; Wheal Coates, St. Agnes Consols, and South Kitty, in St. Agnes parish; Killifreth at Chacewater; South Phoenix and Dunsley Wheal Phoenix, in Linkinhorne; New Wheal Metal (the old Wheal Dog) and Polrose in Breage; Camborne Vean, under the guide of New Dolcoath; East Treleigh Wood at Redruth; Wheal Vincent at Altarnun; West

[1] These two setts had been worked under the name of Creegbraws & St. Michael Penkevil and had been sold up in 1855. Working, then chiefly for copper (on the eastern end of the Poldice and Unity Wood lodes), had been on a considerable scale, Creegbraws possessing 70″ and 24″ pumping engines, another engine for winding and stamps plus three large water-wheels (including one 46′ diameter) and St. Michael Penkevil a 22″ whim and stamps engine. (RCG 20.7.55). The somewhat unusual name of the latter was derived from that of the parish on the Fal where Tregothnan, the seat of the mineral lord (Viscount Falmouth), was situated.

[2] An iron shaft, 22 fathoms deep, was sunk here into the bed of the creek by mooring laden barges over each successive segment at high tide, with a rotary engine erected on the shore at Point.

[3] Carnelloe was re-started in July 1871 by a company whose chief shareholders were Wickett & Co., brewers of Redruth, holding half the mine: there is 'not such a striking geological analogy on the whole north coast of west Cornwall as that of Botallack, Levant and Carnelloe' (MJ 5.7.71). As in earlier trials here, waterpower was used for both pumping and hauling, whilst a short incline tram-road was laid up the cliff to the stamps. (RCG 30.12.71.) The wheel pits can still be seen on the cliff side.

[4] The workings later formed part of Paignton's water supply system.

[5] Wheal Lucy was a celebrated little mine on Hayle Towans, started in August 1871 on a carbona which gave very promising initial returns. About £5000 was spent in equipping the mine with all the latest appliances including a horizontal 20″ engine intended for pumping and working two heads of the then newly invented pneumatic stamps. It was far from being a 'gingerbread' mine, however, and sixty men were employed there in 1871. (MJ 6.8.71). West Wheal Lucy was much smaller, the single shaft there being only 7 fathoms deep and pumped by a portable engine (RCG 8.6.72).

Condurrow; Woodclose & Polgooth, near St. Austell; West Fowey Consols; Penstruthal Consols, and Tresellyn, near Altarnun.[1]

Four major re-workings were East Uny at Redruth, Wheal Busy at Chacewater, South Roskear at Camborne, and Levant. The latter had been taken up by a new company in 1871 and a great deal of money was laid out in re-equipping the mine in the next few years.[2] At South Roskear a group of Glasgow adventurers put up 70″ pumping, 30″ whim and 36″ stamps engines on the old Wheal Gerry section, whilst at Wheal Busy several engines were erected in a major re-working which commenced in January 1872. At East Uny a 70″ engine was put in during the same year, together with a 22″ whim and 36″ stamps, at a cost of £6,000. Other mines of some consequence were Carn Galver, which had been abandoned since 1866, and where a 40″ pumping, 30″ stamps and 20″ whim were erected in 1871, and Trencrom Mine where a 60″ pumping, 30″ stamps and 20″ whim were also erected during the year. Yet another new venture of 1872 was New Cook's Kitchen, separated from the old sett of the same name in June. The new company, in 2450 shares, was largely held by the same adventurers and in October a 45″ pumping engine and 24″ whim were erected to work the mine as a separate entity.[3]

This crop of new companies may be said to have been in part the result of the great decline in copper a few years before, insofar as the greater part of Cornish mining activity had formerly been devoted to the winning of that metal. Now, with copper gone—or virtually so—as a field for mining entrepreneurs, attention was turned to tin on a scale far greater than at any time hitherto. Inevitably there was money to be made in mine promotion, in bal-selling, in such a boom as this, and as in all the earlier waves of mining speculation, a fair proportion of 'Wheal Do 'em's and 'Wheal Rig-me-up's were started. As much money and more could be made from

[1] RCG *passim*.

[2] Although Levant had been a declining concern throughout the 1860's, it was on account of a long standing dispute between the lords and the shareholders that it was being abandoned in 1871 by the company that had worked it so successfully since 1820. Daubuz held 59/150 of the mine and it was made known that some of the lords would agree to no new lease to any company in which he had a share: ' a mine that has been worked half a century, sold a million pounds worth of ore, and given a quarter of a million in profits, is to be extinguished because lords and adventurers cannot agree.' (MJ 18.3.71). All manner of stratagems had been used by both sides during the period of dispute in the previous few years, including the contention—disproved by the calling in of referees—that the company had endangered the safety of the mine by allowing the back of the 40 to be worked away too close to the sea bed. (RCG 27.11.69). This was a reference to an occurrence in 1852 when a panic was caused ' by sea water breaking into a winze above the 40 at 150 fathoms from the Engine Shaft ' (Oats Collection, RIC, Truro).

By September 1871 the bottom 50 fathoms were under water, with pumping stopped and the materials being drawn to surface. (RCG 23.9.71.) When the lease expired on December 2, a new company—prominent amongst them, the Bolithos and Michell, of Trereife—took the mine, paying £3000 for the remaining machinery and materials (comprising a 40″ pumping engine, 33″ stamps, 24″ whim and 20″ man-engine). Like the workings underground, these were all in a very dilapidated state and a considerable sum of money had to be spent in the next few years replacing and renovating the engines, cutting down the main shaft for double skip-roads, substituting wire rope, etc. Not until May 1874 was the mine in fork again to the bottom (266 fathoms below adit) and the outlay then amounted to £17500. (MJ 9.5.74). 200 men and boys were employed at that date underground.

[3] RCG 8.6.72, 26.10.72. The 45″ engine was bought from Balleswidden, having been erected there new in 1845–6.

selling tin mines as from selling tin, as Cornishmen had long known. Amongst the more outrageous puffs put out to enhance the supposed value of a likely sett, one appertaining to Wheal Fortescue, St. Stephen, is worthy of note. Here, in all solemnity, the manager reported that '. . . large rocks of tin, from 1 cwt. to 3 cwt. in weight, almost fit for smelting, have been raised up in cultivating the land.'[1] Carzise, started in 1872, may be taken as a typical modest 'bal-selling' case of the period, although the sett was by no means a worthless one. At the inaugural meeting held in Camborne in May, the sum of six hundred guineas was voted to the two promoters 'for securing leases and other services rendered.' Apart from the profit made on the supply of the 70″ engine (bought from Prosper United at Marazion) erected here, this may be said to have been the lion's share of any profits Carzise was to make. Another more glaring case was Boscaswell Downs which had been bought for £1000 by two men well known in St. Just and Penzance (E. S. 'Banker' Boyns and J. T. Keverne) at a time in the 1860's when it was poor and due to be knacked. In 1872, in better times and with a considerably improved mine, they then parted with it to an up-country promoter named Milstead for £15,000. He in turn promptly brought it to market as the Boscaswell Downs Tin and Copper Mining Association, with a nominal capital of £75,000, securing for himself in the process no less than £44,000 for the mine.[2]

Along with this burst of mine promotion, came electrifying rises in the value of all mining shares. Extraordinary activity ruled in the ranks of the dealers: 'the rise in the price of shares is marvellous' commented the *Royal Cornwall Gazette*.[3] Typical was the case of Dolcoath, where the shares (recently sub-divided by four, into 1432 in 1869) rose from £125 in January 1871 to over £200 by November and to almost £300 one month later. South Frances shares rose from £25 to £75 in the same period and St. Ives Consols 940 shares from £14–£16 in February to £47–£50 two months later. Creegbrawse rose from 30s. to £16 in a matter of weeks, and South Crofty from as little as £8 in 1871 to a peak of £127 in 1872.[4] In March of the latter year it was reported that 'round about Tabb's Hotel, Fore Street, Redruth, from morn to night, the greatest activity prevails'[5] and as this speculation mania in mining shares —much of it by miners and mine agents—grew in extent even the hitherto sacrosanct large room used for the copper ticketing meetings at this

[1] RCG 9.10.69. This resulted in due course in a neat piece of bal-selling when the sett was sold to a new venture, the Fortescue Tin Mining Co. Ltd., in April 1872. Another nearby mine, Terras, where a 50″ pumping and other engines were put up, was the scene of another giant swindle of 1871.

[2] Oats Collection, RIC, Truro.

[3] 9.12.71.

[4] This rise was partly due to the sudden discovery by South Crofty of what was considered to be the extension of East Pool main lode, at much the same time that it was found that the latter, by a dialling error some time before, had worked out some of South Crofty's ground. The encroachment was later found to be only a minor one however, and was amicably settled by a payment by the East Pool adventurers of £2012.

[5] RCG 23.3.72.

premier mining hotel was thrown open for share-dealing and exchange.[1] Other weekly auctions and sales of tin mine shares were also held in Truro, Penzance and Helston, and for a brief period even outside the main mining centres at a weekly auction held in a hotel at Falmouth.[2]

These bonanza times, the benefits of which were being shared by up-country speculator and Cornish adventurer alike, had brought little benefit to working miners. True, they received a better tribute rate when the price of tin was high, but in nothing like the same proportion that the increase affected dividends and share values. There was the occasional rich 'sturt' which achieved the prominence of a mention in the local press—as, for example, when two tributers at St. Just Amalgamated took up £40 in one month of 1871 as a result of taking an old pitch with nothing 'except a layer of spar in sight' but which merely hid a bunch of tin from which they broke within days 1½ tons of tin—but these were occurrences of considerable rarity, a prize to be hoped for in half a lifetime of tributing and of little use to sustain the weekly cost of a miner's home. There was, moreover, a dearth of good miners as a result of the mass emigration that had followed in the wake of the copper collapse. In 1867 and 1868 the outlook, not only for Cornish copper but also for Cornish tin, had seemed bleak to every working miner and many had turned their faces to new fields abroad where there was some future before them. Cornish mining, it should be remembered, was now feeling the full effect not only of one but of a whole successive series of waves of emigration: in the 1820's there had been the men tempted away by the high wages of the first new mines abroad; in the 1830's and 1840's the 'bounty' emigrants whose passage was paid or assisted by colonies anxious for population; then towards mid-century came the Californian gold-rush whilst in the 1850's a host of Cornishmen went out to the world-famous finds in Australia of gold and copper[3] and in the following decade to the even greater copper mines opening up in the Lake Superior district of Michigan. As a result, from 1870 onwards, there were frequent reports in Cornwall of 'complaints of the dearth of labour in the mining districts,'[4] the situation being particularly acute in the St. Just area. Even the new shallow mines, where the management was usually lax and the tutworkers' pay-sheets—from the same cause—more liberal, were hard pressed for men, as witness the advertisement for one hundred hands for Terras Tin Mine in April 1871.[5] At Dolcoath,

[1] RCG 30.3.99. [2] MJ 27.4.72.

[3] By the summer of 1852 an estimated five thousand people a week were leaving England for Australia and there were very many Cornish families amongst this number. These latter went, not from desperation as a decade before, but as a result of the glowing tales of opportunity and the chance of riches in a new land. Newspaper reports of returned Cornishmen landing at Plymouth with 1½ cwt. of Australian gold dust in their baggage played an important part in starting others on the long trail of the gold rush, and small companies were even formed in Cornwall in the early 1850's to find the capital necessary to send small parties of miners out to the gold diggings on a commercial basis. Even by 1853 it was said that 'the tide of emigration to America and Australia from St. Just and neighbourhood still continues to flow rapidly on . . . The old deep mines are beginning to feel the lack of labourers very severely.' (WB 15.7.53.)

[4] RCG 13.7.72. [5] RCG 8.4.71.

in May, the adventurers even considered bringing in five hundred labourers from Ireland, a measure which precipitated near riots in Camborne. There had already been a strike of surface hands at Dolcoath for increased wages the previous autumn[1] and this was followed in January 1872 by a more serious strike at Wheal Basset. Here the men 'refused to go underground and though threatened with a double spale[2] for two days' idleness they left the mine in troops . . . the girls at surface also struck for an increase from 9d. to 11d. per day.'[3] The cause of this disturbance had been the attempt by the mine company to fine the men for not working New Year's Day, but behind this incident, which ended in the miners' and bal-maidens' favour, lay a widespread discontent that the men were not receiving any share in the prosperous times that tin mining was now enjoying. Agitation by the miners was focussed upon what they considered the iniquitous five-week month, a system whereby only twelve months' pay was received in the calendar year, losing them four weeks complete pay per annum.[4] Mass meetings were held at various places within the county, notably at St. Blazey and near Minions. This was a matter which also affected the men working in the surviving Cornish copper mines and no less than a thousand men from Phoenix and the Caradon mines attended the latter, attempts being also made to form a Miners' Protection Society.[5] Tincroft was the first mine to accede to the men's demand to abolish the unfair five-week system of wages—partly because it could so easily afford to make concessions with a divided profit of £123,750 in 1870–72, and partly because the chairman of its committee, William

[1] RCG 23.9.71. Strikes and other forms of industrial unrest were uncommon in Cornwall by the standards of other areas such as South Wales or the North-East, although they were by no means as rare as some writers would have us believe. A serious one had been at St. Just in April 1853 when all the men at Boswedden came out over the victimisation of a pare of men whose leader had played a prominent part 'in a rising some six years since, in consequence of the high price of flour.' (See page 86). The rest of the miners in the parish joined them, and at a meeting attended by between 1500 and 2000 in a field near the village it was agreed that no St. Just man should go underground and that any shopkeeper refusing the men credit 'should never, at any future time, receive the support of the miners in business.' (WB 15.4.53). Demands were made for better tribute rates generally, this having been the alleged cause of the victimisation case which started the strike, and these were acceded to by Levant, Spearn Consols, Spearn Moor, Boswedden, and Boscean, although Botallack, Wheal Owles and Balleswidden adventurers 'expressed their determination not to give way.' (WB 22.4.53). St. Justers were men of an even more independent frame of mind than the miners from other Cornish districts and in 1856–7 those at Balleswidden again struck on the introduction of gas lighting underground, partly from a probable dislike of anything that was new-fangled and partly because it deprived them of the few pence they could make each week on the sale of saved candles. In 1859 yet another strike occurred there due to a cut in the tutworkers' rates of pay. (RCG 4.3.59).

[2] Briefly, spales were fines which could be imposed by a mine manager on any employee for an offence which contravened the mine's byelaws. A list of the latter was normally pasted to the door of the men's dry. These fines, deducted from a man's pay, were usually donated to the Miners' Hospital, Redruth, or other charity. They were the subject of widespread complaint by miners at this time—and not without cause, for instances were quoted where men had been spaled 2s. 6d. for appearing at surface even three minutes before the bell rang for the end of the core. In other mines men were spaled 2s. even for missing their core through sudden illness in cases when the agent had not been sent a message by the miner to this effect. (See RCG 15.10.70).

[3] RCG 6.1.72.

[4] RCG 13.1.72, 27.1.72.

[5] RCG 27.1.72. This was an echo of the attempts six years earlier when there had been bitter strikes in the Caradon, Gunnislake and west Devon mines. At this time miners' meetings were held on Kit Hill, with over a thousand working miners present on one occasion. (RCG 8.3.66, 15.3.66.)

Teague, had himself experienced its injustices in years gone by. Tincroft's example was thereupon followed by East Lovell, and by Dolcoath and other Illogan mines. A County Meeting of mine adventurers was held in Truro and, subject to the men consenting to give up their Maze Monday concession[1] in return, it was generally agreed to abolish the five-week month. In March there were several brief strikes at Botallack and other St. Just mines because the men refused to make this reciprocal concession,[2] whilst at the other end of the county the miners came out *en masse* at Okel Tor where the old pay system was still continued. In June there was another strike at Ding Dong amongst surface workers over the hours to be worked[3] but thereafter, after these concessions on wages, hours, tribute rates and above all, the hated five-week month, these disputes faded.

Certainly the adventurers could afford to be generous at this time for the dividends of 1872 showed how radically the increased price of black tin had benefitted the financial returns of mining. Tincroft paid £48000, Dolcoath £45645, Carn Brea £15500, Cook's Kitchen £9187, Wheal Kitty (St. Agnes) £8590, East Pool £7200, Botallack £3000, North Levant £2800, Wheal Owles £1960 and Wheal Vor £1477. In the very many new mines opening up, the adventurers paid their calls cheerfully enough, confident in the hope, if not the knowledge, that this was only an outlay that would be re-couped ten-fold and more. 'Cornwall was to be the new Eldorado'[4] and a spirit of blithe financial optimism such as the county had not known before pervaded mining affairs. Few investors glanced backwards in time at what had been done previously in these setts into which they joyfully put money, or gave a thought to the disastrous working losses recorded as recently as 1865 in many of them. In fact the great majority of people putting up money for the bubble mines amongst these were gamblers who spared no thought for good or bad mining, or for good or bad mine management. The price of their shares alone was all important and as this hinged upon the dividends, or upon a series of ever more glowing reports from mines still in process of being opened up, there was inevitable pressure on mine committees and managers to maintain these. This was the case even in the large, established mines where it became necessary to pay excessive dividends and if need be to force the working of the mine to maintain them. To justify this, captains were continually pressed by shareholders to see further into a lode than could be seen by either pick or drill, and coping with their reports, making them cautiously optimistic enough to suit all the shareholders, became a major pre-occupation on their time. Thus the wave of market mining at this time did damage even beyond the ranks of the purely bubble mines, for speculation based on the ultimate richness of a sound sett was beneficial

[1] By time-honoured custom and tacit agreement, the first day following the miners' day of rest—occasionally riotous and frequently spirituous—was one on which little work was done. A tributer was virtually his own employer and made up this lost time during the rest of the week, but with 'owners' men' and tutworkers this was not the case.

[2] RCG 2.3.72. [3] RCG 8.6.72. [4] MJ 28.12.72.

but that based purely and simply on the hope of a rise in share values alone did serious harm. To speculators, of course, a profitable deal in shares was as gratifying a harvest as was that of the striking of a rich lode in the same mine to a true adventurer. To the former, moreover, a falsely reported discovery could also be as rewarding as a genuine one in the quick profit it brought, and the number of puffed-up reports and news items inserted in the *Mining Journal* with a view to enhancing the value of a mine's shares were legion at this time. To some extent too the very richness of the existing prizes, such as Tincroft, Dolcoath or East Lovell, carried some seeds of destruction for the hopes of smaller mines. It might be said that ' what renders the larger concern [Dolcoath] a mining Croesus also gives hope and energy to the tiniest adventure '[1] but it was also true that anything less was to leave shareholders dissatisfied and unwilling to accept smaller returns.

Amidst this boom of 1870–72 there were few failed mines to disturb the apparently unbounding prosperity. There was South Grenville, where the 30″ pumping engine was for sale in January 1870,[2] Camborne Vean, sold up in April, and the adjoining Stray Park which was disposed of to Dolcoath a few months later.[3] More important than these was Pendarves United—formed in 1866 from Tolcarne, Wheal Harriet, Tryphena and Condurrow—a group that had risen briefly to the position of the county's third greatest tin producer in the subsequent two years. By 1869 Pendarves United was being worked singularly badly, with all mining carried out by tutworkers and a great deal of lode-stuff being sent to surface that could never pay to be dressed. In one typical month of 1869 1931 tons of this were stamped which gave a mere 9 tons of black tin, only $1\frac{3}{8}$ per cent produce, which was hopelessly low with the working costs of this time.[4] By November the deficit at Pendarves United was £9400, necessitating a call of £36 per 1/256 share and even with the high price of tin the loss by 1872 was £500 a month.[5] After drastic attempts at economy and a revised system of working, it was decided to abandon all except the Tryphena section.[6]

[1] RCG 1.1.70. [2] RCG 8.1.70.

[3] With 60″ pumping engine, 22″ whim and 30′ water-wheel working 48 head of stamps. (RCG 2.4.70). 'About £54,000 has been laid out here.' (RCG 9.4.70).

Stray Park early in the century had been part of Dolcoath sett but was later worked as a separate entity before it was included again in 1870. The mine proved a great disappointment to its adventurers for a considerable sum was vainly expended in the 1860's in sinking 100 fathoms deeper for tin simply because ' we are near Dolcoath and in similar strata.' (RCG 26.2.70). To some extent the two mines had been worked together, with Stray Park more or less drained by its bigger neighbour (although possessing its own 65″ engine) and from 1865 onwards its miners had been allowed to make use of the Dolcoath man-engine. (RCG 18.8.65).

[4] RCG 17.6.69. [5] RCG 17.8.72.

[6] RCG 23.11.72. The proximity of various of the Pendarves sections to Dolcoath had been a potent factor in keeping the adventurers' hopes high. The same was true at Carn Camborne which had been commenced about 1861 largely on the strength of Dolcoath's richness. In April 1870 the adventurers here acquired the old Camborne Vean dressing floors at Pulstrong for £750, equipped with a water-wheel and 48 head of stamps. (RCG 4 23.70, 14.5.70). Two months later there were objections by local residents to a burning-house being erected on the mine, even though use of the big ornamental stack was proposed to carry the fumes clear. (RCG 18.6.70).

Apart from this major casualty in terms of tin, the period was one of continuing expansion, with every mine of any size installing bigger pumping engines to cope with growing depth of shafts, new whims, and more powerful sets of steam stamps. It was an era of unparalleled activity

OUTPUT OF THE THREE LARGEST TIN PRODUCERS IN CORNWALL[1], 1853–81

showing the rise and dominance of Dolcoath

1853	Dolcoath	360 tons	Polberro	282 tons	Lewis	262 tons
1854	Dolcoath	363 tons	Polberro	263 tons	Levant	287 tons
1855	Dolcoath	352 tons	Carn Brea	332 tons	Wheal Vor	313 tons
1856	Wheal Vor	424 tons	Dolcoath	416 tons	Par Consols	316 tons
1857	Wheal Vor	697 tons	Dolcoath	544 tons	Carn Brea	433 tons
1858	Dolcoath	635 tons	Wheal Vor	573 tons	Carn Brea	509 tons
1859	Dolcoath	723 tons	Carn Brea	515 tons	Wheal Vor	381 tons
1860	Dolcoath	805 tons	Carn Brea	653 tons	Providence	321 tons
1861	Dolcoath	864 tons	Carn Brea	536 tons	Par Consols	336 tons
1862	Dolcoath	986 tons	Carn Brea	558 tons	Charlestown U.	399 tons
1863	Dolcoath	1026 tons	Carn Brea	568 tons	Charlestown U.	480 tons
1864	Dolcoath	1030 tons	Wheal Vor	576 tons	Carn Brea	511 tons
1865	Dolcoath	944 tons	Wheal Vor	848 tons	St. Day Utd.	462 tons
1866	Dolcoath	919 tons	Wheal Vor	845 tons	Charlestown U.	509 tons
1867	Dolcoath	848 tons	Wheal Vor	658 tons	Pendarves Utd.	465 tons
1868	Dolcoath	984 tons	Wheal Vor	661 tons	Pendarves Utd.	566 tons
1869	Dolcoath	813 tons	Wheal Vor	576 tons	Botallack	525 tons
1870	Dolcoath	1035 tons	Tincroft	788 tons	Pendarves Utd.	476 tons
1871	Dolcoath	1170 tons	Tincroft	831 tons	Botallack	497 tons
1872	Dolcoath	1285 tons	Tincroft	829 tons	Carn Brea	559 tons
1873	Dolcoath	1045 tons	Tincroft	693 tons	Carn Brea	544 tons
1874	Dolcoath	1121 tons	Tincroft	686 tons	Carn Brea	652 tons
1875	Dolcoath	1241 tons	Tincroft	730 tons	Carn Brea	668 tons
1876	Dolcoath	1263 tons	Carn Brea	852 tons	Tincroft	794 tons
1877	Dolcoath	1405 tons	Carn Brea	1038 tons	Tincroft	796 tons
1878	Dolcoath	1539 tons	Carn Brea	1195 tons	South Condurrow	861 tons
1879	Dolcoath	1780 tons	Carn Brea	1048 tons	West Basset	1000 tons
1880	Dolcoath	1527 tons	East Pool	1268 tons	Carn Brea	1062 tons
1881	Dolcoath	1816 tons	East Pool	1317 tons	Carn Brea	974 tons

in Cornish tin, when the sun shone on adventurers and the roll of ' calling ' mines—other than those in process of being opened up—diminished as never before. At long last Cornwall seemed to have regained her old premier position in world tin production and at a Wheal Owles meeting in 1870 the purser felt safe to boast that '. . . he scorned the idea of a Cornishman being beaten in the production of tin by a Coolie or a Chinaman.'[2] Two years later the *Mining Journal* commented, ' a very few years ago not even the most sanguine believer in the resources and future of the county would have dreamt of such a period of prosperity as now seems to have fairly set in.'[3] This was also the heyday of the little stream-works for there was a colossal waste of tin slimes from the newly developed

[1] Memoirs of the Geological Survey, Mineral Statistics of the United Kingdom, (published annually) 1853–81.

[2] MJ 28.5.70. [3] 6.1.72.

mines with their inefficient and sometimes merely improvised dressing arrangements. Their owners waxed fatter and yet fatter as the value of their produce rose with almost every parcel of tin sold. A revealing word-picture survives of one Red River streamer of this era—formerly, like so many owners of small stream-works, a tin dresser on one of the mines nearby—of whom it was said, 'when he commenced catching tin in the valley, he stuck to it like a leech, up in time in the morning, jumping into any passing donkey-cart to get there more quickly, but the fruits of the river soon produced him a pony and trap and now he has a horse, trap and silver-mounted harness.'[1] There were then no less than forty-one separate concerns working the Red River for tin, with another fourteen on the stream flowing to the sea at Portreath.[2]

During this period of mining boom, the changes in tin smelting in the county had been few but far reaching. In 1869, at the expiry of their twenty-one year lease of the Charlestown smelting-house, Enthovens decided to remove their centre of operations to London.[3] Most of the foreign tin and lead ores that they now smelted were consigned to the Thames and it was an obvious economy to produce metal there rather than after the expense and delay of trans-shipment to and from a Cornish establishment. The Charlestown works was thereupon taken over in 1869 by Williams Harvey, Bolitho and Daubuz in one-third shares. Smelting was continued as before but with the output, costs and management shared between the three. In 1871 a similar arrangement was made between these same three concerns to acquire the Truro Smelting Works which had been run, successfully enough but on no great scale compared to say Carvedras or Trethellan, by John Gatley since 1863. In this case, however, the works was discontinued and as the premises were thereafter leased to a firm of wholesale grocers, the furnaces were presumably dismantled.[4]

In both cases, at Charlestown and Truro, this buying up of existing works had been organised by the chief smelters in the county largely to prevent them passing into the hands of any competitors. The prosperous state of the tin trade made it inevitable that there would be revived interest in smelting as well as mining and that this was the case had become evident in 1869 when two of the partners in the Bissoe works announced their intention to withdraw from the firm and to set up an entirely new

[1] MJ 14.12.72.

[2] The dues on the former produced an income to the Basset estate of £2531 in 1873. *Min. Stats. U.K.*, 1874.

[3] Messrs. H. J. Enthoven & Sons Ltd. still refine lead from foreign sources at their Rotherhithe works, as well as operating a smelter at Darley Dale in Derbyshire.

[4] Bolitho Records, DD/BL, CRO, Truro. Gatley's partner in the Truro Tin Smelting Co., J. H. Bawden, went to Trethellan works to become manager there for Williams Harvey & Co. (RCG 25.3.71).

This marked the end of the chequered career of the Truro Smelting Works, originally set up in 1816 and successively occupied by various companies, interrupted by periods of disuse. From surviving contemporary photographs it is evident that the tall square-based stack of the works—Truro's most outstanding landmark until the completion of the cathedral—survived until about the turn of the century.

works on the Fal estuary at Point, near Devoran.[1] In this venture these two were joined by Edward Michell, who owned the existing lead smelting works already situated at Point.[2] As the Penpoll Tin Smelting Co., this announced that it was open to buy tin ore in January. The following year also saw the re-establishment of the tin-smelting works of the old Union Company which after Tyrie's death in 1863 had found no new occupier during the depression brought about by the American conflict. The new company, known as the Tamar Tin Smelting Co., had offices in Saltash, one of its major shareholders being Michael Loam of Moditonham.[3]

Thus, during the three years after 1869, the works at Charlestown changed hands, one at Truro was dismantled, a new works set up at Penpoll and an old one revived on the Tamar. Another minor change was in the ownership of Trereife at Newlyn, when T. W. Field[4] acquired a share in the company consequent upon the death of R. R. Michell in April 1872.

But for Cornish tin-smelting, and even more markedly for Cornish tin mining this brief era of boom from 1870–72 was to be high water-mark. During this time the tide had flowed more strongly in the adventurers' favour than ever before, and in 1871 the greatest number of mines active in the county at any one time hitherto had produced 16,272 tons of black tin. No-one, smelter or miner, had cause to suspect that an almost unbroken fifty years of decline, depression and dereliction lay before the industry.

[1] RCG 20.11.69. These were Messrs. James and Shakspeare, eminent metal brokers in London since 1844, the first-named (Henry Charles James) having been with Williams Foster & Co. prior to this date. After their withdrawal, the Bissoe Tin Smelting Co. became known from November 1870 onwards as Tregoning & Co. The latter family—originally of Gwennap, but with offices in Liverpool and a tin-plate works in Llanelly—had been the chief partners in the Bissoe works throughout.

[2] Originally built by John Swan in 1827 and taken over on his bankruptcy a few years later by John Penrose (of Tregie) and R. & W. Michell of Truro who owned part of the Calenick Tin Smelting Co. This works, with which the Penpoll works is occasionally confused, was for sale in July 1872 and again in 1873. The two works could formerly be distinguished by the 120′ high stack of the lead smelter situated on the higher ground above Point, and by the 80′ stack of the new tin-smelter beside the works alongside Penpoll Creek on what was formerly known as Lemon Quay. (Whitford Collection leases, DD/WH 185, CRO, Truro).

[3] A notable name in Cornish engineering. At this date Loam also had interests in a foundry at Saltash, specialising in the manufacture of rock drills. (RCG 1.4.71).

About a year later the Tamar Metal Extraction Co. Ltd. was formed by another group to take over what the prospectus termed the Lower Tamar Lead Smelting Works, together with the dumps (estimated to contain 100,000 tons) of the Tamar mines for re-treatment (MJ 19.10.72).

[4] Originally of Redruth. A mine purser and adventurer, later a banker and owner of arsenic manufactories in west Cornwall.

III

THE COMPETITION OF FOREIGN TIN

AUSTRALIA SURVIVED: 1873–1882

'*What Chili has done for Cornish copper mining, Australia bids fair to do for its tin mining.*'

Mining Journal, 1 August 1874

DURING THE BOOM and bustle of 1872, one small item of news in the *Royal Cornwall Gazette*, that 'a competent person, on behalf of the Cornish tin smelters, is to be sent to Queensland to examine and report on the newly discovered tin deposits,'[1] passed without further comment. Yet the opening up of these new sources of tin, and of others shortly to be discovered in New South Wales and Tasmania, was to be an event of great significance to Cornwall and was to have grave consequences for its premier industry.

Throughout 1872, unbeknown to an uncaring Cornwall at large, something like a tin fever had swept Australia. The metal was found, primarily in widespread alluvial deposits but also in lode form, in a 225 square mile area of elevated table land in the southern part of Queensland.[2] At much the same time alluvial tin was discovered in other states to the south and consequent upon each discovery a wave of further prospecting took place, not only around each of these areas but also farther afield. By July 1872 no less than 2240 mining leases for tin had been granted in Australia, whereas twelve months before, the total had been seven.[3] This was hardly surprising, to judge by the richness of one reported new find where four Cornishmen—who had originally emigrated from Redruth to the South Australian copper mines but had now become 'diggers'—obtained nearly 13 cwt. of alluvial tin in one day. This was at Fisher's Gully, 160 miles from Brisbane where, to quote the same report, 'the quantity appears to be unlimited. The difficulty of sending their produce to the seaboard appears to be the only serious question with most of the miners in the district.'[4] All that was needed was a shovel for digging and a pan for washing, perhaps not even that, for in many places it was possible merely to pick up the lumps of almost pure cassiterite lying exposed in the dry river beds.

Since about 1850 a small quantity of Australian tin ore had been sold in London or Liverpool, but in the opening months of 1873 the frequency and size of these shipments increased by leaps and bounds. In 1872 458 tons

[1] 6.7.72. [2] See RCG 27.7.72, 5.10.72. [3] MJ 20.7.72. [4] RCG 31.8.72.

had been landed whilst in the first five months of 1873 956 tons came in, followed by 442 tons in June, 555 tons in July, 518 tons in August and 733 tons in September. The whole of this was bought by Cornish smelters and the ore—all of it alluvial tin, in casks or bags—was trans-shipped on the Thames into coasting vessels bound for Penzance, Hayle or Truro.[1]

By midsummer 1873 this great quantity of newly found tin, which filled the lower holds of every wool clipper from Australia[2] entering the Thames, was reflected in steadily falling prices of black tin in Cornwall. From the beginning of the year the smelters had started to lower black tin prices in anticipation of what would emanate from this new and ominously uncertain source of world supply. Thus was a still largely unsuspecting Cornwall launched into a period of depression that was to exceed in severity anything hitherto known, and that was to make the copper slump of the late 1860's seem a mere bagatelle by comparison. The effect was not of course instantaneous and before the full extent of the Australian ' tin rush ' had become apparent there were more luckless flotations early in 1873: Prideaux Wood, Tretoil, East Polberro, North Dartmoor,[3] the Mid-Cornwall Mining Co., Great Tin Work, Morvah Consols, Crowndale Consols (Tavistock), East Tywarnhayle, Trevegean (St. Just) working the old Wheal Diamond, East Botallack,[4] and St. Blazey Consols. But by midsummer this last trickle of the flooding tide had stopped[5] and the first

[1] All the smelters, from Tamar in the east to Trereife in the west, bought heavily in this first great inrush of Australian tin. Cargoes were even imported and sold by ticket at Chyandour by the Bolithos, perhaps from ' diggings ' in which they held a share themselves, although the greatest quantity was imported by metal brokers in London. (Bolitho Foreign Ores Ticketings Book 1856–73, RIC, Truro).

This new source of ore in 1873 was responsible for the setting up of what was probably the first tin smelting works in England outside Cornwall—the Tin Smelting Co. (Boundy & Owen) at Ripley, Derbyshire, which was active from 1874-76 (Min. Stats. U.K.). In point of fact the whitening of Australian black tin, in Cornwall or at this brief-lived new works in Derbyshire, did not last many years, for the miners were not prepared to accept the unduly high charges made by the English smelters. Australian works were thereupon established and before the end of the 1870's most tin was leaving Sydney and Melbourne in the form of ingots and bars.

Thereafter, most of the foreign tin ores smelted in Cornwall came from Bolivia and regular ticketings were held at Liverpool of the incoming cargoes of barilla throughout the period from 1880–1900. As with home produced ores, the Cornish smelters operated an apportionment scheme to prevent any real competition and in 1878, for example, the agreed proportions of the 6160 tons imported in that year were—Bolitho 1575 tons, Williams Harvey 1575 tons, Daubuz 856 tons, Michell 718 tons, Redruth 718 tons, Calenick 387 tons, and Charlestown 331 tons. (Bolitho Foreign Tin Ticketings Book, 1873–93, RG 1/112, CRO, Truro).

[2] Tin ore (and later ingot tin) made ideal ballast for these wool ships whose main cargo was proverbially difficult to stow. Some of the most famous square-riggers in the world were engaged in this trade in the 1870's; the *Loch Torridon*, the *Patriarch*, the *Windsor Castle*, the *Christiana Thompson*, and others.

[3] On the Rattlebrook, near Lydford. The prospectus of this company, with a nominal capital of £20,000, had been issued late in 1872. The sett consisted of four square miles of ground for which £8000 was to be paid (including £2000 in cash). The property had been ' most favourably reported upon ' by a mining engineer from Llangollen. (RCG 23.11.72).

[4] East Botallack was a sett which had been tried in the 1840's as Wheal Fortune, in the early 1850's as West Wheal Buller, and where a third flotation had been attempted in 1861 under the name (currently fashionable) of West Wheal Metal. The sett actually lay adjacent to Balleswidden but as that mine was knacked and Botallack was then a market favourite choice fell this time in 1873 on the name East Botallack.

[5] What appears to have been more or less the last flotation of the boom was West Appletreewick Mining Co. Ltd., which started sinking a shaft in a small sett on the cliffs near Charlestown in May 1873, only a matter of weeks before the tin market collapsed. (RCG 31.5.73).

signs of an ebb that was to last half a century became apparent. The steadily falling price of black tin began its inexorable toll, just as the steadily rising price had earlier fostered so much hope. At Providence Mines in January there was the shock of no dividend after an unbroken run of ninety-three successive quarterly divisions of profit—totalling £101 12s. 6d. on an initial outlay of £10 6s. 7d. per share—news which brought the share value down from £35 to £12 with a rush; at Balleswidden, long a failing concern, the 80″ pumping engine was for sale in March and the whole mine standing idle.[1] The stoppage was reputedly accelerated by the mine being under-manned, with only enough labour available to work one core a day—the results of long continued emigration in the west of Cornwall.[2] On this score the demise of Balleswidden was said to be welcomed by the other St. Just mines, amongst whom the discharged hands were soon absorbed.

The closing of this great Land's End mine was paralleled at Poldice in Gwennap, where a loss of £1862 was incurred in the first three months of the year.[3] In May it was decided to persevere for one month more, the decision to continue or not resting almost entirely with Sir F. M. Williams who held 1700 out of the 2000 shares, but with the still declining value of tin it was decided to abandon the sett in June, with only a few men retained in the Tolgullow section.[4] This represented the last of the really big Gwennap mines, either of copper or tin, Poldice having been worked for the latter in this final working on an extensive scale and in a most determined way. Great trouble had been experienced with pumping and with 'chokages' in the County Adit, and a new 85″ engine had been started only at the beginning of the year to put an end to these pumping difficulties.

These were the major casualties of the first half of 1873 and with a steadily falling tin standard, reduced in the year by about £24 per ton, it was obvious that worse must follow. Established mines such as Wheal

1 Sales of black tin from Balleswidden had been—

January 1833—October 1838	value	£ 24386 10s. 8d.
November 1838—April 1873 11,828 tons	value	£685098 13s. 7d.
	total	£709485 4s. 3d.

with a further 68 tons (sold for £3892) in the remainder of 1873 from the leavings.

Despite this very considerable output, the mine had not paid much by way of dividends. From 1843 to 1853 they totalled £25375 (reaching a maximum of £3978 in 1845) but thereafter for its remaining twenty years, calls had been made totalling £17558.

Some revealing comments on the state of Balleswidden in 1873 occur in a Report on the mine, 1833–73, printed for private circulation only: 'the pumping and hauling machinery was in the last state of collapse, the big engine shaft alone taking ten men to put it to rights from Saturday to Monday each week, working without cessation.' Railroad Shaft, which was 60 fathoms deep and not connected with the main pumping shafts was kept dry as and when possible by primitive means, 'sometimes hand labour, sometimes horse and bucket' and this, be it noted, even up to the 1860's. One final comment from this Report serves as a fitting epitaph for the once great Balleswidden: 'She was worked by a Committee of Merchants. Verb. Sap.' (Bolitho Records, DDX 143 & DDG RG 1/26, CRO, Truro).

2 At the date of the final report (25.2.73), 352 were employed at Balleswidden, consisting of 233 men, 82 boys, and 37 females. Underground there were 26 men and 6 boys on tutwork, 137 men and 24 boys on tribute.

3 RCG 7.5.73. 4 RCG 28.6.73.

Vor, East Lovell, Trumpet Consols and Wheal Owles dropped from the dividend list—in many cases for ever—whilst Botallack made a call on her shareholders for the first time for many years.[1] Relinquishments (i.e. forfeiting) of shares by those who either lacked the money to pay calls or who had no confidence in a mine's future became frequent, and shut down both Great Wheal Busy and Wheal Rose United in November, neither of these having been thoroughly tried. The sudden slump made the re-working of Wheal Busy particularly disastrous and only seven months had passed from the starting of the big new 90″ pumping engine there to its sale in July 1873. From an outlay approaching £30,000 on this brief[2] re-working, the sixth at least in the long history of Wheal Busy, only about £8000 was salvaged by the sale of the machinery. At Wheal Rose United, 3504 shares out of a total of 10000 had been relinquished and this added one third part to the burden of costs of those who were prepared to continue. The same state of affairs prevailed at Great Work, where the smelters Daubuz and Williams held half the mine between them. In August they relinquished these shares and this brought about the inevitable closing of this once great mine the following month.[3] During 1873 money had been lost steadily and in March, as a means of cutting working costs, the Wheal Grey section had been abandoned despite promising developments. A dispute with the mineral lord over the way the mine was being worked also accelerated Great Work's demise, for the dues had earlier been halved to 1/48 on the understanding that new capital would be put in and more spirited working prosecuted. This was not complied with and full dues were again demanded from early 1873, plus the payment of a considerable sum representing those hitherto foregone.

Amongst the smaller mines, particularly those recently floated, relinquishments occurred on a wide scale, for many shares were held by speculators whose interest was in the market standing of a mine and not its value underground. Under such circumstances the blind and unreasoning confidence of 1871–2 quickly gave way to doubt and from that rapidly to alarm. From this cause, the roll of abandoned mines during 1873 was a long one: West Drakewalls, Florence United, Wheal Damsel, Tresavean, Parbola, Polcrebo, Wheal Lucy, Leeds & St. Aubyn, New Hingston, Belowda Beacon, Wheal Margery, New Wheal Charlotte, East Polberro, East Mulberry, Leeds Tin Mine, New Lovell, Wheal Millett, Grambler, Bosworthen, New Martha, Carzise, Wheal Emily Henrietta, Wheal

[1] Wheal Owles, to the last dividend in August 1872, had given profits of £41800 on an outlay of £6900. Botallack to the same date had given £123,950 on an outlay of £23850. (RCG 9.2.77.)

[2] The company had been formed in January 1872 amid scenes of great joy and celebration in the village of Chacewater, hard hit by the copper recession. (see RCG 13.1.72.)

[3] The actual date of the final abandoning of a mine is not always clear; sometimes mining operations ceased but the pumping engine was kept going for a time in the hopes of finding a buyer or in the hopes that metal prices might recover; sometimes pumping was stopped but tributers kept at work above the adit or natural water level, and the latter appears to have been the case at Great Work for some time at least. Moreover in many large mines, unless the decision to abandon was peremptory, there was usually employment for a few weeks at least working out available ore and cleaning up the stocks at surface.

Daniell, and others. The last named was owned by the same group of shareholders as Great Wheal Busy and the shares of this small but once favoured new venture had fallen rapidly during late 1872, when a long series of relinquishments started, from a peak of £8 to a mere ten shillings by early 1873. Carzise was a similar new mine, a case of bal-selling that has already been cited—which its promoters had unsuccessfully attempted to sell as a going concern in September 1872. The newly started Creegbrawse & Penkevil United—another hopeful flotation of 1871—was also sold up in February 1873, having employed a maximum of 145 persons in its brief period of working, and with an 18″ engine and 16 heads of stamps installed. Such was the bleak outlook for Cornish tin that in the same area, near Scorrier, East Treleigh Wood was sold 'as a going concern' to its erstwhile manager for £50[1] and this despite the 6000 shares into which it was divided having stood at 15s.–20s. each not twelve months before. Another sale during the year was Wheal Kitty in Lelant, advertised in February, with two (33″ and 28″) pumping engines and two (20″) whims. The latter part of the year also saw the sale of some of the machinery at Wheal Buller[2] where most underground work had been stopped about three months before. A loss of £1909 on only twelve weeks working brought this about, although the pumping engine was kept at work and attempts made in the eastern ground to develop Buller as a tin producer. The neighbouring East Uny was wound up in July, £33000 having been called up since the venture was re-started. A new company actually continued to work the mine, but with steadily continuing losses. In the east of the county, one of the attempts to find the long elusive tin of Caradon beneath its proven copper wealth was also abandoned with the closing of West Rosedown, where the 60″ pumping engine and 24″ whim were for sale in February.

The total profit paid out during the year by the twenty-six dividend mines, both tin and copper, of Devon and Cornwall was £87021,[3] less than had been declared by two mines alone in 1872 and the news, in September, that tin ore had been discovered in yet another new Australian locality[4] cast further gloom over the mining districts of Cornwall.

In fact, the worst was yet to come and early in 1874 a further disastrous drop occurred in the tin standard. When the year opened, it was £114, by February it was below £100 and by 6 March it was down to £84, with the smelters anticipating the falling market with every new purchase of black tin that they made. In one week alone there was a drop of 6s. per cwt.—'if prices go much lower it is feared there will be a regular panic and smash, for the prices of materials and labour keep up while prices of

[1] RCG 10.5.73.
[2] Comprising 28″ stamps and 24″ whim. (RCG 15.11.73.)
[3] The three most profitable tin mines were Tincroft (£21000), Dolcoath (£13962) and Wheal Kitty (£7193).
[4] On the Latrobe River, only 63 miles from Melbourne. (RCG 27.9.73.)

produce go down.'[1] The high price of coal, made worse at this time by strikes amongst the Welsh colliers, was an added burden for its cost had risen from about 12s. per ton to as much as 25s. or 30s. within the past few years and this added an estimated £150,000 to the county's mining bill. Next to that of a drop in the tin standard, a rise in the price of coal was the worst of news to Cornish mine adventurers and in 1874 both hit Cornwall with increasing severity. Added to this there were radically increased costs of timber, iron, candles and all other materials.

Throughout the year a long succession of mines shut down or curtailed their scale of operations, with one or more 'knacked bals' named in the *Royal Cornwall Gazette* and *West Briton* each week. In a few cases some were revived again in smaller re-workings in later years but the majority closed forever. In January Wheal Hearle, Polbreen, Polhigey Moor, Pennance (near Redruth) and Carn Galver were abandoned, the last named being bought by Bolithos as 'a going concern' for £1600[2] but with all the materials sold off four weeks later. In February the stoppage was recorded of Fortescue Tin Mine in St. Stephen—the supply of ploughed-up rocks of tin having presumably come to an end—as well as of East Treleigh Wood and St. Blazey Consols. These were unimportant but the decision to stop North Roskear the same month was a matter of some consequence, the end of what had been regarded as an important Camborne mine and one whose pumping was the 'key' to a whole district. The chief shareholder here, with 400 out of the 695 issued shares, decided to relinquish and in March the engines were advertised for sale.[3] In the same month the adjoining Wheal Seton and New Wheal Seton were stopped, the former being another major mine whose future had seemed more or less permanently assured only two years before. Ten tons of black tin a month were being raised but the monthly loss of £200 was more than the adventurers were prepared to meet. Wheal Seton's stoppage and the sale of the engines there in April[4] imperilled West Wheal Seton moreover, long one of the most important of the northern range of Camborne mines. Dams were hastily put in to prevent as much water flow as possible from the abandoned Seton sett but it was obvious that increased pumping charges would result and that an extra engine would be required to supplement the existing 85″ on West Seton. On the other side of the abandoned Seton sett, the smaller East Seton and New Wheal Seton had also to stop operations in April, drowned out by the abandonment of the larger mine. In addition, fears were expressed that the nearby North Crofty would be the next to go under from this increased water, the mine being already in a tottering financial state due to numerous shares being relinquished by those who considered matters hopeless. Elsewhere, in March, Restronguet Creek Tin Mine, at Devoran, was suspended, as also was South Kit Hill

[1] RCG 14.2.74. [2] RCG 24.1.74.

[3] Comprising 70″ pumping, 36″ stamps (96 heads), 24″ and 28″ whims, plus another 16″ engine. £45314 had been called up here since 1861. (RCG 21.3.74.)

[4] Comprising 70″ pumping, 24″ stamps, 22″ whim. (RCG 11.4.74.)

at Callington, Trelyon Consols[1] at St. Ives, and Perran Consols at Perranporth. At the latter, after five years' working, a loss approaching £25000 had been incurred.[2] The same month a more important loss was Wheal Margaret (Lelant) where, even with slight dues (1/36th) and working solely on tribute, £548 deficit was made in the first quarter of the year—'a clear loss of £25 on every ton of tin sold.'[3]

April saw the almost inevitable closing of North Crofty, and in the west of Boscaswell Downs.[4] The 36″ pumping engine was kept going for some weeks in the hope of finding a buyer for the mine as a going concern but the other engines were for sale in May.[5] The same month Providence Mines discharged all except eighteen men, whilst New Wheal Lovell, Spearn Consols,[6] Wheal Vincent (Altarnun) and Hobb's Hill (St. Neot), were abandoned.

As a result of this crop of failed mines and of the glowing reports of the inexhaustible tin of the great Mount Bischoff mine in Tasmania, a panic swept through the ranks of mining investors. The depreciation in the value of Cornish mining shares was certainly frightening. On thirty-three mines in the two years from March 1872 to March 1874 the market value had declined from £2,600,444 to £629,951—the bursting of the speculative bubble with a vengeance. In these same two years, Wheal Jane shares had come down from £42 to £2, Wheal Basset from £110 to £15, Cook's Kitchen £48 to £4, Trumpet Consols £17 to £3, Wheal Margaret £16 to £1. During the first four months of 1874 alone, the shares of Carn Brea came down from £65 to £27, Botallack £70 to £30, South Crofty £35 to £15, and Providence Mines £9 to 30s., whilst in the speculative new mines such as remained 'shares generally could not now be disposed of as a gift.'[7] The words of George Abbot, although relating to the copper bubble of 1824, once again applied; '. . . that indiscriminate and intemperate rage for Mine Investment . . . taken advantage of by designing and unprincipled individuals, to an extent that led to much devastation and ruin, and an almost universal condemnation of every mine operation.'[8] The extent of the speculation and the loss in the 1870's was more widely felt, for the acquisition of shares had become a widespread practice. Commenting on the activities of one bubble mine, it was said, 'The shares have been put all over England. I do not say designedly,

[1] An attempt was made to sell the mine as a whole but the reserve price of £750 was not reached by any bid at the auction. (RCG 9.5.74.)

[2] RCG 7.3.74. The chief shareholder was Richard Hallett, chairman of the celebrated Burra Burra Mine in Australia.

[3] RCG 21.3.74.

[4] Throughout 1872 and 1873 an enormous amount of work had been done here, cleaning up and re-timbering the estimated ten miles of old workings 180 fathoms deep, securing shafts, clearing the adit, installing tram-roads and skip-roads, overhauling and extending the dressing floors, etc., etc.

[5] Comprising 34″ (36 heads) stamps, 32″ (32 heads) stamps and two portable winding engines. (RCG 2.5.74.)

[6] Spearn Consols, which had been formed in 1839 by a combination of Wheal Spearn (started about 1811) with other setts, was sold for £610 to be amalgamated with Spearn Moor. (Oats Collection, RIC, Truro).

[7] RCG 28.2.74.

[8] *An Essay on the Mines of England*, 1833.

Dolcoath, Cornwall's deepest and most celebrated tin mine, about 1910. In this general view can be seen (left to right) the headgears of New East, New Sump, Old Sump, and Wheal Harriet Shafts. The tall square building at the right housed a big vertical air compressor, whilst the rails at left foreground led to the stamps. On the skyline to the south-west is the stack of Carn Camborne.

Looking down the Tuckingmill valley towards Tolvaddon, with Dolcoath dressing floors in the foreground. Cook's Kitchen, South Crofty, Tincroft and Carn Brea lie to the right. The aptly named Red River in this valley, flowing from Newton Moor to the sea at Gwithian, was the chief centre of streaming in Cornwall, with over twenty stream works still active as late as the 1890's.

The site of the new Williams Shaft on the slope of Carn Entral above Dolcoath (seen behind) after the ceremonial cutting of the first turf on 1 October 1896.

Above: *Williams Shaft, Dolcoath, during sinking operations, showing lattice steel headgear and house containing the patent traversing whim engine.* Below: *The engine-house (40″ engine) at Tyrie's Shaft on the North Tincroft section of Carn Brea & Tincroft, about 1914.*

but somehow or other they always seem to find their way into the hands of very ignorant people; clergymen, poor widows, and people of that sort, who are all going to make their fortunes upon the strength of a flaming prospectus. The effect of that is that after a short time the mine is wound up, and everybody cries out all through England against Cornish mining.'[1] In this respect Cornwall was to suffer ever after as a result of this wave of mining promotion and share pushing; some was knavish, some foolish, and some genuine, but all Cornwall suffered from a tarring with the same brush and by March 1874 it was stated that ' comparatively little ' foreign ' capital is now invested in Cornish mines.'[2]

During the remainder of the year 1874, the melancholy spate of abandoned mines continued. In June most of the machinery was sold off from Great Wheal Vor[3] and the news in August that the major shareholder had thrown in his 760 shares virtually meant the end of the mine, even though a few men were retained for a time exploring for side lodes at shallow depth. In July Nangiles was stopped after an abortive outlay stated to be over £60,000,[4] whilst the following month Tremenheere (Wendron) and Wheal Mary (St. Dennis) were abandoned. Mines sold up during August were Rosewarne United with 60″ pumping engine and five horse-whims, North Rosewarne with 36″ pumping engine and two horse-whims, and North Crofty.[5] Amongst the closures of September were Carn Camborne, where £28000 had been expended in vain,[6] and five mines in the St. Austell area—Wheal Bonny,[7] Wheal Tregoss,[8] Terras Mine,[9] Blencowe Consols[10] and Cornubia & Great Beam. Another of the great mines of the past closed in October—Charlestown United, then 225 fathoms deep and with four engines on the sett, a group of mines which with the intermission of only a few months had been worked on a large scale since 1819.

Typical of the way in which all unremunerative work of development and exploration was cut down or stopped altogether in order to reduce costs were the St. Just mines. Here Wheal Cunning United, comprising Boswedden, Wheal Call, Wheal Castle, Boscean and Wheal Cunning itself—a group formed only in the previous February[11]—stopped all save

[1] Report from a Committee of the House of Commons, Stannaries Act (1869) Amendment Bill, 1887.

[2] RCG 21.3.74.

[3] Although a small amount of exploratory work was continued at Wheal Vor in the hopes of discovering, as before, 'a new mine', this marked the end of the major working that had commenced in 1852. Since the concentration of all mining activity at Wheal Metal—but still under the name of Wheal Vor—from 1860 onwards, the dividends to this date had been £88000. In this same period, the chief mineral lord had received £18000 in dues. (MJ 28.3.74.)

[4] RCG 25.7.74.

[5] Including 80″ pumping engine, 28″ (40 heads) stamps, and 24″ whim. (RCG 15.8.74.)

[6] RCG 5.9.74.

[7] Including 33″ pumping engine, 24″ stamps and 14″ whim. (RCG 5.9.74.)

[8] Including 26″ rotary engine and 32 head of stamps. (RCG 19.9.74.)

[9] Including 50″ pumping engine and 22″ rotary engine. (RCG 26.9.74.)

[10] Including 70″ pumping engine, 36″ stamps, and 18″ whim. (RCG 26.9.74.)

[11] Boswedden & Wheal Castle adventurers had started to open the Wheal Cunning section in 1871

the shallow parts of the last named section, and sold off four engines.[1] Wheal Owles, in March, stopped eight of her eleven engines in the same manner, abandoning the old Wheal Owles, Cargodna, Wheal Boys and Wheal Grouse sections and intending to confine operations only to the newly developed cliff section of Wheal Edward and West Wheal Owles. The first call was then made since 1842, when £2000 had been required to put an engine (bought from the Wherry Mine) on Wheal Grouse. At Botallack, where heavy calls were made during 1873 to make up the deficit on working, work was suspended at Crowns and Wheal Hazard in February 1874 and even the future of the most productive section, the Higher Mine, was for a time in the balance.

To cut down working costs, an attempt was made to reduce wages, both by the direct implementation of a lesser wage or tribute and by an attempt to return to the recently abolished five-week month. As a result, a number of strikes occurred in many mining areas of Cornwall and west Devon, commencing at Tincroft in October 1873 when the daily pay for surface labourers was cut from 2s. 10d. to 2s. 6d.[2] In March, a few weeks before the mine closed, the men at Boscaswell struck against an attempt to re-introduce the five-week system,[3] whilst the same month more serious strikes and disturbances took place at South Condurrow and Wheal Basset from the same cause. Rosewarne United and Wheal Uny miners also came out in sympathy and a large number marched through Redruth ' with shouts and acclamations.'[4] By October, however, it was reported that wages had declined by some 20 per cent, although it does not appear that the five-week system was revived anywhere[5] in view of the widespread feeling of the men against it.

The amount lost by investors in Cornwall in 1874 will never be known but the total declared dividends came to £49121, something like one-fifth of the 1872 figure of £236,742. Of the nineteen dividend paying mines, the two richest were Dolcoath (£9773) and Tincroft (£9000). The former, like a few other mines, decided to stock much of the output of tin and hold it against the rise that surely had to come. The richness of the mine continued unimpaired and in January over seventeen tons of tin ore had been raised that was so pure it had been taken to the smelters unstamped.[6] Another prosperous concern of the year, owned largely by Sheffield shareholders, was the little Park of Mine, south of St. Columb. This, only 45 fathoms deep and destined to be very short-lived, nevertheless sold 231 tons of black tin in the year, all of exceptional richness, and

and then bought Boscean in February 1873 for £3000, new shares being issued at the same time and the whole ensemble re-named Wheal Cunning United.

[1] At Boswedden and Wheal Castle some work was continued (after an attempt to dispose of them as a going concern in 1874) but these also were to be abandoned in August 1876. The four engines advertised (RCG 21.2.74) were a 37″ pumping at Wheal Call; a 36″ pumping, 24″ stamps and 20″ whim at Higher Boscean.

[2] MJ 11.10.73. [3] RCG 14.3.74. [4] RCG 21.3.74.

[5] The strikes and disputes over it went on the longest at Devon Great Consols, and at the nearby Okel Tor.

[6] RCG 24.1.74. Almost all Dolcoath black tin was smelted at Mellanear (Hayle).

averaging about one cwt. of black tin to the ton of lode-stuff. As at Dolcoath, some had been raised that contained 70–80 per cent of black tin which was sold to the smelters unstamped.

The two years 1873 and 1874 had weeded out the first heavy crop of failures; the mines that were ageing, that were badly managed, that had shareholders lacking in confidence, or that never should have been started or re-started in the brief and now forgotten boom.[1] The low tin prices of 1875 continued the process. By the middle of the year, tin was down to £75 a ton in London and Cornishmen watched in vain for any sign, however slight, of a decline in Australian shipments. In 1873 output from there had rocketed to 2990 tons, to 5800 tons in 1874, to 7210 tons in 1875 and to 8392 tons in 1876. Making matters worse was the return to something like normal conditions in the Straits where British military intervention in 1874 had rapidly brought law and order. Under settled political conditions, aided by the coming of the ocean steamship and the opening of the Suez Canal in 1869, Straits tin had made great strides in the next few years. Financially, 1875 was as disheartening a year for Cornwall as 1874 had been, with dividends of £47019 in all, as against far, far greater losses. Mines stopped during the year were New Rosewarne[2], Rocks Tin Mine (St. Austell),[3] North Wheal Jane (Kenwyn),[4] North Wheal Metal (Breage),[5] Polberro,[6] South Phoenix (Linkinhorne),[7] and the final remaining part of Wheal Buller.[8] More important, however, than any of these was the demise of St. Ives Consols, St. Just Amalgamated and Crenver & Wheal Abraham United. At the latter, strenuous and extremely costly efforts had been made by more than one set of shareholders to emulate Dolcoath's finding of rich tin beneath the worked-out copper zone and after the value of the red metal declined so radically this offered the only hope of success to the Crenver company. In September 1872 the directors had expressed 'their emphatic opinion that the property is one of large present and still larger ultimate value as a tin mine . . . every fathom gained downwards is in the right direction' and in January 1874, on the strength of the tin ground opened up, the erection of 36 heads of stamps was authorised. Unfortunately, despite the optimism, the outlay, and the determined working, success eluded the company and in March 1874 winding up was escaped only narrowly by the last minute raising of a further loan to enable working to continue for another six months.

[1] The false prosperity of 1870–72 had carried with it the seeds of its own destruction. Old setts that had never paid and new ones that could never pay had been feverishly taken up, and now the disastrous after-effects of this were endangering legitimate mining.

[2] With 30″ pumping engine. (RCG 16.1.75.)

[3] With 20″ rotary engine. (RCG 20.2.75.)

[4] With 28″ pumping engine. (RCG 1.5.75.)

[5] With 32″ pumping engine. (RCG 1.5.75.)

[6] From 1865 to this date Polberro had sold 1248 tons of black tin, realising approximately £80,000. From 1852–64 sales (by Royal Polberro Consols) had been 2920 tons, realising £190,644. (Acc. 547, CRO, Truro.)

[7] Sold for £1050 as 'a going concern' (RCG 2.10.75.)

[8] Comprising 40″ pumping engine, 22″ and 10″ whims. (RCG 9.10.75.)

With no improvement at the end of this time, the materials were sold off in November. St. Just Amalgamated came to an end rather more abruptly, being shut down peremptorily in April[1] without there being money enough in hand even to pay the miners' wages.[2] The closing of St. Ives Consols, which half a century before had been Cornwall's premier tin mine, was described as 'a terrible blow to the district,'[3] although it might have been something less than this to the shareholders who had borne heavy losses for some time. In August it had been necessary to make a call of £3 17s. od. per share to clear up debts over and above a mounting monthly loss of £200. As a precautionary measure preparatory to stopping pumping, dams were put in to protect Rosewall Hill & Ransom United and early in September the engines were stopped. The next month the mine was for sale as a going concern at an asking price of £5000.[4] Later, the mine was bought by Messrs. Bolitho and Williams for £3000, reputedly for the value of the machinery and the tin leavings on the old dressing floors[5], and it was later re-worked by them on a small scale.

One of the few bright features of the year was the agreement in October on the amalgamation of the important Lelant mines, Wheal Mary, Wheal Kitty and Wheal Margaret (the latter two having already been amalgamated one year before). All three were standing unworked and with water in them to about 80 fathoms from surface, as a result of disputes over pumping between T. W. Field, the Trereife smelter who owned most of Wheal Mary, and Bolitho, who owned most of Margaret and Kitty.[6] Under the new name of Wheal Sisters, this important group of mines, with eleven engines in all and something like five hundred people employed, was thus saved and continued to be one of the largest of the Land's End mines in active production.

In the Wendron area particularly, mining was at a very low ebb by this time, with the majority of mines closed or suspended and the local financial depression was made worse by the failure of the Helston Bank late in 1875. One of the Wendron mines briefly in the public eye was Great Lovell, one of the few still being worked. Here, the 44 fathom level was being driven a long way away from any shaft or other means of ventilation and although the conditions were known to be very bad, the only attempt made to supply air to the end of the level was by the primitive mode of burning furze. The men working there were 'persistently cautioned to be careful' and one September morning when the night core did not come to grass, the worst was at once presumed. In the

[1] The St. Just Amalgamated engines (40″ pumping, 36″ stamps, 24″ and 26″ whims) were for sale in July. (RCG 24.7.75.)

[2] RCG 1.5.75. With miners paid several weeks behind at all times, it was not uncommon for them to be owed money when mines shut down. Under Stannary law, labouring miners were nominally prior claimants for such debts to them, although in practice secret mortgages and other loans on a mine's effects had precedence. This was one of the major injustices removed by the Stannaries Law Amendment Act of 1887 (see page 195).

[3] MJ 6.1.77.

[4] RCG 16.10.75. With 60″ pumping engine, 26″ (52 heads) stamps, 22″ and 20″ whims.

[5] RCG 18.12.75. [6] RCG 16.10.75.

level, fifty fathoms from the shaft, the three men were found, one dead and the others taking days to recover from their near asphyxiation from bad air.[1] The nearby East Lovell at this time was making calls, with the new Tregonebris section suspended in January 1876 and the engine—

VALUABLE MINING SETTS, MACHINERY, AND PLANT, IN ST. IVES, CORNWALL, FOR SALE.

MR. A. BERRYMAN has been instructed to OFFER FOR SALE, BY AUCTION, on the Mine, on Wednesday, the 27th October instant, at Two o'clock in the afternoon, as a going concern, the WHOLE of the VALUABLE and EXTENSIVE MINING SETTS, called

ST. IVES CONSOLS,

Situate in the parish of St. Ives, in the county of Cornwall, together with the WHOLE of the excellent PLANT and MACHINERY thereon, including:—

ONE 60 in. PUMPING ENGINE, with TWO 10 ton BOILERS.

ONE 26 in. STAMPING ENGINE, with two large fly-wheels, wrought iron axle and sweep rod, with ONE 10 ton BOILER.

ONE 16 heads AXLE, and THREE 12 heads ditto, together 52 heads, with lifters, &c., complete.

ONE 22 in. WINDING ENGINE, with cage attached, and ONE 8 ton BOILER.

ONE 20 in. WINDING ENGINE, with cage attached, and ONE 10 ton BOILER.

ONE 6 in. horizontal ENGINE, with vertical multitubular BOILER, with an 8 fm. 4 in. working lift attached.

9 fms. 19 in. pitwork; 27 fms. 12 in. ditto; 70 fms. 8 in. ditto; and 107 fms. 7 in. ditto, with H pieces, doorpieces, and working barrels, &c., complete; 20 fms. wood rods, 30 fms. 10 in. ditto, 30 fms. 9 in. ditto, 30 fms. 8 in. ditto, and 70 fms. 6 in. ditto, with strapping plates, &c., to match; 2 balance bobs, balance rods, &c.; 50 fms. 2½ in. round iron rods, at surface, with joints, &c., complete; about 800 fms. of underground railroad, with tram wagons, &c.; all the tin floors, together with the excellent and lofty woodsheds covering the same; about 900 fms. ⅝ in. chain, kibbles, &c.; 21 large centre head buddles, 6 square ditto; 124 self-acting and other tin frames; Brunton's patent calciner, 12 ft. bed, with water wheel and driving gear; 3 plunger lifts for pumping slime; 4 water wheels for driving dressing machinery; a large number of tin kieves; and all the necessary dressing tools requisite for a large mine.

Also, the RICH TIN LEAVINGS throughout the mine, being the accumulations of many years, during the greater part of which this has been one of the largest tin-producing mines in the county. The setts are held upon favourable terms, are large in area, and comprise rich and productive mineral ground. The water charges are easy, and the mine is favourably situated for the supply of material.

ST. IVES CONSOLS has paid in profits to the adventurers about £140,000; and, coupling this with the fact that a large proportion of the ground included in the setts remains unexplored, a rare opportunity is now offered to an enterprising company for mining investment.

For further information and to view, apply on the Mine; to Capt. JOHN GILBERT, Capt. APLIN, the Purser, or to the Auctioneer, 28, Clarence-street, Penzance.

Dated 11th October, 1875.

Sale notice of St. Ives Consols, Mining Journal, 16.10.75

a 50″ started in December 1871—'hooked down.'[2] Calvadnack Mine was shut down during May 1875 and the next month all underground operations ceased at Basset & Grylls. At Balmynhear, the winter rains had closed the mine during December 1875, the mine having already changed

[1] RCG 4.9.75.

In view of the grossly deficient ventilation at Great Wheal Lovell, a prosecution was brought against the manager of the mine under the Metalliferous Mines Act and a fine of £10 was subsequently imposed. (MJ 16.10.75).

[2] This meant the virtual end of East Lovell, although small scale mining activity continued there until April 1888, with only eighty tons of black tin produced in the final ten years. A total of 1749 tons had then been sold in all, for £115,638, on which dues of £5982 were paid. (East Lovell Cost Books, RIC, Truro).

An interesting echo of East Lovell came in a legal dispute in 1895. Nothing had been done on the sett for years and the lease had expired in July 1891 but when in 1895 the mineral lord's agents sold off the remaining engines (50″ and 30″ pumping, and 18″ whim) as they stood in the houses for scrap, the company was belatedly galvanized into an unsuccessful action to recover damages. (See RCG 22.8.95).

hands in April 1874 when the original company decided to sell up.[1] Alone in the whole Wendron district, Medlyn Moor Mine showed any signs of weathering the storm.[2] In Breage, South Great Work and West Great Work were well nigh at a standstill by midsummer 1876. The former had been kept going on the proverbial shoestring after narrowly escaping being stopped in April of the previous year when the manager gave his promise to the shareholders that, if the mine were continued, he would not make another call upon them,[3] whilst the latter managed to avert being compulsorily wound up in the Stannary Court in April 1874 only by rapidly re-constituting itself as a limited company.[4] This procedure was one that at least a dozen mining companies had recourse to during 1873 in order to escape the dreaded results, in uncertain expense and delay, of having matters wound up in the Court.[5]

In St. Just, Botallack was improving somewhat after the fears and rumours of a complete stoppage, whilst some activity above adit had recommenced at St. Just Amalgamated. The loss at Levant, where un-watering and re-opening the mine was proceeding apace, was not as heavy as might have been expected. At Wheal Owles, the policy—possible only in a small, rich and tightly knit costbook company—was to continue raising tin and then stock it against a banker's advance, awaiting a better price. By the beginning of 1876 the mine had 350 tons of black tin in stock and a debit balance due to the bankers of £14,460, and this state of affairs, of ' a full hutch and a heavy banker's debt ' was to continue for some time.[6] The policy of stocking tin had been tried by St. Ives Consols but if sales had, perforce, to be made on a market that had declined yet further the loss incurred was only made worse, and this had reputedly been a prime factor in closing this latter mine.[7] Ding Dong, where one half of the shares were held by those indefatigable promoters of mining in the far west, the Bolithos, was far from paying her way, whilst Rosewall Hill & Ransom was sold in March 1876 as a going concern for £1300.[8] At Providence, there had been some resumption of work but the mine was reported to be ' kept afloat rather as a charity than as an enterprise.'[9]

Other losses of 1876 were Penhale Wheal Vor, closed down in February

[1] Balmynhear at this time had a 45″ pumping engine and 24″ (48 heads) stamps.

[2] A sett which had been tried before in the early '50's under the name of Wheal Medlyn. The machinery here, chiefly comprising a 23″ engine and 21′ water-wheel, were for sale. RCG 16.6.54.

[3] RCG 17.4.75.

[4] RCG 18.4.74.

[5] MJ 27.12.73. An extreme example of the legal and other costs incurred in winding up the affairs of a mine in the Stannaries Court was Okel Tor, where £1891 was incurred in this way. (RCG 30.11.77.) In another case, at Sithney & Carnmeal United, a mine which was started in 1860 and wound up in March 1866, the shareholders were called on for £5 10s. 0d. per share in the Stannary Court four years later, in February 1870. (RCG 12.3.70.)

[6] RCG 1.1.76.

[7] RCG 1.4.76.

[8] RCG 1.4.76. Two months later, however, the new company was wound up and all mining stopped. (RCG 24.6.76). Under the name of Goole Pellas, a small part of the mine was later re-started by a Mr. Brown of Buckfastleigh (see RCG 15.6.77) and from July 1876—April 1881 the sale of 572 tons of black tin for £17160 is recorded. (Goole Pellas Tin Sales Book, RIC, Truro).

[9] MJ 6.1.77.

although later again to be re-worked; East Grenville, abandoned in March[1]; New Fowey Consols, where the tin potentialities of the old Fowey Consols had been given a very half-hearted trial, in July; North Metal & Harriet (Breage) and the once rich Bottle Hill Mine at Plympton in August; Trevarrack (Lelant) in September[2]; Park of Mine, in October; and New Pembroke (St. Blazey) in November. At the last named mine the company had opened up considerable quantities of tin ground beneath the old Pembroke copper workings but at the ruinously low price of the metal market the loss was too great to keep the mine open. Another mine sold up was Pednandrea, acquired by a new company in September for the sum of £2500 after a heavy loss on the re-working since 1851 of over £100,000.[3] Also for sale during the year was Penpoll Tin Smelting Works, at Point[4]—evidence of the appalling trade conditions in Cornwall—whilst even a few of the proverbially profitable stream works on the Red River and elsewhere had been thrown idle and were either up for sale[5] or were being worked by trustees on behalf of creditors.

The decline in the number of working mines in Cornwall and Devon in the period 1874–76 was dramatic. In 1873 there had been 230 mines in existence and out of these 47 closed down in 1874, 48 in 1875, and a further 37 in 1876. Thus 132 mines were abandoned out of 230 in this three year collapse, leaving 98 working in 1877.[6] The effect upon Cornwall of this rapid decline in the scale of mining can well be imagined, producing a depression as severe as had occurred in the late 1860's with the decline in copper. Then there had been the existing tin mines to which some working miners could turn, whereas now there was nought but emigration to fall back upon. A number were absorbed by the china-clay works of the St. Austell area, but for the great majority the only hope for the future lay abroad. Few of the men could afford to take their families with them and the latter remained, half starving and dependent upon uncertain remittances sent home, or on the parish and the County Distress Fund,

[1] With a loss of about £40,000. (Fortescue Records, DDF 388, CRO, Truro).

[2] September 1876 also saw the closing of New Trumpet & Lovell United in Wendron, one of the many Lovells. This had been one of the numerous small mines of the 1870–72 boom, with a career typical of many such. From the initial meeting in July 1870 to the winding up in September 1876 £6655 was expended with only two sales of tin (of 6 tons 14 cwt.) totalling £301 to show for it. (New Trumpet & Lovell United Cost Book, RIC, Truro).

[3] In this twenty-five year period, the expenditure at Pednandrea had been £451,851. Sales of ore had realised £357,896 and apart from dividends of £1423 in 1871 the luckless adventurers had paid calls amounting to no less than £101,804. Although the mine was sold for £2500 a further call was still necessary to clear off the arrears of unpaid calls amounting to £5392 plus the existing debit balance of £1802. (MJ 30.9.76).

[4] RCG 26.2.76.

[5] Sales included Excelsior Tin Stream Works 12.12.74, Gwellyvellan 21.3.74, South Nancemellyn 4.4.79. (RCG)

[6] R. Hunt, *British Mining*, 1887. Many of these were mere paper concerns and a more realistic appraisal of the declining number of true working tin mines occurs in a memorandum prepared for the Bolithos in 1876 by their chief mining agent. In this, a total of 144 mines is given as being at work producing tin in 1862: of these 99 had since stopped, leaving 45. A further 55 new or newly re-started mines had been put to work since 1862, of which 18 had since stopped. The total thus left in 1876 was 82 working mines out of 199 that had formerly been active. (Bolitho Records DD BL/30 CRO, Truro).

for most of the necessities of life. Families were split up and children boarded out elsewhere by the Distress Committee whilst the men who had remained, or who sought money for an emigrant passage, walked twenty and even thirty miles a day endeavouring to get work. Whole villages such as Chacewater were kept alive by remittances from men gone abroad whilst at St. Just in 1878 it was said that ' first rate houses may be had at a yearly rental of £2 and many are inhabited by tenants whose only payment is a promise to keep the house in decent repair.'[1]

To add to the economic difficulties of Cornish mining at this time, the physical problem of pumping cropped up again in the exceptionally wet winter of 1876–7. In almost every mine, engines were worked at their fastest, adding crippling coal bills to the all important monthly costs. In several mines the workings simply could not be kept in fork and in January 120 men were discharged from Wheal Grenville where the small engine was completely over-powered. At South Condurrow, 40 were discharged as a result of the bottom levels being drowned out, whilst in Breage both West Godolphin and Penhale Wheal Vor were temporarily stopped from the same cause. Everywhere there were fears for adits, Peevor being partly drowned out by collapses in a branch of the County Adit and the Wheal Basset adventurers having to drive expensive ' side-lyes ' to by-pass bad chokes in their adit. Penhalls was flooded out to adit level for two weeks, and an extra monthly contribution towards the cost of pumping there was made by the adjoining Blue Hills Mine.[2] As a result of these hindrances to production, a new 80″ engine was thereafter ordered by Wheal Grenville, where a new company had taken over[3], and a new 65″ for South Condurrow, whilst in April the West Godolphin company bought the existing 60″ left on Wheal Vor (Wheal Metal).

Unfortunately there was still no sign of any reversal or even any arresting of the fall in tin prices and the question now being asked was how long the depression would, or could, continue. Operations were curtailed at Wheal Coates in March, with 60 hands discharged; St. Agnes Consols was completely shut down in June and in July the materials at East Basset[4] were for sale, as was also the Providence Mines' machinery in August.[5] In September the engines at Spearn Moor[6] (St. Just)—which included the recently acquired Spearn Consols—were also advertised, followed by the sale of the little Carnelloe Mine's water-wheel and other

[1] RCG 26.4.78. A great number of St. Just men emigrated to the newly discovered diamond mines of Kimberley. When fire swept the great De Beers mine there in July 1888, one of the first places to know of it was St. Just, informed of the disaster and of the deaths of three St. Just miners by telegraph. (RCG 19.7.88).

[2] At St. Aubyn United, the engine shaft itself collapsed and ran in on account of the incessant rain: ' the shaft tackle, or poppet heads, have gone into the shaft, as also one of the legs of the shears; the other leg being left to bear the whole strain fell over and broke.' (RCG 19.1.77).

[3] Wheal Grenville's last dividend prior to this date had been in September 1872 and the loss since then £23375. (Fortescue Records, DDF 388, CRO, Truro).

[4] Including 60″ pumping engine, 36″ (32 heads) stamps and 24″ whim. (RCG 6.7.77).

[5] Including 40″ pumping engine, 30″ stamps, 23″ whim and 20″ man-engine. (RCG 31.8.77).

[6] Comprising 26″ pumping engine, 30″ (24 heads) stamps, 26″ and 18″ whims. (RCG 7.9.77).

materials in November. Such was the state of mining that earlier, in June, the Creegbrawse sett and the surviving account-house materials etc. were all sold at auction 'for a £5 note, stock and block.'[1]

ENGLISH TIN, AND HOW TO COMPETE SUCCESSFULLY WITH AUSTRALIA.

USE DYNAMITE.

Invaluable for BLASTING the HARDEST and WETTEST ROCK; **SAFER to** USE; and EFFECTS a GREAT SAVING of time and money.

Pamphlets free by post. An experienced man sent underground to **give** instructions when necessary, free of charge.

Apply,— STEPHEN WILLIAMS, CAMBORNE.

Few now realise that Australian tin was Cornwall's greatest fear and competition in the 1870's, as is evident in this advertisement for dynamite in the Mining Journal, 23.10.75

The occurrence of the year which made the greatest news, however, was at West Basset where a deficit of £25000 was revealed in June, made up of a series of debts that had been systematically concealed from shareholders. This had not been done with any nefarious intent, but to shield the company from calls that might bring about its closure. This was the sum stated in the press[2] but it actually amounted to an adverse balance of £10337 (including £7274 loss in the previous five months' working) plus £20995 bank overdraft, obtained against stocked tin. In explanation to the shareholders it was stressed that the outlay had been considerable since West Basset had changed hands in 1871: £15000 had been expended on the New Stamps, £12000 on re-modelling and extending the Old Stamps, and £12000 to open out the mine on the Great Flat Lode and to put it into first-class order. Almost £40,000 had thus been spent in this way in a programme which seemed both economic and desirable with tin at its 1871–2 price.[3] Despite this long explanation however, it was as well for the West Basset committee that suppressed accounts were not a penal matter under the existing Stannary law.[4] The shares, during the first shock of the disclosure, went down to a mere penny apiece, but by October they had recovered to a more normal 30s.–35s.[5]

With black tin below £40 a ton, few indeed were the mines that could now pay their way, let alone pay dividends. The smelters, acting in even more determined unison than ever before, combated the heavy shipments from Australia and the Straits in 1878 by depressing even further the price they would pay for Cornish output and a great many parcels of black tin were bought at prices from £35 down to as little as £27 10s. 0d. per ton. The calls made during the year (1878) in thirty mines were estimated by

[1] RCG 12.10.77. [2] RCG 29.6.77.

[3] West Basset, Book of Reports and Balance Sheets, Trestrail Collection, CRO, Truro.

[4] In a cost-book concern, a balance sheet in the form in which it is known today was not presented to shareholders. Although the Stannaries Act of 1869 provided for the keeping and presentation of regular and correct accounts, it was not until a further Amendment Act in 1887 that manipulation of accounts in this way became an offence under Stannary law.

[5] RCG 26.10.77.

the *Mining Journal* at £133000, whilst dividends in the thirteen profitable mines totalled £34107. South Condurrow paid £7041, Dolcoath £4296, Wheal Eliza £4096, East Pool £3040 and Wheal Peevor £2250. South Condurrow's results for the year were almost identical to 1877, when 649 tons of black tin were sold (costing £32 2s. 0d. per ton to produce) which realised £27846, leaving a profit of £7082 in that year, after deduction of dues amounting to £1056. Amongst the calling mines of 1878, the case of Wheal Sisters was typical, with a production of 190 tons of tin in sixteen weeks' working which made a loss of over £1300 for her twelve shareholders to bear.[1] Wheal Owles was similarly placed, and in July £62 10s. 0d. per share[2] was called. The mine's future hung in the balance and the purser met strong criticism for his huge stocks of tin. But the calls here were far outweighed by those at Tincroft and Carn Brea where, as at West Basset the previous year, accumulated liabilities were suddenly revealed in September to the dismayed shareholders. At Tincroft £6000 was needed to meet the deficit but at Carn Brea no less than £20,000 was required.[3] Again, there had been no fraudulent intent, the day of reckoning having been deferred time and again in the hopes that metal prices would rise and the deficit be wiped out. Captain William Teague was himself the owner of one third of both mines and had his share to pay.

The mines stopped during the year included Wheal Unity Wood, where the 70″ engine was for sale in March; Trumpet Consols, abandoned in June; Cape Cornwall Mine, where the 30″ pumping engine and 24″ stamps were for sale in July; and Wheal Sparnon, wound up in August. Ding Dong, reputedly the oldest mine in Cornwall, also fell victim during the year, having sold 3473 tons of black tin since January 1850 for £225,162.[4] Two other important mines in a precarious state were Wheal Basset and South Crofty, and in October it was decided to stop all activity at the latter except the driving of the bottom (205 fathom) level by boring machine, with the whole costs not to exceed £200 a month.[5] In the smelting field, a casualty of the year was the Bissoe works, shut down in July by Tregoning & Co., caused primarily by the death at this time of J. S. Tregoning, the principal partner. There was a rumoured intention of re-establishing smelting in Llanelly where the company's tin-plate works was continued, but if this took place it did not continue long.

In October, the *Royal Cornwall Gazette*[6] reviewed the existing state of mining in west Cornwall after the havoc of the previous four years. In St. Just, four mines only (Botallack, Levant, North Levant, Wheal

[1] RCG 17.5.78.
[2] viz. £5000, the mine being divided into 80 shares.
[3] £20 for each of the 1000 shares held. To soften the blow it was made payable in four £5 instalments over a period of a few months.
[4] Report dated 1924, DDX 104, CRO, Truro. Ding Dong sett covered 504 acres and at the time the mine was abandoned it was calculated that there were nine miles of levels from adit down to 80 fathoms depth and a further mile under that down to the 125.
[5] RCG 18.10.78. [6] 11.10.78.

Owles) survived; in Marazion, St. Erth and St. Ives, only Wheal Sisters; in the Helston and Breage area, nothing except 'a small mine or two in the last stage of consumption'; in Wendron, Medlyn Moor alone; in St. Agnes, Wheal Coates, Wheal Kitty, Penhalls and Blue Hills; in the Truro area Wheal Jane and Killifreth, whilst in the Redruth-Camborne area, the ranks were now sufficiently thinned to little more than twenty names.[1] To complete this picture, in east Cornwall there was only Phoenix United left that was of any importance and here too operations had been curtailed to the shallower levels, with the pumps stopped in the deep eastern part of the mine, and with the Duchy dues given up to assist the mine in keeping going.

Despite this fall in the number of working mines, the actual production of ore had scarcely fallen, due to the dire necessity for increased output from the survivors to meet diminished prices and partly to the ripping out of many mines in their last few months in a final desperate bid to meet working costs at any price. In fact, output in 1878 (15045 tons of black tin) was higher than it had been in 1872 (14266 tons), yet the value of this produce had been more than halved—from £1,246,135 to £530,737. The major factor which brought about this increased output in the surviving mines and which also brought a considerable reduction in labour charges was the re-introduction of rock boring machines. Dolcoath and Tincroft had tried these in the late 1860's but Doering, their great pioneer in Cornwall, was reported to have been laughed at by other mine adventurers for his pains and even to have lost a great deal of money in endeavouring to convert the conservative Cornish miners from the old slow hand boring.[2] By the 1870's, however, the new machines had proved their worth in other countries and attention was turned anew in Cornwall to their use as a way of cutting down the cost of driving levels—appallingly expensive as the one time copper mines in soft killas (clay-slate) were now working tin veins in unyielding granite. Moreover, long continued emigration had skimmed off many of the best men whilst the heat of the deep workings sapped the strength of those left. In hard ground, in the days before machine drilling, a two-man team—one wielding the hammer and another turning the borer—might take three hours or more to drive a hole twenty inches deep, using a dozen or more steels.[3] With the quality

[1] West Seton, South Roskear, South Crofty, Cook's Kitchen, New Cook's Kitchen, East Pool, Wheal Agar, Dolcoath, Tincroft, Carn Brea, North Pool (on a very limited scale), West Tolgus (copper), Pednandrea, Wheal Peevor, South Tolcarne, South Condurrow, Wheal Grenville, West Frances, South Frances, West Basset, Wheal Basset, Wheal Uny and East Uny. The last named had been bought at this date by Wheal Uny adventurers (for £1250), the 70″ pumping engine there having been maintained by them for some twelve months.

[2] RCG 29.11.78. Patents for various types of rock drilling machines were particularly numerous at this time, over sixty being granted in Britain alone from 1850–75.

[3] The exceptional speeds attained in hand boring contests held from time to time between mines bore no relation to normal work underground. In one such contest attended by 1,000 people in August 1888 five teams competed and a Tincroft 'pare' bored 13″ in hard granite in under seven minutes, at a striking rate (with the hammers slung fully round) of 91 per minute. (*Trans. Mining Asscn. of Cornwall, 1888*).

of metal then available a borer might last twenty minutes at most before it became too dulled for further use and had to be set aside to be re-forged. Thus hand boring was, above all, slow whilst machine drills felt neither the heat of the deep workings, nor the fatigue and pain of 'beating the borer.' Their introduction was fast becoming a necessity and not, as formerly, a luxury, and Dolcoath was the first mine to give them further trial, installing one in the 314 fathom level in November 1875 with another of different make also tried at Wheal Agar. Six months later the results at Dolcoath were said to be successful apart from the necessity for frequent stoppages, the machine having 'to be brought to the surface about every twenty-four hours to be repaired.'[1] South Crofty followed suit with one in June 1877, the success of the Dolcoath one being confirmed in December 1877 when a new and more powerful air compressor was ordered to enable more similar machines to be worked. In 1878 at the same mine it was reported that a level could be driven 10 fathoms a month at a cost of £17 per fathom as against 1½ fathoms a month by hand at £26 per fathom. West Basset, South Roskear, and Carn Brea each put one in during that year, with another at Botallack in August 1879.[2] These machines were of various makes and types, Eclipse, Barrow, Mckean, Darlington, Jordan, Beaumont, McCulloch, and others, and during this trial period were operated by contractors and not installed by the mine companies themselves as permanent equipment. Not all were satisfactory, and the one tried at South Roskear in 1878 had a hard time with a greenstone dyke compared to those working elsewhere.

In conjunction with mechanical boring, a more powerful explosive was needed, both to keep pace with the machine's capacity and to shatter the rocks with more force ready for tramming out and stamping. On this account dynamite, first manufactured in 1867, was so markedly superior that the old black powder rapidly passed out of use.[3] The new explosive required a smaller bore-hole, was impervious to wet conditions and could be used in greater safety since a stray spark would not cause it to explode. Although monopolistically dear, dynamite almost halved the cost of driving levels in hard ground compared to blasting by common powder. Moreover it made less smoke,[4] although the acrid fumes and gases it produced were still a serious problem.

[1] RCG 13.5.76. The necessity for frequent repairs was commonplace in all early rock-drills. A few years before this, during the construction of the Mont Cenis Tunnel in the Alps, opened in 1871, two hundred machines were kept in hand in the repair shops to ensure that sixteen were maintained in constant service.

[2] Here it drove one level 22 fathoms in ten weeks, as against 58 fathoms in twelve months by hand labour. (RCG 21.11.79).

[3] The chief gunpowder works in the county had been the Kennall Gunpowder Company near Ponsanooth, the Nancekuke Powder Mills in Illogan, the St. Allen Mills near Truro and the East Cornwall Powder Company's Mills at Herodsfoot, near Liskeard.

[4] One captain reported to the Mining Explosives Committee (formed in 1879 to hold comparative trials) that after gunpowder he had gone into smoke so thick he could not see a candle 'two feet from my nose.'

Thus, by economy of production and by improvements in technique, metal mining survived the 1870's. Only by sheer staying power, by the stout-heartedness of some adventurers and the providential richness of a few mines, was it not annihilated. In 1870 the idea that mines could stay open with tin at 1878 prices would not have been believed, such was the change wrought by introducing new machinery and new ways. On average it was costing £20 per ton less to produce black tin in 1878 than eight years before, although from a long term point of view few mines were able to open out sufficiently far ahead of the productive stopes. The cutting out of wasteful working and the economies that had been made were nowhere better shown than in the numbers employed: in 1873 26500 persons had been employed in Cornish mines, whilst by 1878 this was down to 13730. As output was more or less maintained at the 1873 figure it will be seen that productivity was doubled per person employed, a revolution indeed in Cornish mining.[1]

This halving of the population employed in mining brought distress as bad as any that Cornwall had yet seen and miners emigrated in a steady stream. The exodus of the seventies was largely to South Africa, to the newly opened diamond fields near Kimberley, whilst others went to colliery districts in Scotland, the Midlands, Lancashire and elsewhere. Conditions in some west Cornwall mining districts were so bad that the County Distress Committee was revived to deal with the worst cases. In 1879 £6809 was collected into the Committee's fund, over half of this being subscribed by Cornish people in Australia, chiefly from Adelaide, Kapunda and Ballarat. To some extent it was justice that this aid should come from Australia since it was from there that the malaise affecting Cornwall emanated. Writing of the mineral specimens from the new continent displayed at the 1878 Paris International Exhibition, one Cornishman wrote despairingly of 'the solid nature-rolled masses of brown stream tin, from the size of a walnut to boulders as large as a man's head . . . No one having at heart Cornubian interests could fail to be deeply impressed with the witnesses to the indubitable wealth in tin of sundry distant regions which have poured their produce so largely into the commerce of the world as to occasion, in part at least, that depression and grievous trial which at present afflicts a large portion of our western population '.[2]

In 1879, however, after five years, there were the first signs that the worst was perhaps over and that a return to something approaching relative prosperity might be ahead. Two or three mines stopped during

[1] This decline in the total numbers employed was particularly noticeable in the case of children, partly from the use of improved and semi-automatic dressing equipment and partly because very young children could no longer be employed after the passing of the Factory and Workshop Act of 1878. To big tin mines this was a matter of some concern and at Dolcoath, for example, sixty youngsters had to be ' dismissed ' as a result of this Act. (See RCG 1.8.79).

[2] *Transactions Royal Geological Society of Cornwall*, 1878.

the early part of the year but none of great significance.[1] South Crofty had been suspended completely, even including pumping, in February but was re-started on a limited scale in May,[2] whilst at Wheal Basset a new group of shareholders decided to stop the old mine and devote all attention and expenditure to the North Basset section instead. Here, the intention was to open up the Flat Lode running east from South Condurrow to which Grenville and South Frances were now sinking,[3] and a new 80″ engine was ordered, plus a 26″ whim. Two other engines erected in the year were at New Cathedral Mine, Gwennap,[4] and at Wheal Violet,[5] west of Camborne—the first mines to have been started for several years.

With the smelters raising prices steadily from October onwards, Dolcoath announced the intention of selling 200 tons of stocked tin to clear the existing £10,000 overdraft, on which the bankers' annual charges were over £1000, whilst Wheal Owles decided to follow suit with their 400 tons when once the price reached £60 per ton.[6] For the first time for several years Cornish mining, in 1879, almost paid its way, with calls of £55000 in thirty concerns more or less equalled by the seven dividend paying mines of the year.[7]

The financial results of 1880 as compared to the previous year admirably demonstrated how the prosperity of Cornish mining hinged upon the price per ton of black tin. Below a certain price mines made a loss, and frequently a loss that could not be met even by increasing output, for there were standing charges, such as pumping, that were irreducible. Above a certain price, they made a profit and any subsequent rise over and above this brought a more than proportionate rise in the latter. On this basis, with black tin up from an average of £40 to £50 a ton, dividends of the year more than doubled, to £118768. The long catalogue of abandoned mines came to a halt and instead a number were re-started, commencing

[1] Including Treleigh Wood (with 60″ pumping engine, and 20″ stamps) and the little Combellack Mine in Wendron (worked by a 14″ horizontal pumping and stamping engine): West Godolphin, with its four engines (60″ and 30″ pumping, 14″ horizontal whim and 15″ horizontal stamps engine) was also sold in February but was to be re-worked. The same was true of Penhale Wheal Vor, where the engines (60″ and 40″ pumping, 30″ stamps and 20″ whim) were for sale in October.

[2] RCG 14.2.79, 9.5.79. When pumping was about to be stopped, East Pool offered a monthly sum towards the cost of keeping the South Crofty engine at work, but without success. Above the 160, the two setts were holed into one another.

[3] The stopping of Wheal Basset meant water troubles for South Frances, where dams were put in during January. These included the first cast-iron one ever used in Cornwall, although they had long been used in collieries in the north. (RCG 7.2.79). Subsequently it was decided to form a committee from Basset, West Basset, West and South Frances to keep Richards' Engine at old Wheal Basset at work.

[4] Soon after the mine was forked there was a serious accident in January 1881 when old workings were holed into and eight miners drowned. (WB 27.1.81).

[5] RCG 20.6.79. Wheal Violet, later better known as Violet Seton, was a venture, financially single-handed, on the part of a Lieutenant-colonel Fludyer, a colourful character who had served in the Grenadier Guards and who forsook his native Rutlandshire to devote his energies—and a great deal of money—to its working. According to one report (MJ 24.8.89) he had vainly expended £80,000 at Violet Seton but throughout a whole series of vicissitudes the mine was to be worked on a quite considerable scale until it was closed on Fludyer's death in the 1890's.

[6] RCG 17.10.79, 7.11.79.

[7] East Pool £12640, South Condurrow £11024, South Frances £10010, Wheal Peevor £5025, Dolcoath £4296, Wheal Eliza £3584, Mellanear (copper) £3500. Total £50079.

with West Kitty late in 1879. St. Just United, West Wheal Peevor[1] and Wheal Coates were resumed in January; Wheal Union, at Redruth, and Carn Camborne in February; Pendarves United, Morvah Consols and Wheal Elizabeth in March; South Penstruthal (Gwennap), Great Rocks (St. Austell) and West Godolphin in September; East Botallack in October[2]; whilst in December companies were formed to re-start East Uny as a separate entity, and to re-work Basset & Grylls, as well as Wheal Boys and Prussia & Cardrew United, near Redruth.[3] Allied to this new wave of activity was an extension of operations in the existing mines, with boring machines reported as introduced at Levant[4], West Frances and Cook's Kitchen in March, Wheal Owles in May and at South Frances in July. An attempt was also made to institute a system of regular tin ticketings at Redruth on the same pattern as the copper ticketings then dying a natural death but this was not a success. The smelters saw in this a possible challenge to the existing *status quo* already endangered during the year by the re-establishment of the Penpoll Tin Smelting Co. set up in the old works which Messrs. James & Shakspeare had endeavoured to sell four years before. The new company was promoted by the adventurer William Teague[5] in conjunction with Arthur Strauss & Co., metal brokers of London, and an overt price war immediately broke out between them and the old ' ring.' Establishment was successfully achieved, partly on account of both Teague's and Arthur Strauss' extensive mine holdings, but for most of 1880 it was a case of ' Penpoll *versus* the rest.'[6] A feature of the year was strikes at various of the Camborne mines, amongst surface workers. There was strong feeling that although tin had risen greatly and profits with it, wages were still at the starvation levels of the worst of the depression, and men with families could no longer live on fifty shillings a month. The strike originated at Dolcoath in January and quickly spread to Cook's Kitchen, Tincroft and other mines adjoining. A march through Camborne by between 400–500 men, some armed with cudgels, showed the widespread discontent and a general rise in rates of surface pay was rapidly granted by the adventurers thereafter.[7]

[1] West Peevor was started largely as a result of the contemporary success of Wheal Peevor which had been more or less the only survivor from the 1872 boom. In the best four years (from 1878–82) Peevor sold 2,287 tons of black tin—at an average price of only £43 per ton—and paid £27,716 in dividends during this time. Earlier however, from 1872–6, calls of £22650 had been necessary, and losses began again after 1883. (*Trans. Mining Association of Cornwall*, Vol. 1. 1885).

[2] East Botallack was later sold to the New Balleswidden Co.

[3] In most of these mines, the existing machinery was still *in situ*, having been unsaleable in the previous few years of depression.

[4] This rock drill tried at Levant was later said to have been a failure. A critic of the mine said ' It had cost them £10,000 and the adventurers had never had a shilling return . . . the man who would work the small lodes in St. Just mines with a boring machine was the vilest idiot alive. It would knock up every mine in the parish.' (MJ 21.6.84).

[5] Teague had started at Tincroft in 1854 and by 1863 had become managing agent. The mine was exceptionally badly mis-managed by a London company and during this time he put all his savings into it, gradually acquiring 2620 of the mine's 6000 shares. By the end of 1862 his holding in the mine was worth £52,000 (MJ 14.3.63) and he later rose to be one of the best known figures in late nineteenth century Cornish mining.

[6] RCG 4.6.80.

[7] RCG 30.1.80, 27.2.80.

The minor revival, for such it now was with black tin about £60 a ton, continued throughout 1881. Kit Hill Mine and Wheal Jennings[1] were resuscitated in January—the latter under the name of Parbola—with Basset & Buller Consols formed in February to re-work the old East Basset, Copper Hill and Wheal Buller setts. In April a 36″ pumping engine was started at New Kitty in St. Agnes, and an attempt was made to re-start Polberro by a limited company in July. At Tregembo, an engine was erected in September, whilst some work was also proceeding at Great Work by a handful of miners employed by William Teague. Other flotations of the year were Great Polgooth United, Owen Vean & Tregurtha Downs, and three small mines in the Breage area—New Great Wheal Vor, Great East Vor and West Vor & Leeds United.[2] The first named was largely a fraudulent company promotion which later achieved notoriety in the law and criminal courts, whilst the latter three all issued highly coloured prospectuses.

By this date Cornish copper mining was a dying industry, very much subsidiary to tin in its importance to the county. In 1881 the latter realised £697,444 (12893 tons) whilst copper ore brought less than one third of this—£190,057, from the sale of 52556 tons. Arsenic was in fact becoming an article almost of as much commercial value as copper ore, produced in the form of arsenical soot when the stamped tin ore was calcined preparatory to further dressing. After 1880 it was to become a by-product which was an important addition to the revenue of many mines, in particular South Crofty, East Pool and Tincroft.[3]

In the established mines, the endeavours to prune working costs continued, whilst at the same time efforts were made during this period of newly found prosperity to re-equip mines to give increased output without greater cost. Formerly the richest part of the lode had alone been removed, the old miners following its heaves and its underlie quite haphazardly. Now, an era of what might be termed bulk mining came into being, in which the whole lode—leader, capel, and the entire mineralised part of the standing wall—was broken out and raised. Indeed many mines worked for their last few years on the produce of old levels alone, on workings that an earlier generation of miners had considered exhausted. The two modes of mining might be exemplified by a comparison of the Wendron district before it declined in the 1870's, where small rich lodes were hand-worked and rock boring machines were never used, with the Camborne

[1] RCG 7.1.81. In the worst of the '70's, Wheal Jennings, owned by a Mr. Richard Jennings who had oil interests in Pennsylvania, had all but paid costs. Worked entirely by tributers, it was not sunk by so much as a foot in seven years, yet sold a regular four or five tons of tin a month during this time. (RCG 31.8.77).

[2] RCG 16.12.81.

[3] The first rise in the value of arsenic, then in demand for killing the 'potatoe bug' (the Colorado beetle) and later for controlling boll weevil on cotton crops, came in 1875 when the price per ton of crude arsenic (i.e. unrefined arsenic soot) rose to £10. (See RCG 6.3.75). The demand prior to this date, when its main uses had been for hardening shot, in flint glass manufacture, and the making of sheep dip, had been relatively small.

Goold's Engine Shaft (with 80" engine) at Grenville United Mines, just before the first World War. Wire rope from the winding engine in the foreground.

Michell's Shaft (left) and Engine Shaft at the old East Pool, about 1910, from the west. This was the site of the great underground collapse in the workings in 1921 which necessitated sinking the new Taylor's Shaft to the north.

The extensive dressing plant of East Pool at Tolvaddon in the Tuckingmill valley. Here, in the 1920's, 8000 tons of ore-stuff a month were being treated, using eight heads of pneumatic stamps and seven Brunton calciners. The main arsenic stack (one of two still standing), which served no less than 42000 cubic feet of condensing chambers, is a little to the right.

Levant Mine, St. Just: a general view of Skip and Engine Shafts. The arsenic works and stack are visible on the cliff edge above the whim engine-house, and the stamps engine-house on the right above the dressing floors which extended down the slope hidden beyond. The newer houses for the man-engine and compressor lay slightly inland, out of sight. The count-house, extreme right, is now demolished, its dressed granite masonry re-built into a private residence at Marazion.

St. Just United Mines: a general view of Bailey's Shaft shortly after re-working had commenced in 1881. Somewhat unusually, the whim engine was sited behind the 36″ pumping engine and hauled over the latter's house. A third, stamps, engine was added in the left foreground some years later. Beyond can be seen the stack and buildings of Cape Cornwall Mine.

mines of the 1880's, equipped with fast winding whims, double skip-roads, boring machines, enormous stamps batteries and the like. The difference between the two was responsible for the decline of the tributer as more and yet more underground work came to be set at a fixed rate per fathom. The changes necessary to bring an old established mine into a modern state were considerable. Shafts and small levels, in a few mines less than five feet high and two feet wide, had to be enlarged and, where crooked, straightened in order to take tramming roads. The latter had been widely used for decades in copper mining, with its massive ore and bulk output, but only rarely in tin and the cost of enlarging a level to say 7′ x 5′ was considerable. Mechanical boring and high explosives alone made it possible and in 1886 Captain Josiah Thomas of Dolcoath said that despite the current richness of the mine, it would hardly be paying costs if gun-powder and hand boring were still used. Compressed air, both for boring and for its beneficial effects upon ventilation, was alone enabling the deep mines to work at a profit.[1] Ventilation and high temperatures had long been major problems in deep tin mines, owing to the irregular and cramped workings, together with the long exploratory headings continually being driven. The compressed air used in the machines was also doubly valuable in being fresh from surface, the hand power fans and other ventilating devices of the less progressive mines being mere air circulators which did not introduce fresh oxygen. This was now of extra importance, for, although dynamite made little smoke in comparison with common powder, its fumes after the frequent blasts were extremely injurious. William Teague, in 1887, recorded that from dynamite fumes in a poorly ventilated mine it was 'not unusual for men to come six days to work and vomit their breakfast every morning . . . and to come up to surface bringing up their dinners just as they had carried them down.'[2] Yet dynamite, for all its injurious effect, was now essential and the attempt in 1880 to renew the patent covering its manufacture was strongly opposed by the Cornish mining interest. A large number of mines contributed towards a fund opposing Nobel's application for renewal.[3]

With greater output of ore from underground, better means of hauling to surface were required and by the mid-1880's chains and kibbles had

[1] Examples of the greater speed of development with boring machines have already been given. In another case at Cook's Kitchen, in the early 1880's, using rockdrills and a 12″ compressor, a handful of men drove three ends 16 fathoms in a temperature of 90° in the same period of time that it would formerly have taken sixty unaided miners.

[2] *Trans. of Mining Assocn. of Cornwall*, 1887.

Even with rock drills in use, the working conditions in the deep Cornish mines of the 1880's were harsh to a degree. Tin workings in granite had been reckoned cool by the standards of the infamous Hot Lode in the Gwennap copper district but this was in the 1860's at little more than 200 fathoms depth and twenty years later, with many tin mines almost double that depth, air temperatures of 90° and even 100° were not regarded as exceptional. For a graphic account, in detail, of mining under conditions of great heat, much of it applicable in a lesser degree to Cornwall, see Eliot Lord, *Comstock Mining and Miners*, 1959.

[3] The explosive cost 2s. a pound in Britain as compared to 1s. a pound in Germany, calling forth the comment from the *Mining Journal*—'if everybody connected with dynamite has not made his fortune, it has not been for want of knowing how to charge.'

largely given way to wire ropes and skips.[1] The installation of skip-roads into the four main hauling shafts was completed at Dolcoath early in 1882[2] and most other mines rapidly followed suit. The capacity of the stamps batteries now put a limit on the output of most mines and these were enlarged or additional sets installed to treat the growing tonnages hauled

ANNUAL OUTPUT OF TIN IN CORNWALL AND DEVON, AND AVERAGE ANNUAL PRICE 1873–94 [3]

	Black Tin (tons)	Metallic Tin (tons)	Average Price per ton (black tin)	Average Price per ton (metallic tin)
1873	12885	9972	£78	£133
1874	14039	9942	£56	£108
1875	13995	9614	£53	£ 90
1876	13688	8500	£44	£ 70
1877	14142	9500	£41	£ 73
1878	15045	10106	£35	£ 66
1879	14665	9532	£40	£ 72
1880	13737	8918	£49	£ 91
1881	12900	8600	£54	£ 97
1882	12470	8077	£57	£107
1883	12890	8419	£51	£ 97
1884	13622	8709	£44	£ 85
1885	12934	8470	£46	£ 89
1886	12780	8378	£55	£101
1887	12740	8406	£62	£113
1888	12451	8080	£62	£117
1889	12411	8067	£53	£ 97
1890	12944	8414	£53	£ 98
1891	13029	8470	£51	£ 94
1892	12890	8418	£51	£ 97
1893	12343	8023	£47	£ 89
1894	11507	7480	£38	£ 73

to surface each year, Dolcoath having three sets of steam stamps in operation at this time, for example, with 48, 120 and 60 heads. Stone breakers also came into use to supplement the stamps, although these were slow to be adopted in preference to ' the muscular arms of our maidens.' The only aspect of production which had failed to keep pace with this technical revolution in Cornish mining was tin dressing and a great deal of cassiterite stamped out of the ore was certainly being lost to the mine companies at this time. On the chief streaming river in the county, the Red River which flowed from Bolenowe Moors past Dolcoath and Tuckingmill to the sea at Gwithian, there were still no less than twenty

[1] For an account of the technical development of Cornish mines at this period, including the attention given to winding engines, stamps, man-engines, etc., see D. B. Barton, *The Cornish Beam Engine: Its History and Development*, 1965.

[2] Prior to this all ore had been raised at Dolcoath in one-ton kibbles, with each whim engine working two shafts. In this fashion a mere 6 tons per hour were being hauled to surface in 1869 and far more rapid winding was overdue long before 1879 when Dolcoath's first skip-road was put in.

[3] From *Memoirs Geological Survey, The Geology of Falmouth and Truro, etc.* 1906. Output from Cornwall only from 1882 onwards.

works active in 1879, even after the decline in their numbers during the 'Australian' depression. In that year one-eighth of all Cornish tin came from stream works such as these, with four of the Red River works alone selling 35 tons a month.[1] Determined efforts and a great deal of money were to be laid out by the mines in an attempt to cut down the loss of revenue draining into the Red River and others but the stream works, for the rest of the century, were to remain far more prosperous than the mines that gave them their chief sustenance.

Surprisingly enough, the period of depression from 1873 onwards, with its cutting of all working costs to a minimum, did not result in any rise in the accident rate in Cornish mining. From 1874–83 the death rate by accident in the two south-western counties was 1.64 per 1,000 employed as compared to 2.24 per 1,000 in the collieries of the United Kingdom. Most of the serious accidents occurred either when raising men from underground by skip or from the use of the new high explosives which miners treated with scant regard for safety.[2] As in the old days of gunpowder, the Cornish miner was remarkably, almost fatalistically, inured to the numerous blasting accidents that took place and he was, by temperament, far too individualistic to take much account of official regulations.

In 1882 the three richest mines were East Pool, Dolcoath and South Condurrow. In February the first named had £10,000 balance in hand after twelve weeks' working, the bottom (200 fathom) level having 'one end, for ten feet wide, producing stuff which will average 2½ cwt. of black tin to the ton.'[3] A month later the mine was raising 40 tons a week, with a profit of £13880 declared in April.[4] At Dolcoath the engine shaft, then 364 fathoms below adit, was worth £250 per fathom, whilst South Con-

[1] Thirty separate stream works are listed in the official returns for the period 1872–81 in addition to 25 on the Red River and its tributaries. The location of some of them may now be considered surprising, including Carbis Beach, Porthledden Cove and Letcha foreshore below St. Just, Lelant stream, Loe Pool, Gwithian, Chapel Porth and Short Horn Beach, near Par. In St. Agnes, Trevaunance and Trevellas beaches also produced quantities of tin, as did the tidal sands at Perranporth. These latter, however, worked the tin ore exposed and disintegrated by the action of the sea against the outcropping lodes in the cliff and they were minor affairs compared to the big Red River works. At the Reskadinnick works, one of the largest, 210 persons (mostly children) were employed, with over three quarters of an acre of settling pits, buddles, frames and other dressing equipment. From Tuckingmill itself down to Gwithian a succession of twenty works found profitable employment on the banks of the stream, with £40000 worth of machinery extending a distance of 1½ miles—'one of the sights of Cornwall, and a sight unequalled in the world.' (RCG 12.10.77). Even where the river finally discharged onto Gwithian beach, incarnadining the sea offshore for miles, a few men earned a living bringing back the sand in carts to be re-processed for its final tin content in a few buddles driven by a single small water-wheel. (See RCG 13.6.74).

Other centres of Cornish streaming in 1882 were in the valley below Wheal Vor and Wheal Fortune where four works were active, and the Cober Valley where another four were at work from Lower Town to Loe Pool. Another important works was at Relubbus, for sale in 1881 (RCG 12.8.81) with eight water-wheels, three scoop wheels, eight heads of stamps, 100 ragging frames, 57 dead frames, etc., etc.

[2] Examples of this were at Dolcoath in May 1886 when a miner was killed as a result of carrying a lighted candle and sticks of dynamite in the same hand, and at Carn Brea in June 1882 when three were killed in the 250 fathom level as a result of storing explosives and detonators together in close proximity to burning candles. Even the companies themselves were lax in their application of official safety regulations on the use of high explosive and legal action was brought against two mines in August 1889 for allowing men to carry over four pounds of dynamite loose and not, as was required, in sealed canisters.

[3] RCG 3.2.82. [4] RCG 28.4.82.

durrow was productive enough to put in another engine on a new western shaft to open up the Grenville lodes without noticeably reducing her dividends. New ventures of the year were New Fowey Consols floated in January and with a 36″ rotary engine (and 12 head of stamps) started in July; East Blue Hills, in St. Agnes; North Metal, in Breage; and New Terras Tin Mining Co. Ltd., in St. Stephen, in June. The only suspension was South Dolcoath, in December,[1] whilst an attempt was made to sell Wheal Sisters as a going concern in July,[2] the mine still not paying its way despite the relatively high price of tin.

THE THREAT OF THE STRAITS: 1883–1894

'*... at Banca, the tin is near the surface, and the expense of raising it is so small that if it should be brought into unrestricted competition with the Cornish tin, the latter must be driven out of the market.*'

West Briton, 2 May 1828

UNFORTUNATELY AN END to this slight revival in the fortunes of Cornish tin in the two years after 1880 came in 1883 when the price of the metal again commenced a downward trend. Indeed, before the decline had fairly set in it was being questioned whether, at £60 per ton, Cornish mining could generally be made to pay. Wheal Agar, Carn Brea, West Basset and Tincroft were all making serious losses, although well run and tolerably well equipped. In May 1883, the loss at South Frances was £2398 in only four months, whilst in July West Frances was carrying a deficit of £10637. There was some danger of the latter being stopped and dues were halved to 1/60 as a means of helping the mine.[3]

Despite the somewhat bleak outlook, or perhaps before it became apparent, a few further small ventures made their appearance, including New Great Work, New Trumpet, Polcrebo, Wheal Castle Mines Ltd.,[4] Calvadnack & Wendron United, East Tregembo and North Blue Hills. Most if not all of these were on a small scale, the last named confining its brief period of fitful activity to re-opening the Goongumpus adit, although with a portable engine started there in June. Of more significance was the final stoppage of St. Ives Consols, or the part of it that had been re-worked since the main 1875 suspension. In the previous seven years, output had been 489 tons 11 cwt. of black tin, sold for £21170.[5] Botallack was also

[1] With 50″ pumping engine and 23′ water-wheel. (RCG 19.1.83).

[2] The group then had 60″, 40″, 28″, 36″, and 26″ pumping engines; 30″ (32 heads) and 24″ (20 heads) stamps; three 24″ and three 20″ whims—a total of thirteen engines.

[3] RCG 13.7.83.

West Frances, having come into the dividend list in 1869 (see page 137) paid only £7680 in dividends thereafter up to October 1872 before relapsing again into a calling state. By April 1877 an overall loss had been made here of £52175. (Fortescue Records, DDF 388, CRO, Truro).

[4] To re-work Wheal Castle at St. Just. An engine was started in November 1883 but working ended again in October 1886.

[5] DD BL 1/31, CRO, Truro. The maximum output was 72 tons in 1877.

in low water from this time onwards, being advertised as a going concern in September.[1] The main news of 1883, however, came from Dolcoath where a 'fine' of no less than £40,000 was demanded of the company by the mineral lord for a renewal of the lease. Rumours of the possibility of this had been in the air for some months but the announcement created a furore in the mining districts against what were considered rapacious lords. Relations between G. L. Basset and the Dolcoath company had deteriorated since 1877 when he disposed of his holding of 147 shares in the mine.[2] From 1880, he had increased the dues to the full rate of 1/15 from the 1/20 which had been accepted up to this date under the variable terms laid down in the lease of 1866 granted by Sir J. F. Basset.[3] Moreover this same heavy rate of 1/15 was required under the new lease, whilst originally no less than £40,000 'bonus' was requested, in almost ultimatum terms in which arbitration was ruled out before it could even be suggested. These were terms for lease renewal that shook the foundations of Cornish mining and chilled the enthusiasm of speculators not only in Dolcoath or in other Basset mines but in every Cornish venture. The effect upon Dolcoath shares was electric, with £60,000 clipped from their market value within ten days. The precedent for this action was the case of Devon Great Consols where, over twenty years before, the Duke of Bedford had exacted a renewal fine of £20,000.[4] At Dolcoath it was pointed out that Basset was paid a clear £8000–£9000[5] each year plus a high rent of £100 for water and £100 for every rood of land used by the mine. Local feelings were strong, and on the March day that the shareholders met on the mine to consider the matter, no Dolcoath miner went underground. Instead a procession wound its way through the streets of Camborne awaiting the outcome. There was a very real chance that the company would throw their hand in and during the four years of the lease still remaining would pick the eyes out of the mine rather than submit to such 'blackmail.' Indeed, had it seemed possible to exhaust Dolcoath's richness in that time, this would have been done. Fortunately however, reason won the day and the new lease, to run from August 1883, was agreed upon after a levy of £25000, a much lesser sum than had originally been demanded.[6] An interesting echo of this Dolcoath case came a few months later, when a new lease was granted to Tincroft by

[1] With four pumping engines, two sets of steam stamps and three whims. (RCG 7.9.83).

[2] Dolcoath was then divided into 1432 shares. Originally, in 1799, it was in 186 shares but this number was doubled in 1858 and quadrupled to 1432 in 1871. In the dark days of 1846, when Dolcoath's market value was £4000, Lady Basset had held 32/186, singly the largest adventurer when once the Foxes had withdrawn. (See W. Herbert Thomas, The History of Dolcoath. *The Cornishman*, July–October 1893).

[3] RCG 4.6.80.

[4] In another less well known case at Phoenix Mine, Linkinhorne, in 1862, a fine of £5000 for lease renewal was also involved. (See RCG 13.6.62).

[5] In 1887, for example, his dues from Dolcoath were £9034, from East Pool £4170 and from Wheal Basset £510. (RCG 27.1.88).

[6] W. Herbert Thomas, *op cit.* Although arbitration was ruled out by Tehidy, Bolitho intervened as peacemaker between the two sides to find mutually acceptable terms. (See RCG 5.8.87).

Lord Robartes. Here the dues were to be on a sliding scale, 1/40 when tin was under £55, and 1/26 when over that price. The *Royal Cornwall Gazette* comment upon this was—' Carn Marth is higher than Carn Brea. The Basset family monument on one is become a byeword. It would be a fitting tribute to Lord Robartes to erect a monument on Carn Marth, and thus hand to posterity the story of the contrast.'[1]

The Dolcoath shareholders also had the shock, during 1883, of the Mayne forgeries, which came to light in May. In this affair, a clerk at the mine, John T. Mayne, was found to have issued no less than 203 bogus shares in Dolcoath, to the value of some £12,000. This had been carried out over a period of four or five years, showing only too plainly that the financial management of the company was lax and that no proper register of shareholders was being kept: ' That such a state of affairs could have existed for so long a period without being detected seems almost incredible.'[2] This, occurring at Cornwall's premier mine, shook the confidence of outsiders in cost-book concerns yet again and a few weeks later the suggestion was made that the Committee should make good the loss ' and keep up their credit as businessmen.'[3] In September, however, it was decided that the fraudulently issued shares should be ' thrown over the book ' and absorbed by the existing shareholders. Mayne received a sentence of seven years' penal servitude after pleading guilty, whilst Dolcoath shareholders received the unwelcome news of the legal fee of 750 guineas for the cost of his prosecution.[4] Dolcoath was fortunately rich enough at this time to withstand these reverses and throughout 1884 the bottom of the mine improved steadily on a tin lode far exceeding anything ever seen in depth in Cornwall before. In March, it was reported that ' the whole bottom of the mine (the 375 fathom level) is one long run of rich tin ground,' whilst a few weeks later it was said that ' the present generation will not see the tin taken away.'[5] East Pool, too, was continuing very prosperous and in fact surpassed Dolcoath in terms of dividends for the year. These totalled £76661 for the whole county,[6] although calls far exceeded this sum for more mines were making losses than were either paying dividends or meeting costs. At Wheal Uny the deficit was £2221 on four months, at South Tolcarne £2400 on five months, whilst at Cook's Kitchen—where the smelters Daubuz and Bolitho were the main shareholders—a massive deficit of £20,068 had to be cleared off.[7] Botallack was still working, but at a loss, Levant just paying costs, whilst at Pednandrea it was only ' thought desirable to give the mine further trial for a few months.' The stoppages of the year were Tregembo, in

[1] 21.9.83. [2] RCG 11.5.83. [3] RCG 8.6.83.

[4] RCG 18.1.84. This charge, for such a brief trial, was so exorbitant that the company would pay only 550 guineas and no action was taken to claim the balance.

[5] RCG 14.3.84, 4.4.84.

[6] The seven major tin mines paid—East Pool £22200, Dolcoath £21385, West Kitty £8400, Wheal Agar £6750, South Condurrow £7041, Wheal Eliza £6123, Wheal Grenville £3000.

[7] ' The shareholders present [at the meeting] seemed too much cast down to make any strong remarks, but a deep feeling of disappointment prevailed.' (RCG 4.1.84).

April[1] and Penhalls which was sold the same month for £2000. The adventurers here had been faced with paying off the existing deficit of £12,000 in February[2] but decided to continue the mine for a few months longer, the dues having been entirely given up.[3] Since the re-working had commenced in 1858 the overall loss had been about £20,000, sales having been 3601 tons of black tin. Polrose (Breage) was sold up in June[4] whilst Wheal Coates closed down in November and the Mulberry Tin Works (Lanivet) in December.[5] The same month Wheal Jane stopped after a long and gallant struggle.[6] The mine had all but closed in 1874 but managed to weather the storm by the most rigid economies. In the succeeding ten years dividends of £2560 were paid, as against calls of £31,000. The dues in this time had been £3926,[7] adding fuel to the burning question as to why mineral lords should reap their profit even when adventurers were paying calls and labouring miners were receiving starvation wages. Certainly some reform of the dues system was overdue, for even when mineral lords were pressed to help a mine, they only remitted part of their quota solely out of grace—' given up during pleasure.' Wheal Jane was regarded as a particularly glaring example of arbitrary mineral ownership for the company were prepared to continue work above adit to give employment to the men, and then to re-commence pumping when the price of tin recovered, but permission to stop pumping was refused and Wheal Jane was thus thrown idle, never to be worked successfully again.

The plain truth was becoming apparent that Malaya could make a profit with tin at a price less than the bare cost of production in Cornwall, and this, moreover, despite a not inconsiderable export duty on Straits tin, amounting to 25 per cent of its value in the 1870's and only gradually reduced thereafter to something like 15 per cent. Output from the Straits had risen enormously, from 3014 tons in 1877 to 7900 tons in 1878, and to 10985 tons in 1879. By 1884 one state alone, Perak, with an output of 10190 tons, was rivalling Cornwall. The effect of this growing output was made worse by Australia's still considerable production. Even though the best of the virgin alluvial tracts had now been skimmed, there was still

[1] With 60″ pumping engine and 24″ horizontal engine working stamps and whim. (RCG 4.4.84).

[2] RCG 15.2.84.

[3] The Penhalls machinery, comprising 50″ pumping engine, 36″ (64 heads) stamps and 20″ whim were advertised for sale MJ 7.6.84.

From 1879 onwards Penhalls had been worked jointly with Blue Hills, managed by the same committee (under the chairmanship of J. C. Daubuz, the Truro smelter) and with meetings of both companies held together either at Penhalls account-house or at Carvedras. Penhalls was a losing concern at this time, going steadily poorer and worked to a large extent on 'assisted tribute' and some of the minor economies effected—such as the appointment of an engine-man to serve also as the mine's engineer at an extra 5s. od. per week, dispensing with any night agent, the selling of all available scrap iron lying about the sett, ' one burning-house man to be replaced by a woman or child '—are evidence of the state of the mine's finances. In September 1882 tenders had been sought for an engine and air compressor to work two rock drills but this was not proceeded with. (Minute Book of Penhalls & Blue Hills Mines, Stannary Court Records, CRO, Truro).

[4] With 40″ pumping engine, 32″ (60 heads) stamps, and 16″ horizontal whim. (RCG 13.6.84.)

[5] With 20″ engine and four water-wheels, including one of 50′ diameter. (RCG 26.12.84).

[6] The machinery comprised 60″ pumping engine, 30″ (80 heads) stamps, 20″ and 24″ whims, 12″ and two 9″ horizontal engines. (RCG 5.12.84).

[7] RCG 29.2.84.

an enormous expanse of 'wash-dirt' available and, using Chinese labour, this could still be most profitably worked. In addition, Tasmania had also become a producer of importance, with almost 6000 tons of tin exported in 1878—much of it from the 'tin mountain' of Mount Bischoff.[1] 1885

OUTPUT OF PRINCIPAL CORNISH TIN MINES 1884–88

	1884 tons	1885 tons	1886 tons	1887 tons	1888 tons
Dolcoath	2423	2555	2838	2366	2239
East Pool	1573	1512	1440	1233	1179
Phoenix	552	—	611	786	733
Levant	316	452	456	492	436
Carn Brea	464	392	535	578	760
Cook's Kitchen	175	86	157	179	155
Great Work	109	44	—	—	—
Killifreth	153	181	154	115	154
Blue Hills	53	98	118	94	105
Botallack	231	271	318	338	284
Levant	72	83	100	136	129
Pednandrea	162	134	151	153	154
St. Just United	270	157	98	67	50
South Condurrow	546	440	439	458	445
„ Crofty	169	92	99	125	125
„ Frances	433	399	417	318	343
Tincroft	456	504	543	574	476
West Basset	479	449	336	297	326
„ Frances	394	343	385	415	422
„ Kitty	372	349	318	313	318
„ Seton	203	189	242	208	219
Wheal Agar	583	637	642	675	514
„ Basset	267	395	379	370	416
„ Eliza Cons.	483	338	323	287	317
„ Grenville	463	595	589	639	481
„ Kitty	175	181	159	195	168
„ Owles	245	133	109	97	82
„ Sisters	359	301	282	241	226
„ Uny	243	65	—	16	78

brought a somewhat better price for tin but the melancholy crop of failed mines in Cornwall continued. The last remnants of Great Work were offered for sale hopefully as a going concern in January[2] and in the same month the little Combellack Mine in Wendron.[3] South Roskear closed in May[4] and during the rest of the year, New Great Wheal Vor, North

[1]Although first discovered in December 1871, and opened as a mine about a year later, the true extent of Mount Bischoff's capacity in terms of tin only became apparent after 1884 when a tram-road fifty miles long was constructed from it to the coast. By 1904, the mine—a whole mountain, worked open to the day—had produced 65,000 tons of black tin, second in the world only to Dolcoath in terms of tonnage. See *Trans. Inst. of Mining and Metallurgy*. Vol. XIV, 1904–05.

[2] With 60″ pumping engine, 30″ stamps and 32″ whim. (RCG 16.1.85).

[3] With 14″ horizontal engine and 16 head of stamps. (RCG 9.1.85).

[4] With 70″ pumping engine, 36″ stamps, 28″ whim, 16″ air compressor, and 36″ stamps or whim. (RCG 15.5.85).

Treskerby, East Botallack, Kit Hill Mine, and the luckless Cathedral Mine in Gwennap. In April Wheal Uny, after a call of £3 per share (£9000 in all),[1] was shut down but with the engines left *in situ* in the hopes of finding a buyer for the mine, whilst West Poldice changed hands in February for a reputed £2450,[2] to be worked thereafter under the name of Tolgullow United. Another casualty of the year was Wheal Prussia which had been re-started in 1880. This, adjoining North Downs, was an old copper mine that had been under water for something like forty years and presented considerable problems to re-open. A 70″ engine from Wheal Sparnon was put in and by March 1883 unwatering was complete, with about 80 men then employed. The fall in the price of tin disheartened the company at this time and an attempt was made to sell the mine as a going concern.[3] Although this was unsuccessful, activity was continued on a lessened scale for a further two years until the summer of 1885, the calls having then amounted to £25500.[4] In the year a total of ten mines in Devon and Cornwall paid dividends amounting to £93515.[5] There were now 84 working mines in the two counties although by far the greatest proportion of the output came from only about thirty of these (see page 188). Amongst these the number that had been predominantly, and at one time exclusively, copper producers was very marked. Amongst the remainder activity was in many cases confined to cleaning up waste tin on the dressing floors or to a few tributers picking over burrows and stripping out shallow levels for the tin-stone left standing. This, moreover, even though in 1884 the names of over a hundred defunct mining companies had been expunged from the register—there having been, on paper, an official record of 240 mines in the two south-western counties as late as 1882.

Amidst the depressing welter of closed mines it was heartening for Cornwall to see the rise and progress of at least one rich new mine. This was West Kitty, in St. Agnes, floated in November 1879 to re-work what had formerly been Wheal Rock.[6] This ground had never been successfully worked to this time, even though the Wheal Kitty Main Lode made rich right up to the boundary of the sett, but by one of those

[1] This was intended as the final call on Wheal Uny and the end of the company but the matter was not in fact terminated. Months later, the unlucky adventurers received a demand from the Stannary Court for a sum approaching £3000 for further liabilities and the costs of winding-up. This occurrence may be said to have been the final nail in the coffin of the Stannary Court and the decision to transfer its legal affairs into the normal County and other courts.

Tin sales by Wheal Uny to this date (1885) since the re-opening in 1851 had been £416,786 (RCG 5.8.87).

[2] RCG 20.2.85.

[3] RCG 27.7.83. With 70″ pumping engine and double 12″ horizontal whim.

[4] Wheal Prussia Ledger, RIC, Truro.

[5] East Pool £36610, Dolcoath £23295, West Kitty £11400, Wheal Grenville £7500, Wheal Agar £6500, South Condurrow £4952, Wheal Basset £793.

[6] Earlier again, in a brief working in the 1860's, the name West Wheal Kitty had been used. It was re-started as Wheal Rock in the early 1870's and for a time thereafter had also been known as St. Agnes Consols, being sold up under the latter name in 1877. (RCG 15.6.77).

reversals of fortune which form such a recurring feature of metal mining history, the new 1879 West Kitty was rapidly to become a 'prize.' In 1881 the mine only had nine heads of stamps available, yet in sixteen weeks' working sold £1000 of tin.[1] The dressing floors had perforce to be sited at the head of Trevellas Coombe, and here sixteen more stamps were installed late that year, worked by a 30′ water-wheel. The following year further stamping power became necessary to keep up with West Kitty's output and a 30″ steam stamps was put in. By July 1888 2147 tons of tin had been sold for £126,813—all from the one 'Wheal Kitty' lode—and dividends of £55,500 had been paid.[2] This was a financial record which showed what could still be done in Cornwall by a young mine, and following West Kitty the adjoining setts of Trevaunance United and Polberro were re-started by companies which, although nominally separate, had the same financial management and offices.

The general depression of the 1880's had brought changes not only in mining but also in the ranks of the smelters. The successful revival of the Penpoll works in 1880 had been the first major breach of the old smelting ring that had existed for so long. Teague and Strauss brought radical changes, for they were not prepared to join in the existing system whereby tin was bought quietly by private contract at prices below fair market value. Their success, despite the opposition of the smelting fraternity who were so solidly entrenched in the day-to-day management of many big mines, led to the re-establishment of the Tamar works in 1884[3] and to the re-starting, under limited liability, of the old Redruth Tin Smelting Co. in 1886. The following year another and entirely new works was set up, on what had formerly been the East Basset dressing floors at Seleggan, near Carnkie, by the Cornish Tin Smelting Co. Ltd.[4] These newcomers inevitably took business away from the existing old-established houses[5] and the smelting empires of Daubuz, Williams, and Bolitho began to crumble. The Angarrack works, owned by the latter, was the first to succumb, being closed down in 1881. The output here in 1875 had only been 2050 blocks (7378 cwt.) and, although it had risen slightly to 2370

[1] MJ 2.4.81.
West Kitty black tin was of exceptional quality and in the next few years usually brought the highest price per ton of any sold. In 1891 the recently re-started Great Hexworthy Mine on Dartmoor ousted West Kitty from the position of sending the finest quality ores to the ticketings, with some parcels sold at £60 and more per ton.

[2] MJ 7.7.88. The initial calls in 1879–80 had amounted to 12s. per 1/6000 share.

[3] RCG 29.2.84. This was a revival, by a new limited company, of the works re-started in 1871 which had shut down in 1876. It disappeared again from the list of bidders at the tin ticketings in 1890.

[4] The partners in this were Alfred and J. C. Lanyon (Redruth merchants and mine adventurers who had earlier had a share in the Redruth smelting works in the 1870's), John R. Branwell of Penzance (a flour miller and merchant related by marriage to the Battens), and John Jose of Mellingey (Perranwell) whose wealth had come from Peruvian copper. Most of these were major shareholders in the currently rich East Pool, where at first almost all the new works' ore emanated. Smelting commenced in April 1887. (MJ 9.4.87).

[5] The progress of one of the newer works, that of the Redruth Tin Smelting Co., established in 1862, was typical. Here, in the ten years after 1867 the output doubled, although it is true that some of this increase was from foreign and not Cornish ores. Statistics of the tonnage smelted, the cost,

blocks in 1878 (8899 cwt.),[1] this output was now no longer economic. In 1883 Daubuz' works at Treloweth, one of the oldest and most important of the Cornish smelting-houses up to about 1840—also passed from the scene[2] and this was followed by the closing down in 1884 of Bolithos' Charlestown works.[3] In the early 1880's Wheal Eliza had been the latter's single largest source of black tin bought for smelting each year[4], whilst a quantity of Bolivian ore had also been smelted to supplement Cornish supplies. This closing of three smelting works left six of the old-established ones still existing, namely Trereife, Chyandour, Mellanear, Calenick, Trethellan and Carvedras, as well as the four newcomers, Penpoll, Tamar, Redruth and Cornish. These latter continued to prosper at the expense of the older 'ring' especially when the practice of holding regular public tin ticketings was established at Redruth. The institution of these was described by the *Mining Journal* as 'one of the most important events that has taken place within living memory in connection with our tin mines.'[5]

and the coal used in the years 1867–1883 are recorded in a Memorandum Book of the Redruth Tin Smelting Co., now at CRO, Truro. (DDX 39/1/2)—

	Black tin smelted (tons)	Tin metal produced (tons)	Coal used (tons)	Smelting cost per ton of metal
1867	984			
1868	1021			
1869	1339			
1870	1092	669		£3.11.2
1871	1081	657		£3.12.0
1872	958	570	886	
1873	1331	795	1380	
1874	1278	753	1275	
1875	1028	597	1015	
1876	1306	762	1358	
1877	1725	1022	1739	£3. 2.4
1878	1819	1070	1819	£2.18.8
1879	2226	1310	2150	£3. 3.3
1880	1613	925	1721	£3. 7.0
1881	1285	730	1283	£3. 9.9
1882	1306	735	1269	£3. 7.2
1883	1564	860	1546	£3. 5.5.

[1] Bolitho Records, (DD BL 1/30), CRO, Truro.

[2] The premises were advertised to be let by tender under the name of The Lamb and Flag Smelting House, St. Erth (RCG 6.4.83). In 1888 they were in use again as a butter factory and creamery, promoted in that year by the Bolithos (See RCG 10.5.88). Later again, when the new creamery was erected near St. Erth station, the old works became, as they still remain, the transport depot of the dairy company. The Lamb & Flag public house, adjoining the smelting works, also remains today.

[3] All Bolithos' black tin purchases could now be smelted at Chyandour or at Calenick and the Charlestown works, like Angarrack, was largely superfluous to requirements. The works was still run by the Charlestown Smelting Company although entirely owned by Messrs. Bolitho since 1879, after having been three years under joint ownership with Williams Harvey & Co. after Daubuz had withdrawn.

[4] Another staple of Charlestown smelting works to the last was Minear Downs, a vast open-work lying to the east of the celebrated Carclaze, which had been worked off and on since about 1870. This area contained innumerable fine veins of tin, many 'thinner than a sheet of paper' and the system of working was to remove the ground from the slope of the hill in the dry summer months, stamping and washing it in winter. In this economical fashion working in very soft ground, 'although the produce of the mine or quarry is only from 4 to 6lbs. of tin per ton of stuff, it is a paying concern'. *Proc. Miners' Assocn. of Cornwall & Devon*, 1870.

[5] 1.1.87. Ticketings for tin had been held before this date but in name only, for the smelters' bids were almost identical for each parcel of tin offered by a mine: 'if every crucible was wholly independent of the other, the figures speak volumes for our practical metallurgy.'(RCG 21.11.74).

Acceptance of the ticketing system came only gradually and the newcomers, particularly Strauss who had instituted it, met stiff opposition from the Bolithos, Daubuz, and Williams, as regards the many mines in which they were major shareholders.[1] The death of Strauss' mining partner, Teague, in March 1886 and the acquisition of his shareholdings by Messrs. Harvey & Co. would have been something of a set-back had the ticketing system not been accepted by this date.

Continuing technical development in the mode of working the mines was evidenced throughout the 1880's in various ways. Dolcoath introduced a traction engine for hauling tin to the smelters in January 1885; East Pool tried, with great success, the first Holman compressed air winch for underground hauling in winzes, etc., whilst the same mine started to put in a man-engine during 1887, and both Cook's Kitchen and Dolcoath extended theirs in depth to enable the men to do more work underground in the course of a shift. Double skip-roads had now come into use wherever possible, or single skip-roads in shafts too small for a double road, to cope with the increasing output of ore. The use of boring machines also continued to expand and the bigger mines now started acquiring them on their own account instead of hiring them on contract terms from outside firms as hitherto.[2] Carn Brea had four machines at work in 1886, Cook's Kitchen two, and Dolcoath three. At the latter mine a new compressor was ordered in June of that year capable of supplying the hitherto unheard of number of eight or ten machines and at the same time the first tentative suggestion was made that it might be worthwhile using these boring drills for stoping as well as for driving, rising or sinking.[3]

There was evidence of what was being done underground by rock drills and high explosive at this time at West Frances and West Basset. At the former, to open up a rich shoot of tin,[4] a diagonal rise on the course of the lode was put up from the 174 fathom level to meet a vertical shaft newly sunk 35 fathoms from surface. This was finally holed in April 1886—a feat not only of technical mining skill but also of surveying, for it had involved the correct measuring and dialling of 880 fathoms of levels and shafts in the process, a triumph 'which practical miners only can fully understand.'[5] At the latter mine a no less interesting achievement which 'far exceeds what has ever been accomplished in the mining world in this direction',

[1] Bolithos had important holdings in Levant, Wheal Sisters, Carn Brea, Cook's Kitchen, South Condurrow, Tincroft, Wheal Jane and elsewhere; Daubuz, in Blue Hills, South Frances, West Basset, Carn Brea, Cook's Kitchen, Tincroft, and elsewhere, whilst the principals of Williams Harvey owned a large slice of Dolcoath—whose tin was always sold by private contract to the latter's Mellanear works—West Frances, and shares in most other large mines.

[2] This had formerly been necessary when there were so many different makes, of varying reliability and power. Holman Bros., for example, ran a separate Boring Machine Company until 1889 or later. (See RCG 29.8.89).

[3] RCG 13.7.83.

[4] This discovery had been the saving of West Frances, which tottered on the brink of closing in July 1883. The mine was the last to drop onto part of the not always predictable Flat Lode which sustained this whole district and even as late as March 1882 it had been said that 'the attempt made here to develop the new flat lode has been a failure.' (RCG 24.3.82).

[5] MJ 1.1.87. For a description and illustration, see also MJ 17.4.86.

was the bringing down of 1500 tons of ground 'at one blast' using Bickford Smith's new instantaneous fuses and £50 of dynamite.[1] This, it was said, would have taken several years to achieve by manual labour and the proceedings were marred only by the consequent rush of air from the fallen ground which swept the agents and other men from off their feet at the end of the level and caused some injuries, including a broken arm.

The year 1886 saw an improvement of about £8 in the price of black tin, partly due to a rise in the price of metal and partly to a reduction, as a result of greater competition, in the smelter's profit margin. By June metallic tin was over £100 per ton and a heartened Cornwall was disturbed only by the rumour, quickly dismissed as unfounded, of the discovery of large stanniferous deposits in North America, in Dakota. The rise continued in 1887, with Cornwall more prosperous in a modest way than for many years past. A few mines were re-started, including Polgooth, Parka Consols, Tregembo and West Grenville. At the last named, formerly worked under the name of Trevoole, a 70″ engine was erected on a shaft intended to cut the western extension of the Flat Lode. This was now proving the mainstay of the whole district from Wheal Uny at Redruth three miles westward to beyond Troon, with Grenville currently the richest of the mines along its length. South Tolcarne alone was not paying costs and after a heavy call in May 1886 the adventurers decided to throw in their lot with South Condurrow[2] and the two setts were thereafter amalgamated. At West Frances an engine was being erected on the new shaft to cope with increased pumping,[3] whilst West Basset had commenced work in the old North Frances section—a sett that had been abandoned in 1865 with a loss on the working of £30,000.[4] In St. Agnes the prosperous state of West Kitty heartened the Polberro and West Polbreen adventurers to find the money for pumping engines to be put in, whilst Wheal Uny and East Uny were again re-started in March 1887 after having been idle and full of water for almost two years.[5] In the east, Phoenix United acquired the moribund West Phoenix[6] with its 70″ and other engines. This cost £2005 and was an acquisition that gave a new lease of life to the company which, for some years past, had confined

[1] RCG 3.12.86. In West Basset and other mines on the Great Flat Lode there were in places enormous stopes many fathoms wide.

[2] South Condurrow had been one of the most consistent dividend paying mines after 1870, totalling about £80,000 by the end of 1886. (MJ 8.1.87).

[3] At this same time, to secure better ventilation in West Frances, a level was holed into Wheal Grenville. The use of a mechanical ventilator—rare in Cornish mining—had been necessary in the former for some time for it was a hot and, to this date, a particularly badly ventilated mine. A similar type of ventilator was also used in 1887 in the 368 at Cook's Kitchen where the air temperature was a constant 100°. (MJ 29.1.87).

[4] RCG 27.1.65.

[5] East Uny had been bought by a local syndicate (headed by Colonel Fludyer) in March 1886 for £1000, the value of the machinery. Two months later it was sold for £1250 to the company formed to re-start Wheal Uny, and thereafter the two mines were worked together. (East Uny Account Book, RIC, Truro).

[6] RCG 18.2.87. There was already a section of Phoenix United which had earlier, and somewhat confusingly, been named West Phoenix.

operations to the shallower levels of the old mine. In the west, the Levant company had, after fifteen years, some return to show for their heavy outlay and with a dividend of £1 per share was said to be 'well on the way to paying back the £30,000 called up.'[1] In the bottom (278) level, far beneath the sea, the lode was 5′ wide and almost 3′ of it solid copper, worth £70–£80 per fathom even at the depressed ore prices of the day.[2]

Dividends for 1886–7 showed a welcome rise[3] particularly in the latter year when a total of £147,703 was paid out compared to only £95,000 in calls. Amongst the major mines still barely paying their way, even with better prices for tin, were Carn Brea, Wheal Owles, Botallack, Blue Hills, Pednandrea, Tolgullow United, Wheal Sisters, New Cook's Kitchen and West Seton. Profits were made in some of these during various periods of working, only to be swallowed up by equal or even greater losses during others. Two things kept such ventures, with their overall losses, going—the ever present possibility and hope that a lode would be struck that would rival Dolcoath, and the mercantile interests. The latter, although paying heavy calls, recouped or more than recouped these by their trading profits with such a mine. With the requirements in materials of a large mine amounting to £10–20,000 each year this was a factor of importance, particularly in the money values of those days before twentieth century inflation. By the 1880's downright manipulation of a company's affairs so that the lion's share of profits went into the costs and not into dividends was a thing of the past, but the merchant 'in-adventurer' influence was still very strong indeed—as many a mine agent who sought to see his mine get a fair chance found out to his cost.

Under the cost-book system, with no reserves and with profits divided up to the hilt in the current period in which they were earned, even tolerably rich mines gave widely differing annual results, with poverty and near abandonment alternating with riches and unbounding optimism on the part of the adventurers.[4] West Frances was a case in point, with shares down to 6d. in 1883 when the mine seemed finished but selling for £14 each—after having been sub-divided into three in the interim—'or £42 for what was 6d.' in 1887.[5] Those who held Cornish mine shares

[1] MJ 19.3.87. [2] MJ 11.2.88.

On copper, Levant was to pay about £8000 in dividends in the next two years before lapsing into a state in which profits and dividends alternated almost year by year with losses and calls. The mine was both deep and hot and 'it took three men to do two men's work. They had to pull all their stuff to one shaft and all the timber had to be sent down it also.' (MJ 12.1.89). By 1892 the cost of tramming and filling was very great and out of the 370 employed underground (335 men and 35 boys) only 270 (240 men and 30 boys) were actually engaged on breaking ore or in opening ends. (MJ 6.2.92).

[3] In 1886 the nine Cornish dividend mines were—Dolcoath £45162, East Pool £33280, Wheal Grenville £12000, Wheal Agar £9300, West Kitty £6000, Levant £4375, South Condurrow £4185, Phoenix United £3500, Wheal Basset £768. In 1887 the twelve mines were—Dolcoath £43852, East Pool £30400, Wheal Grenville £16500, Wheal Agar £12000, West Kitty £10500, Phoenix United £9000, Levant £6875, South Condurrow £6125, West Frances £5076, Wheal Basset £3380, Tincroft £3000, Wheal Eliza £1024.

[4] Not so the agents, for their living, in a failing mine, depended upon cautious optimism at all times—a widespread practice which brought Cornish mine reports eventually into disrepute.

[5] RCG 15.7.87.

wanted iron nerves indeed to withstand such losses and gains as these. Up-country speculation in Cornwall, however, had never recovered from the bursting of the bubble of the 1870's and the shortage of investing capital was made worse by the flow of much available money at this time into foreign rather than home mines. The cost-book system of working mines was mistrusted outside the county, or at least its workings only imperfectly understood, whilst on the other hand the limited liability concept had a far worse record since its inception in 1862. In 1888 it was reported that out of over five hundred limited companies registered in the Stannary Court in these twenty-six years not more than three or four were still in existence[1] and the financial loss had been enormous. The reasons why limited liability had been a failure to this date were threefold; it was the favourite method whereby a mine company could be promoted simply as a market proposition; it could be promoted as a purely speculative venture appealing to distant shareholders with little or no knowledge of metal mining; and finally, the margin of actual, as opposed to nominal, capital was usually far too low. In the latter instance, matters were invariably made worse by the difficulty of obtaining further capital from a large number of small shareholders who were indifferent to the long term development of a mine, and doubly so when there was no further risk involved to them. By contrast, under the cost-book system, a small group of determined and wealthy shareholders could work wonders, providing they remained united.

In 1887 an attempt was made to overhaul and bring up to date the scope of the Stannaries administration to meet the changing conditions of mining and mine financing. Eighteen years had elapsed since the passing of the Stannaries Act of 1869 and there were many aspects of this that were out-dated or requiring revision when this new amendment act was passed in 1887. Copious evidence was given to the House of Commons Committee that reported on this Bill, drawn not only from adventurers and managers but also from working miners who had been unofficially deputed to give evidence at earlier meetings in the various mining districts. On the financial side one of the major proposals was the establishment of a registry of mining leases, a provision intended to stop the more pernicious cases of 'bal-selling' in the feverish surges of interest which formed a recurring feature of Cornish mining history. In evidence one case was quoted in which the lease of a sett had been obtained for £21—the cost of the lawyer's fee, the marking out of the limits, and the relevant maps. Then £100,000 in promotion money was paid for that lease. This, although particularly bad, was an isolated case and the Bill was intended more to remove some of the worst injustices which Cornish working miners still had to suffer. One of these was the delay in payment of wages under the Cornish system, under which men might be paid two months

[1] RCG 7.6.88. It now seems doubtful whether these figures were correct.

or more late,[1] and even though the risk had gone of losing these entirely if a mine company failed, it still meant that the only alternative was for a miner to apply to the agents for ' subsist ' in the interim. For this it was necessary to plead poverty and even then the granting of it rested with an agent or manager who might be either friendly or unfriendly, the whole system being repugnant even to the lowliest labouring miner with any pride. By this date many Cornish miners had served a stint or two in American or South African mines and they could see that there was now no valid reason for perpetuating the shortcomings of the Cornish wage system which had originated many years earlier under very different conditions. ' Send up the stuff one skip, and send us down the money the next ' was the comment of one miner giving evidence, a Redruth man who had spent some years on the Californian goldfields and in the Michigan copper mines.

A political undercurrent existed in this Bill, in which the Radical Colonel Fludyer (see page 176) played a prominent part and it was made amply clear by the evidence given that working miners had more than enough cause for complaint at some of the treatment they received.[2] They could be dismissed at a day's notice by any agent of whom they fell foul, and without cause. Any complaint about unfair sample-taking, or at not being allowed to have any check-weight of the amount of ore-stuff they had raised, was liable to result in dismissal, and woe betide those who complained at still having to pay—at company prices—for dynamite, candles, fuses and the breakage and wastage of tools. In one specific case, quoted by Colonel Fludyer, three engine-men had been summarily dismissed by the manager of South Condurrow because one of them had refused, on grounds of safety, to work two engines overnight (a pumping engine and a whim engine, situated a quarter of a mile apart) single-handed; this, moreover, after periods of service as engine-men on the mine of from 14 to 26 years. Such instances, rare perhaps but undeniable, made it apparent that the brunt of foreign competition was falling on the shoulders of the working miner year in and year out, and only on the pocket of the investing adventurer in bad times. The miners, more particularly those who had worked abroad, were in favour of a union to defend their interests but they saw the difficulty in starting one. 'Who is going to be the starters . . . ?' asked the man who spoke for the St. Just district, only too well aware of the discrimination that would be shown thereafter by agents and managers. In fact, despite a growing awareness amongst

[1] Long before this, there had been complaints at the long delay, occasionally a full three months, in the payment of miners' tribute pay. In 1872 it was reported that in ' one of the best mines in the Redruth district the system of payment is such that the tin may be dressed, smelted, manufactured into tinplates, and be seen shining on the tops or roofs of the Canadian houses, before the tributer is paid for raising it.' (RCG 3.2.72).

[2] In one case there was a complaint, in an unspecified mine, of delays in men being hauled to surface after core, with preference being given sometimes to the raising of ore instead. Occasionally miners had to wait as much as three hours in this way if they were not prepared to climb the alternative ladder-road.

A rare photograph of Phoenix United Mines, Linkinhorne, at about the time when working ended in 1898. In this view, looking east, the road from Minions to Sharp Tor can be seen in the foreground crossing the tramroad which connected with West Phoenix.

Killifreth Mine, near Chacewater, sometime before working ended in 1897. In the foreground the original 50″ pumping engine put in during 1873 and in the background the 80″ from North Treskerby which was erected here in 1892–3 on Hawke's Shaft. Between them is a whim drawing from the latter. All three houses can still be seen, Hawke's having been re-built to take an 85″ in the final working of the mine shortly before the first World War.

Blue Hills Mine, Trevellas Coombe, St. Agnes. One of the last of the old school of Cornish mines, closed down in 1898 after an unprofitable period of working lasting almost thirty years.

Cornish miners of the protection they might receive from some form of union, this was still a quarter of a century away.

In the next few years there were to be several signs that there was dissatisfaction amongst working miners. In 1891, for example, there were several instances of tampering with pitwork at Dolcoath and Killifreth, in endeavours to cripple these mines by damming back water from the pumps underground or by dropping metal bars into them, whilst there were cases of damage to dams sealing off old workings at East Pool and other mines. At South Frances, West Basset and Wheal Basset there was a series of unexplained fires, damaging, and in one case totally destroying, pumping engines etc., ending in the destruction by arson of the Basset count-house. Two complaints on the part of the men were that in the event of injury no mine doctor went underground to the casualties—no matter how serious their condition—but waited at surface for them to be brought up by their comrades as best they could[1] and also that the system of working by contract tutwork was unfair. Under this, the men were paid at so much per fathom driven or stoped and they had a disincentive to do any more than their contract as they would otherwise suffer a reduction in their rate when the next contract was set. They could lose but not gain, and such a system was injurious to both sides in the long run. Once it might have been easily remedied by putting men working productive ground onto tribute but in an age of specialisation in big mines, with rock drills, steam stamps and huge tonnages milled, the old tribute system could no longer be applied.

Another clash between the rival interests of landlord and tenant occurred in 1887 between the Tehidy estate and the West Seton adventurers. The mine had been an extremely rich one for copper and following the example of Dolcoath, as so many times elsewhere, the company ploughed back a great deal of their copper profits into continuing the mine for tin. One of the conditions of the new lease under negotiation for the mine in 1887 was that the company should purchase coal imported solely through Portreath, Basset's own privately owned harbour which had recently been made ' open ' by the ending of the Portreath Company's formerly sole rights to its use. This insistence on coal via Portreath, or alternatively on higher dues to be paid by the mine company henceforward, was of little enough significance in itself had it not represented a threat—and the thin edge of a possibly very wide wedge —to the rival port of Hayle. Harvey & Co., who owned the latter, held shares in West Seton, as did the Bolithos who supported Harveys in this issue, and for a time it became a head-on clash between the two sides

[1] RCG 20.4.93. This was stated in evidence at an inquest on miners killed by a fall of ground at Wheal Basset.

Earlier, when attempts were being made in 1872 to form a Miners' Protection Society it was said that at Phoenix Mines there was only one doctor who could attend any accident, and this with 801 employees. Even he lived at a distance and it had taken over six hours on one occasion before he attended an accident. (RCG 27.1.72).

which attracted a great deal of press comment throughout much of the year. The outcome was that West Seton coal henceforward came via Portreath, after a period during which the possibility of stopping the mine was imminent.[1] The whole incident revived the old animosity between the mineral lords and the adventurers. It was pointed out, not for the first time, that most mine leases depended for their equity upon the goodwill of the lord alone and the vexed question of strained relations on this basic foundation of good mining was even raised in the House of Commons at this time. Certainly some measure of legislation was required, for of one mineral toller (lord's agent) it was said that 'a dozen such men would kill Cornish mining.'[2]

The star mine of 1888 was undoubtedly Dolcoath, which paid its highest ever dividend of £58750 in that year.[3] On the occasion of declaring 70s. per share of this in February, a champagne dinner was given to the attending shareholders by the chairman George Williams. The mine had been worked to capacity throughout the year, with the limit on output determined solely by there being only three main hauling shafts in the eastern section. On each of these a whim engine was worked from early Monday morning until late on the Saturday night,[4] frequent interruptions to hauling ore having to be made when sending down the enormous quantities of massive timbers required to make the main lode workings safe from collapse. During the year work was started sinking Stray Park Shaft, not so much because it was thought necessary or desirable but because it was required under the terms of the lease. An attempt had been made earlier to sell this old sett for a rumoured £25000 but the Tehidy estate made it clear that they were not prepared to allow a new company to hold it. The presumed value of the ground adjoining Dolcoath and its fabulous main lode[5] was, however, responsible for the floating early in 1888 of Dolcoath West Limited—actually a revival of another company of the same name that had been promoted in 1885 or 1886—to resuscitate the old Camborne Vean, then 282 fathoms deep. This however appears to have been a market rather than a mining venture.

East Pool's long period of prosperity continued in 1888, marred only by the dispute with Wheal Agar over pumping. This had been a bone of contention between the two mines for years, East Pool with a 50″ and Agar with a 70″ each considering that they bore more than their fair share of the inter-dependent pumping costs. Over the years, this dispute had grown more and more bitter, as was later to become apparent. Dams had

[1] RCG 9.3.88. [2] MJ 12.1.89.

[3] The other dividends of the year were—East Pool £31200, West Kitty £13500, South Condurrow £12858, Wheal Grenville £11250, Levant £7812, Phoenix United £9900, Tincroft £4500, West Frances £8294, Wheal Agar £9450, Wheal Basset £3072, Wheal Eliza £2564, Wheal Kitty £644.

[4] RCG 28.4.89. Except for shaft or engine maintenance, Sunday work was an extreme exception in Cornish mining, and when an attempt was made in 1867 to institute seven day working on the dressing floors at Condurrow Mine a strike almost resulted. (See WB 8.2.67).

[5] In the bottom of New East Shaft, it was worth £500 per cubic fathom, a richness at this depth 'of vast importance to the future of Cornish mining' (RCG 9.8.87).

been put in as a precautionary measure between the two in the intercommunicating levels but during a particularly acrimonious period of the dispute in 1873 there were threats that the plugs would be pulled out of these and duplicate dams were thereupon put in. The disagreement came to a further head in 1888, for Wheal Agar now had a big 90″ engine for pumping bought from East Wheal Rose and they held what they considered the whip hand. The Agar adventurers decided to try and force a monthly payment for their pumping costs from East Pool and when this was not successful they announced that pumping operations would ' be suspended until such time as the East Pool people choose to deal equitably with Wheal Agar.' This drastic course of action brought Agar shares down radically for a time but it had the desired effect, for it was agreed at length, after long negotiations and arbitration, that East Pool would pay £300 p.a. to Agar for her share of the pumping cost.[1] Actually East Pool was usually short of water rather than the reverse, for its 128 heads of stamps and its dressing floors at Tolvaddon drew their supply from the inadequate and far from clean[2] Tuckingmill stream. In 1887 the shortage was such that the company contemplated erecting an engine on the old Wheal Emily Henrietta to raise additional water for dressing purposes.[3] In addition, the distance between the mine and Tolvaddon, something over a mile, was an added burden which East Pool wished to lessen if at all possible and in July the mine manager went to Sussex to inspect and report upon the suitability of an electric ' Telpher ' line as a replacement for the numerous horses and carts in use.[4] Wheal Agar declined East Pool's offer to share its costs and use but in May 1889 the company decided to go ahead alone. Although the contract then specified the line's completion by 13 August, work was not started until November and from a series of delays it was still incomplete in April 1891.[5]

Meanwhile 1888 was by no means a successful year for Cornish mining as a whole. On the towans at Hayle, an engine was being erected to re-try the Wheal Lucy sett,[6] whilst yet another attempt to re-start Great Work

[1] RCG 5.9.89.

[2] In terms of mine waste. In 1884, for example, an estimated 284,703 tons of sands and slimes were discharged into the Red River by the mines along its course. (*Trans. Mining Association of Cornwall*, Vol. 1.). Not included in this were the sands and tailings from Tincroft where there was insufficient water to carry them away. About 10,000 tons a year was dumped here, an accumulation that was to be re-treated with some success just before the first World War.

[3] It was artlessly suggested to the Tehidy estate that they might pay part of the cost of this on the strength of the suggestion that the extra water would benefit also the Red River streamers, thus increasing their dues. (RCG 27.1.88).

[4] RCG 12.7.88. This was at Glynde, near Lewes, where the pioneer Telpher line was in use carrying chalk. Briefly, it consisted of an overhead mono-rail on which trains of skips were hauled by small electric ' locomotives.' As a system it showed great promise but gave considerable teething troubles which destroyed its commercial chances.

[5] RCG 9.5.89, 10.10.89. The original specification was to have three 5-ton trains on the line at once, capable of carrying 180 tons of tinstuff to the stamps daily.

In April 1891 it was reported that ' The Telpher Co. have been very persevering and appear to be gradually getting their line into working order,' (RCG 16.4.91) but it is not certain whether this equipment was ever put into full service, partly due to installation faults as well as to reputed difficulties in carrying the line over the turnpike road near Tuckingmill.

[6] RCG 14.3.89.

was made in November, but these were outweighed by the stopping of West Polbreen, Tolgullow United, Tregembo and West Godolphin. It was also decided to cease operations at St. Just United in May, a decision confirmed at a special meeting held in June when it was reported that the mine was unpayable and that 'the mainstay, the Cream-pot Lode, had failed.'[1] The re-working here, from its commencement in January 1880 had been a determined one, a 36″ pumping engine and a 24″ whim engine being installed. In 1881 a 25″ stamps engine had been acquired from the moribund Wheal Cunning with 32 heads of Cornish stamps which replaced the two heads of Sholl's patent stamps which had been found unsatisfactory. All these and another 26″ whim were thereupon advertised for sale in November 1888.[2]

Also stopped in August was Wheal Sisters in Lelant, where, when the company's affairs had been finally brought to a close in 1890 the Bolithos had contributed £27341 3s. 6d. in calls towards the working of the mine since 1876. This moreover represented only about one third part of Wheal Sisters during this period.[3] In Wendron, New Trumpet Consols' machinery was sold up in November, last of the Wendron mines to survive. At Wheal Uny, re-started in March 1887, the problems of unwatering and cleaning up the mine were considerable and a loss of £3045 was made in the sixteen weeks to July 1889. Both 70″ engines, on Uny and East Uny, had been used to re-fork the mine, and the water was only down to the 120 after seven months of pumping. The pitwork was in bad condition and many of the levels were weak after being under water for two years and it was not until March 1888 that the deepest shaft (Hind's: 214 fathoms below adit) was cleared of water for sinking to re-commence.

The dividends of 1889 were considerably less than those of the previous year, totalling £73085, Dolcoath with a profit of £21737 paying more than double that of any other mine. An innovation of the year was the opening of the National Explosives Company's works on Upton Towans, near Hayle, the first of its kind in Cornwall. It became the chief supplier of high explosive to the west Cornwall mines and was successively enlarged in size until by 1906, when it was engaged on ordnance work as well as mining and quarry explosives, no less than 700 persons were employed. At much the same time, a rival but smaller dynamite works was set up near Cligga, west of Perranporth, but this was not as successful and was to pass into the hands of Messrs. Nobel, of Glasgow, in 1893. In Cornish smelting an important change in one of the foremost companies had been the withdrawal of George Williams, the surviving principal partner in

[1] Oats Collection, RIC, Truro.

[2] RCG 13.11.88. This last (26″) engine was the rotary on Cape Cornwall which sett had formed part of St. Just United.

[3] DD BL/31, CRO, Truro.

Like St. Just United, Wheal Sisters was partly owned by Messrs. Bolitho and the final meetings of both companies were held at Chyandour. An unsuccessful attempt had been made in June 1888 to sell locally 2000 Wheal Sisters shares at £1 each, to keep the mine going and to raise money for the introduction of boring machines. (RCG 5.7.88).

Williams Harvey, in 1888. For some years there had been attempts to dispose of the Williams' holding in the company, and in 1884 Michael Williams had come close to selling the Trethellan and Mellanear works outright for £20,000 to a London company then engaged in working North Treskerby and other mines. In 1888, however, the company was acquired by Frank Harvey, Gilbert Pearce and Richard Pearce, in partnership.[1] At first it was intended to re-name the company Harvey Pearce & Co., but presumably to avoid any loss of goodwill the name of Williams Harvey was retained. The new company acquired both Trethellan and Mellanear works but in December 1890 it was decided to discontinue smelting at Truro and the Trethellan buildings and wharf were assigned to Harvey & Co. for use as trading premises.[2]

A feature of the late 1880's was the unaccountable fluctuations of the Cornish mining share market. This had always been, inevitably, an unpredictable one, with ' bears ' and ' bulls ' attempting to push up or pull down the price of certain shares, but at this period it had what the *Royal Cornwall Gazette* colourfully termed ' hot and cold fits, and generally the up and down.'[3] Such transient occurrences as the bursting of a boiler or the breaking of a stamps axle had a quite disproportionate effect upon the daily price of shares but it would appear that a more important and widespread factor in unsettling the market was share-jobbing by agents and others connected with mines. Such people had early and sometimes prior knowledge of the varying fortunes of the mines with which they were connected and in 1889, on the occasion of the unexpected cutting of a new lode at West Frances, news of this was sent by telegram to the Mining Exchange to prevent any buying-in by those first to know.[4] A serious accident to a mine or its machinery had, of course, a serious effect upon its market value. At Wheal Agar, for example, when the bob of the new 90″ engine broke in February 1889, the share value of the mine came down from £230,000 to £66,000 in a matter of weeks,[5] whilst the same occurred with the shares of Tincroft a year later due to the similar breakage of the engine bob at much the same time that there was a serious fall in the rich levels underground. An unusual accident at Dolcoath, but one which had surprisingly little effect upon its share prices, was an underground fire in January 1890 but fortunately this had only a temporary

[1] The first named was the principal of Harvey & Co., of Hayle. Gilbert Bennett Pearce, also of Hayle, was a director of Dolcoath and had been manager of the Mellanear works since the 1870's, whilst his brother Richard lived in Denver—a well known American metallurgist and connected with Colorado's largest smelter. Both were natives of Camborne, and had worked in their youth at Dolcoath count-house and assay office—Richard being also laboratory assistant at Cornwall's first mining school (at Truro) in the 1850's.

[2] Harvey Records, DD 852/10, CRO, Truro.

[3] 5.12.89.

[4] RCG 22.8.89.

In the Comstock mines of Nevada during the 1870's, to achieve this same end, and prevent undue speculation, the miners were customarily confined at work for days underground—in some cases for as much as a week—when the lode was expected to be struck. See Eliot Lord, *Comstock Mines and Miners*, 1959.

[5] RCG 22.8.89.

effect upon the mine.[1] Dolcoath was now the bell-wether of Cornish mining, her management looked to for a lead in anything of a crisis affecting the mining interest of the county. The continuation of Dolcoath lodes eastward had long been the mainstay of Cook's Kitchen, Tincroft and Carn Brea, although the great cross-course—corresponding approximately to the line of the Tuckingmill stream—marked the limit of its greatest riches. This quirk of geology was not, however, apparent in the 1880's and in 1889 it seemed that Carn Brea, by the discovery of 'a wonderful course of tin in the western part of the mine' was going to rival her.[2] Amid great excitement in the share market, Carn Brea shares more than doubled in value to £80 apiece and although the mine was never to outclass Dolcoath, its dividend in 1890 was Cornwall's best.[3] Both Carn Brea and Tincroft had spent a great deal of money in bringing the machinery and equipment up to date, with new whims and skip-roads put in, plus new pumping engines.[4] This had been required for years past but with no outside capital available it had only become possible with the richness of the mines themselves.

In only one aspect of production, the dressing of tin, were the big mines still behind the times. The stream works in the 1880's sold 14290 tons of black tin, value £572,532, and in the next decade 9922 tons for £326,092, most of which was money lost to the mining companies. With the cassiterite becoming of a finer and finer grain as the mines increased in depth—often becoming so finely disseminated in the rock as to be visible only to the expert eye—the proportion of 'slimes' increased considerably and it was this grade that ran so largely to waste or, to be more accurate, into the streamer's more capable hands.[5] In 1890 the five major mines (Wheal Basset, West Basset, Wheal Uny, Pednandrea and Wheal Agar) whose waters drained into the upper reaches of the Portreath river and its feeders, sold 1247 tons of black tin for £67745, whereas the stream works

[1] This occurred amongst stull timbers in the 254 fathom level near the Man-Engine Shaft and was thus not in any danger of spreading to the main eastern workings. Two shaftsmen out of the party of miners who went down to find the source of the smoke which came in increasing volume from the shafts were later found unconscious but recovered, and crowds of people gathered on the mine despite pouring rain when the news spread through Camborne that Dolcoath was on fire. Smouldering is perhaps a more accurate description than fire, for dense smoke and fumes were its concomitant rather than flames. Two days after its discovery the fire was virtually out but for some time more no men could work anywhere in the mine. To aid the natural ventilation through the workings, 'all taps at the bottom were also opened in order that the great air-compressor might clear out all the poisonous gases.' (RCG 16.1.90). A fire had also occurred earlier, at Tincroft in 1881, in the 195 fathom level, but this too did nothing worse than stop the mine for a few days. (RCG 1.7.81). In each case, and in the other subterranean fires that have punctuated Cornish mining history at infrequent intervals, the probable cause was a candle stub left burning on a timber stull by some careless miner.

[2] RCG 26.9.89.

[3] Carn Brea £25500, Dolcoath £17038, West Kitty £14100, East Pool £9240, Tincroft £9000, Wheal Grenville £4500, South Condurrow £3219, Wheal Owles £2750, Levant £625, Lovell £540. Calls of about £70,000 were made as against these dividends totalling £87112.

[4] At this date the largest shareholders in the two mines, each divided into 6000 shares, were Bolithos and Harvey & Co. The latter held 939 in Tincroft and 408 in Carn Brea; the former 961 in Carn Brea and 450 in Tincroft. (Bolitho Records, DD TL/BOL, CRO, Truro).

[5] In 1884 the Red River stream works sold 1154 tins of black tin. the mines along it 7332 tons. It was also estimated that a further 190 tons a year were carried into St. Ives Bay despite the successive ministrations of the streamers along the river's course.

along its banks on the lower reaches sold 428 tons worth £17126. The former were all working at a loss, yet the latter made a profit. Moreover the streamer had no lords worrying him to sink shafts or pay dues on profitless output—' his bread was certain and his water sure.'[1]

In a modest way Cornish mining paid its way in 1891, based on the *Mining Journal's* figures that dividends in eleven mines[2] totalled £91459, whilst calls in another nineteen mines came to some £70,000. Yet, in relation to the capital invested in the industry, its turnover and the numbers employed, this credit balance was a poor one and it was obvious that Cornish mining could not last on any scale with tin at its ruling price.

An important mine to disappear in 1891 was Pednandrea, still a considerable mine to the last, with 266 persons employed (138 underground, 128 at surface) a few years before this date. Since being taken over by a new company in 1876 the losses had been continuous, amounting to no less than £48700 from 1880 to 1889.[3] There was, quite simply, no money available for anything except day-to-day renewals or repairs and from the point of view of machinery the whole mine was in a ruinous state, especially the stamps upon which so much was dependent for maintaining the all-important sales of tin. In 1890 alone there were four major incidents and breakages at the mine; the boiler-house roof falling in, the stamps axle breaking, the cylinder of the stamps engine splitting, and the bursting of one engine boiler. As a result, the losses in the last few months of the mine's career were even worse than hitherto, amounting to £7700 from July 1889 to the initial decision to stop all work in February 1891.[4] The Bains[5] and Harvey & Co. each held a thousand out of the mine's 6,000 shares and the losses incurred were becoming too much even for merchant adventurers.[6] A final unsuccessful attempt was made to persuade Redruth residents to take up shares, on the score that Pednandrea ' helps the rates, flushes the drains,[7] in addition to being the means of circulating a much larger sum of money in the town than the current loss.'[8] No one

[1] MJ 23.8.02.

[2] Carn Brea £27000, Dolcoath £19975, East Pool £12640, West Kitty £9900, Tincroft £9750, Killifreth £3900, Levant £2812, Grenville £2250, Lovell £1620, Wheal Owles £1000, South Condurrow £612.

[3] Pednandrea Cost Books, (RIC, Truro).

[4] Pednandrea Journal No. 2. (RIC, Truro).

The loss on Pednandrea from the re-starting in 1850 to the closing in 1891 (excluding 1877–79, during which time the loss is not known) was thus about £158,000 (see also page 169). This hitherto unpublished information on the financial career of this famous old Redruth mine reveals it as a prime example of a merchant's mine.

[5] Redruth merchants, shipowners at Portreath, and by this date proprietors of the Redruth Tin Smelting Co. D. W. Bain had been the manager of the port of Portreath in the days when it was leased and run solely by the Foxes and Williams.

[6] Colonel Fludyer was also a major shareholder in the mine and at one time chairman of its committee of management.

[7] RCG 11.6.91. The water pumped from Pednandrea flowed through the surface gutters of the town's main street. In the previous workings it was far from clean, being thick with dirt and red as rust, and some years after Pednandrea was re-started in 1850 it was insisted that the dressing floors which caused this contamination should be removed to the nearby Wheal Sparnon.

[8] It was hardly surprising that so little local enthusiasm could be raised for putting money into Pednandrea, for it was known only too well locally that there had been losses and calls throughout the working.

rose to this bait, however, and in July 1891 the mine was completely stopped when there was no offer forthcoming at a final attempt to sell it by auction.

In the west, all the hands at North Levant, comprising 60 men and 16 boys, were paid off in November, whilst another major loss in the year was West Seton where, in February, the liabilities were such that ten shillings per share was being offered to anyone who would take them over from their existing holders.[1] The mine had made repeated calls on its shareholders since the early 1870's and had all but closed down in 1888. Working, however, was continued but an influx of new capital was desperately needed in order to replace the old machinery[2] and enlarge the badly equipped and totally inadequate dressing floors. With low tin prices and no new shareholders willing to come in, there was insufficient revenue to develop the mine adequately or to re-equip it after its great copper heyday,[3] and after a final call of £4000 in March 1891 work underground was abandoned and the machinery sold off in April for its scrap value of £3250.[4] Another tottering mine that seemed about to follow Pednandrea and West Seton was South Frances, where pumping troubles were particularly acute. In 1887 a second big 80″ had been installed as a 'shammel' engine to pump the water from Pascoe's Shaft back to Marriott's engine[5] and the coal costs of these two were very considerable. Most of the water from the adjoining West Basset sett found its way into the mine and it was stated that every £20 expended in pumping at the 50 fathom level there would save £120 similar cost in South Frances.[6] The latter also urgently needed a greater number of stamps and larger dressing floors, yet could not afford any outlay for this, whilst West Basset had more than sufficient stamping and tin-dressing capacity. The solution was obvious—and indeed had been advocated ten years before this, in 1880—and amalgamation between the two setts was agreed upon, in a new company to be called South Frances United.[7] This took effect from February 1892, with the West Basset shareholders receiving 1500 of the 6000 new shares, and the first undertaking of the newly constituted

[1] RCG 26.2.91.

[2] West Seton achieved the unenviable distinction of having two separate boilers burst within twenty-four hours at one period in the 1880's. (RCG 9.5.89).

[3] The mine had paid great dividends in its copper days but was unsuccessful in terms of tin. From the first dividend in August 1854 to June 1869, 90,994 tons of copper ore and 226 tons of black tin had been sold for £566,020 and a profit divided in this time of £190,000. (MJ 24.6.69).

[4] RCG 16.4.91. [5] RCG 19.8.87. [6] RCG 3.12.91.

[7] This meant the disappearance of West Basset as a separate mine. After the 1877 debacle (see page 171) a series of calls had been made, amounting to about £14000, and by April 1880 the debit balance had been cleared and small dividends commenced. The mine was particularly rich at this time and in 1879, for example, had sold 1000 tons of black tin. Three rock drills were in use and in 1882 operations were started by the company in the old North Frances sett. By 1883 the mine was again poor, with consistent losses (ranging from £1180 in 1885 to £5472 in 1883) thereafter. A great deal of low-grade ore ground was worked, almost entirely on tutwork, and the average earnings underground in the 1880's were only 55s. 0d. per month. Not until May 1888, moreover, was the use of ladder-roads superseded, when a new skip-road was put in at Thomas's Shaft 'for lifting the men out of the mine by steam power' (from the 180 fathom level). (West Basset, Book of Reports and Balance sheets, Trestrail Colln., CRO, Truro.)

company was to lay a tram-road between South Frances and the main West Basset stamps at Carnkie.[1]

Another major amalgamation of 1891 occurred amongst the smelters, with the formation of the Consolidated Tin Smelting Co. Ltd. in July. This was a combination of the Bolithos, with their Chyandour and Calenick works, Daubuz with Carvedras, and T. W. Field—successor to R. R. Michell—with Trereife. The capital of the new company was £250,000 and their first action was to close down the Calenick works and to confine smelting operations entirely to Carvedras, Chyandour and Trereife.[2] Stocks at these three works when the amalgamation took place on 31 December 1891 were 57474 cwts. of black and refined tin, then worth about £170,000. As had invariably been the case, tin smelting was still a much more profitable venture than mining and the Consolidated company's profits for its first year of operations were £15266.[3] The consolidation was an obvious move to enable these companies to effect economies of production and management as well as marketing—the latter, with its necessity for Liverpool or London offices and an American agency, being particularly costly. It was due also to the effect of the McKinley tariff in the United States, which militated against British tin-plate exports and which had already caused havoc amongst the Welsh works.[4] This same tariff, designed to build up America's own tin-plate industry, was also responsible for a contemporary revival of interest in the tin deposits of Dakota in the Black Hills near Custer City. This second flotation was bordering on the fraudulent, with everything on a scale calculated to impress investors: the capital was to be £2 million and mention was made of lodes yielding 300–650 lbs. of tin per ton of rock, of an output of 12000 tons of metal per year, from what was described as the most important mineral discovery in the world since the finding of gold in California or Australia. Cornwall would have been troubled indeed at this news if it had been genuine, but to mining men in Cornwall the over-blown terms in which the whole promotion was couched were sufficiently revealing.[5] Moreover, back in 1884, these Harney Peak mines

[1] One minor disadvantage of the amalgamation of separate mines evidenced itself the following year when the West Basset tributers came out on strike in protest at the discrimination they considered was being shown them by the South Frances manager. 390 had stopped work in sympathy throughout both sections of the mines—and this number marched through Redruth to hold a meeting in front of the Mining Exchange. Tempers ran high and there were threats to burn the tutworkers' clothes if they did not support the strike. Within a few days the men were back at work however, although the results of the stoppage are not recorded. (RCG 24.11.92.)

[2] RCG 2.7.91. The Calenick works was advertised for sale, together with the crucible works alongside, the following year. (RCG 17.11.92). It was also stated that the Trereife works might be closed down, but this did not take place for some years.

[3] DD BL/30, CRO, Truro. This profit was made on the smelting of 6753 tons of black tin from Cornish sources, plus 517 tons of foreign ores, chiefly from Bolivia.

[4] This added 10s. to the cost of a box of tin-plates, then sold at about 14s.—an import duty of approximately 70 per cent.

[5] This was not the first time that a so-called ' tin mountain ' had been allegedly discovered in North America. In August 1867 there had been momentary alarm in Cornwall following rumours that lodes larger than anything ever seen before had been found in the Ozark Mountains in Missouri. But the panic was momentary and in October it was said of the so-called Missouri mountain of tin that ' as good might be found on Portreath beach.' (RCG 17.10.67.)

in Dakota had been discovered and first promoted as an American venture to provide ' all the tin that America would ever want for the manufacture of its own tin-plates '[1] and they had again been taken up by a London company in 1887–9. Bolithos had quietly and promptly had one of their American agents visit the Harney Peak district to inspect and report, as well as to send samples to Chyandour for analysis. This was followed up by a visit from one of the Bolithos' mine agents, a Captain Curtis, who was sent from Cornwall to make a confidential report to his Penzance principals,[2] from which it became apparent—to Bolithos, if not to the world at large—that there might be tin in Dakota but it need not unduly concern Cornwall as regards its economic possibilities. As it was, in 1891, the Harney Peak flotation came about largely as a result of renewed interest in America's tin producing capability, hand in hand with an opening-up of the region by railways as well as by the economic boost given to it by the McKinley tariff barrier. This time, Captain Josiah Thomas of Dolcoath—reckoned to ' knaw tin ' better than any man alive —was called in to make a public report and was paid one thousand guineas to visit the Black Hills. His findings brought a collapse yet again of Dakota's hopes and an end to American tin mining: the metal was there all right but in a complex ore and nowhere in the whole sixty square miles of the district likely to yield more than 30 lbs. per ton—too low a grade to compete in world terms.

In Cornwall the same mining authority was coming in for some criticism at this time. Under his managership, New Cook's Kitchen was being worked in a very languid fashion and its sale of £392 of tinstuff in 1892, the result of sixteen weeks' working of the mine, was described as ' playing at mining or Micawber-like waiting for something to turn up.'[3] Yet it could be argued that even with great capital outlay and spirited working, results were little better. Levant, during the last quarter of 1892, sold £10,000 worth of minerals (£8890 black tin, £1110 of copper ore) yet made a profit of only £146. A member of the committee bemoaned that they were ' between two mill-stones. The upper one was the merchants, and the lower one the landlords, and they were grinding down the out-adventurers exceedingly small.'[4] This was by no means an isolated case and it summed up all too accurately the feelings of many who had money invested in Cornwall and who were getting tired of what was later to be termed ' philanthropic ' mining. It might be still worthwhile for the merchants, the smelters and the mineral lords to keep a mine afloat but as

[1] MJ 26.10.01.

[2] RCG 3.5.88. Captain Curtis's adventures in the Black Hills, secretly reporting for Messrs. Bolitho, but posing as a British tourist, would make interesting reading had any record of them been preserved. (See also MJ 22.9.88).

[3] RCG 16.6.92. Moreover rock-drills were in use at this time.

[4] RCG 5.1.93. Another comment from this same quarter was that ' it was worth nobody's while in England to work a bal for that profit.' (MJ 7.1.93). This state of affairs was common in many mines in the latter days of Cornish mining, selling perhaps £40,000 of ore in a year yet clearing only a few pounds profit.

a commercial speculation or investment it was no longer worthwhile to an outsider.

Yet if there were complaints as to the inadequate financial returns for adventuring in mines at this period in the early 1890's, what might be said for the men who worked in them. Wages underground were perhaps eighteen or twenty shillings a week[1], and liable to be reduced peremptorily if black tin prices came down, and this for work under conditions of indifferent ventilation, great heat and ever present danger. The latter factor came into prominence in 1893 when two serious mining accidents occurred. Early in January Wheal Owles was inundated by a massive inrush of water and twenty miners were drowned. The mine was then a pale shadow of its former self and activities were confined solely to the East Cargodna section near the cliffs, Wheal Drea having been given up in 1884. When the disaster occurred forty-one men and boys were underground and only twenty-one of these made their escape from the 65 fathom level where the great inrush from the old Wheal Drea workings came in. The speed with which the mine filled—from the 120 fathom sump to the 30 fathom 'sea level' in twenty minutes[2]—showed that this was not a mere 'holing' but more the collapse of a whole bar of ground standing between this level and old water-filled workings alongside it. The flooding was so swift that many survivors thought the sea had broken in. It was decided later that errors in dialling in the past, chiefly in not making allowance for annual magnetic variation, were the root cause of the disaster, for the new level in question was being driven away from the old Drea workings according to the plans.[3] A sad feature of the disaster was that there was insufficient pumping power on the mine with which to drain it to recover the bodies. The sole remaining Wheal Owles engine, a 36″ rotary which was pumping in the Cargodna section by means of flat-rods, could never fork the mine, let alone the workings that now connected with it, and the outlay necessary for a 60″ engine and new pitwork, and the period of pumping thereafter, was said to be £4000—far more than the mine adventurers were prepared to put up.[4] Thus, to add

[1] In 1899 the average pay for unskilled Kaffirs and other natives working in the Rand mines was 12s. per week, plus free food and 'compound' lodging. (T. A. Rickard, *Man and Metals*, vol. 2, 1932).

[2] MJ 14.1.93.

[3] About ten years earlier there had been another minor holing, from the Drea section into the old Boscean, although fortunately without loss of life. Some years before this there had also been a similar occurrence at Boscean when two miners were killed (MJ 28.2.57). From about 1888 onwards the agents at Wheal Owles had been aware of the constant dangers of holing. Even where old plans existed they were unreliable from this cause, for old-style dialling was an art rather than a science. In 1894 it was said that 'from 25 to 50 years ago mine captains would consider that if they read their bearing with a quarter degree it was nice enough; surveyors now often read as fine as 10 seconds which is the 360th part of a degree.' (E. W. Newton, The Evolution of the Mining Theodolite, *R.C.P.S.*, 1894).

[4] RCG 26.1.93. A disaster fund was started for the men's dependents and reached a total of £3293. Several of the relatives of the victims were only dissuaded with difficulty from bringing actions against the adventurers for alleged negligence, whilst one was awarded £25.

About twelve months before, a somewhat similar tragedy had occurred at New Great Work, Breage, when two miners were trapped and drowned by an inrush of water in the 30 fathom level. Here it was known that flooded workings were close at hand, from the dangerous pressure of water coming through a fissure in the rock, and the men were under strict instructions to blast the end and

to the distress in the miners' home village of St. Just, the victims of this disaster, Cornwall's worst 'holing into a house of water,' remained in the water-filled levels of the mine. Not only in Cornwall but throughout the world, miners have an abhorrence, tinged with superstitious dread, of leaving the bodies of men killed in workings underground. Feelings in St. Just were that no matter what the financial cost some attempt should be made to give those who had died in Wheal Owles a Christian burial. And yet when all was said and done, after a hopeless attempt by one man to raise a fund with which to start pumping out the mine, a brief burial service held at the mouth of the water-filled Cargodna Shaft on the mist-wreathed cliffs was all that was possible.

The second disaster, at Dolcoath in September 1893, was of a different kind—a serious fall of ground in the fabulously rich 412 fathom level. This occurred at a time when seventeen men were at work replacing some of the main timbers securing the ground and only nine of these, warned by the sounds of cracking timber and the first trickles of dust and stones, had time to make their escape. Of the eight entombed, one could still be heard alive twenty-five hours after the collapse but could not be reached, and the only survivor was a twenty year old miner from Troon who was released after forty terrible hours. Rescue work was greatly hampered by the grave danger of further falls and not until October 12 was the last body recovered, a full twenty-two days after the disaster.

The place where this run of ground took place was near New East Shaft in a section of ground about 30′ wide and 72′ high, which had been so rich that no supporting pillars, or arches of ground, were left. Massive timber stulls, of 24″ square American pine, had been used to make the excavated area safe but it was evident that this method was not sufficient to bear the enormous pressure of something like a hundred fathoms of more or less loose ground above.[1] Dolcoath had now been worked for copper and for tin on a massive scale for a century at least and during this time an enormous amount of veinstone had been removed, leaving big gunnises which rendered the whole area weak. As early as 1865 there had been one big collapse when thousands of tons fell away and when 'some persons went so far as to say that Dolcoath would never be a mine again.'[2] In fact, however, the occurrence was said afterwards to have been beneficial to some extent for much of the fallen ground was found to be of sufficient value to be worth stamping, whilst the 'immense rocks—some probably of hundreds of tons weight—[which] fell away, which are now

not to use picks or borers on any account. Only two men were in the mine at the time of the accident on Christmas Eve 1891 and the first intimation of something wrong was when another miner found the main shaft full of water to the adit level. Two points of interest emerged at the inquest upon the men; that efforts to obtain existing plans of the adjoining old sett from the mineral lords had been unsuccessful, and that the older of the two deceased (48 years of age) had been to America five or six times in his mining career. (RCG 21.1.92.)

[1] W. Herbert Thomas, *op. cit.* In the ten years prior to this accident, sixteen persons had been killed at Dolcoath.

[2] RCG 9.11.65.

resting across the gunnis' were deemed to be useful props against the likelihood of further similar falls. Mercifully this accident occurred between cores and no-one was trapped.

Dolcoath, at the time of this second great run of ground in September 1893, was at her peak. In terms of richness the 412 was better than ever, with the lode over 36′ wide and worth £600 per fathom and in 1892 one 'pare' of men had broken fifty tons of ore from it in the space of three weeks. The mine was an impressive undertaking at this time, with no less than twenty engines at work in all: an 85″ and two 65″ for pumping, plus one smaller one for raising water for dressing purposes; three engines driving Cornish stamps plus one driving Californian stamps[1]; one non-condensing engine driving 'the small stamps on sand'; four steam whims; two air compressors; the man-engine; two workshop engines; two stone-breakers and one working the machinery on the acres of dressing floors. A total of 1336 persons were employed, 745 at surface and 591 underground. Amongst the latter the most highly paid were the tributers, about 120 in number, who got £5 per month on average. At the other end of the scale the 'bal-maids' received 1s. 0d. or 1s. 2d. per day, unless they were engaged, as was more usual, on piecework when they got 4d. or 6d. per ton: 'even for strong girls who have the knack of striking rocks at the right spot and breaking four or five tons a day to a size suitable for the stamps, the wages are decidedly well-earned.'[2] This, moreover, was for work carried out in all weathers in the open or at best in the scanty shelter of wooden 'spalling' sheds.

Apart from the Dolcoath and Wheal Owles disasters, 1893 was notable for the closing of the last of the Redruth mines when Wheal Uny and East Uny were closed down in February.[3] The small but once promising East Blue Hills was also abandoned in March, sales of black tin having been £17907 to this date.[4] The sett was thereupon acquired by Blue Hills, which also received permission in May to work some of the old Penhalls,

[1]A development of the normal Cornish stamps, with heavier, rotating heads. So called from their first use in the Californian gold-fields.

[2] W. Herbert Thomas, *op. cit.* Statistics of Dolcoath's output and dividends to 1887 according to this writer were—

1799 to December 1869	Total sales		£3,234,693 . 19 . 6
1868 to 1887 inclusive	Black tin sold (31820 tons)		£1,829,356 . 7 . 11
	Copper ore sold		3029 . 4 . 3
	Arsenic and sundries sold		6253 . 7 . 4
	Total		£1,838,638 . 19 . 6
	Working costs	£1,271,292 . 3 . 2	
	Lord's dues	106,282 . 8 . 1	
	Premium for lease	25,000 . 0 . 0	
			1,402,574 . 11 . 3
	Paid in dividends		£436,064 . 8 . 3

[3] RCG 9.2.93. The engines (70″ pumping, 36″ (80 heads) stamps, 25″ whim at Wheal Uny: 70″ pumping and 22″ whim at East Uny) were for sale 13.4.93 and realised only £1425 at their auction.

[4] RCG 16.3.93. At the prices then ruling, about 450 tons.

which was partly drained by the Blue Hills engine. In the Camborne area, two failing mines on the brink of stopping were West Frances and Cook's Kitchen. The former was so poor that all dues were given up for five years whilst at the latter a loss of £2649 was announced in July on only sixteen weeks' working. Early in 1894 there was the disturbing news that 'all work that is not distinctively remunerative is to be stopped' at South Condurrow, a mine that had been one of the most consistent dividend payers in the period from 1880 onwards.[1] Following this came the ominous news of another big fall of ground at Dolcoath[2] and of a loss there, announced in July, of £1528 for the previous quarter.[3] The first tentative suggestion of the need for a new vertical shaft was made at this meeting and in October a further loss of £2529 was announced. That all was far from well at Dolcoath became apparent to the world at large with news of yet another serious fall underground in December.[4] Shares in the mine had depreciated from £70–£80 to less than half this figure by January 1895 and the catalogue of the mine's misfortunes continued with a fire in the same month that burnt out the count-house and destroyed all plans of the mine. In February Captain Josiah Thomas retired and a fortnight later the mine committee announced a loss on the quarter of no less than £5100. Even more serious, it was revealed that the last collapse in December had gone worse in the succeeding weeks and had extended upwards a considerable distance. This rendered a large portion of the richest section of the mine unworkable and grave criticism was expressed, not least by the mineral lord, at the way Dolcoath had been worked; 'outside experts have said for years that the mine would collapse if the foreign way of working wide lodes was not adopted at Dolcoath. The fear of the miners of such a collapse has, it is said, prevented them giving their full energy to their work.'[5] The whole eastern part of the mine was in a risky state and it was evident that masonry arches, as used in Mexico and elsewhere, should have been put in to secure the ground rather than timbering. In fact it was decided after the December fall to put in massive concrete arches above the 412 but before this could be done the further runs took place and the whole section, despite its richness, had to be abandoned.[6]

Thus, in a matter of months, was Dolcoath brought from being a mine allegedly too rich even to be unduly concerned by the low price of tin, into a losing concern in which economy was becoming a watch-word. And, as the undisputed leader of Cornish mining, Dolcoath's troubles had a depressing influence upon the industry as a whole, with the technical

[1] MJ 1.1.94. At the same meeting, the Chairman dismissed the usual optimistic reports of the agents (whose livelihood hung upon the continuation of the mine) with the words 'Advice which came from Cornwall should be taken *cum grano salis*, for the Cornish people loved mining and looked upon it as being the right thing for people to invest their money in mines and perhaps even to lose it.'

[2] RCG 26.4.94. [3] RCG 12.7.94. [4] RCG 4.10.94, 27.12.94. [5] RCG 4.4.95.

[6] To prevent any possible similar trouble in the area of New Sump Engine Shaft, concrete arches were put in there during early 1895. (MJ 23.3.95).

Cast-iron supports, chiefly consisting of old pump columns reinforced, had earlier been tried at Dolcoath for securing weak ground, but evidently without success. (See RCG 7.2.79).

reasons for its rapid transformation from evident riches to evident poverty only imperfectly understood, if at all, by the investing public at large. Even in Cornwall and amongst mining men, hopes were fading that the county could continue its stand against foreign alluvial tin and the *Royal Cornwall Gazette's* comment that ' large shareholders are getting sick of philanthropic mining '[1] summed up the mood that now pervaded even the ranks of the smelters and entrepreneurs who were the mainstay of the industry.

[1] 22.11.94.

IV

THE FIGHT FOR SURVIVAL

THE PRE-WAR CRISES: 1895–1916

'The history of the past will never be repeated in the working of our Cornish mines, and the time for old-fashioned methods and old-fashioned patience and perseverance has vanished for ever.'

Mining Journal, 3 August 1901

THE CAUSE OF the depression that had descended upon Cornish tin by 1895—an intensification of the recession which the industry had endured for two long decades—was not far to seek. In the late 1880's and early 1890's the alluvial tin of the Malayan peninsula began to attract the attention of European capital and during this period, with increasing technological development,[1] output rose even more swiftly than it had done in the preceding years under Chinese ownership and control.[2] The prophetic *obiter dictum* of Sir Stamford Raffles, voiced to a junior member of the Bolitho family on the occasion of his visit to Wheal Vor in the 1820's[3] that 'you will live to see the day, my little boy, when the mines of Banca will eat up the mines of Cornwall '[4] had, at length, come true.

In the Straits area production doubled between 1887 and 1894 alone, and one of the most important factors in bringing this about was the depreciation in the value of silver following the abolition of its use as a monetary standard in Europe. The European value of the Straits silver dollar had declined from 3s. 8d. in 1879 to 2s. 1d. in 1894 and thus the mine proprietor there was able to produce and sell his tin that much cheaper.[5] His men were paid in silver but his tin was sold for gold and the ratio of value between the two had altered out of all proportion as a result of 'bimetallism' in the world's monetary systems. This was perhaps the greatest single factor in bringing Straits tin to its predominant position at this time, allied to the inherently easy mode of working alluvial ground,

[1] The introduction and growth of hydraulic mining (by sluicing and gravel pumps) was the most important step forward, increasing output from the Straits enormously. The tin dredge was a later innovation (see page 263).

[2] Despite this, 80 per cent of all tin production in Malaya was still in Chinese hands in 1912, and even by the outbreak of the second World War this proportion had only fallen to about 30 per cent.

[3] MJ 9.5.63.

[4] RCG 1.1.70. Raffles, founder of Singapore and of British influence in that area, visited Cornwall in 1823, shortly before his death.

[5] See MJ 24.11.94. 'Straits tin producers get the equivalent in silver of £130 per ton, the Cornish tin producers for the same tin £60 per ton in gold currency.' (RCG 21.2.95.)

Carvedras Smelting Works at Truro, shortly before it was closed down in 1898. This photograph was probably taken in 1891–2 when a new lease was granted and the works passed to the ownership of the Consolidated Tin Smelting Co. Ltd., of Penzance. An earlier photograph shows a single tall masonry stack rising from the centre of the furnace house, not these individual square furnace stacks as re-built. Above it is the old timber viaduct of the Great Western Railway at the eastern end of Truro station. Seen from a passing train at night the flames from the stacks must have been impressive.

The works of the Cornish Tin Smelting Co. Ltd., at Seleggan near Redruth, looking towards Carn Brea monument. Established in 1887, this was the largest tin smelting works in the county in the twentieth century and was also the last to survive, closing in 1931.

Mellanear Smelting Works at Hayle. **Built** *in 1837, this was Williams Harvey's chief works in Cornwall and after being eclipsed by the company's new Mellanear works at Bootle (Lancashire) was revived from 1915–19 to cope with the war-time smelting in England of Bolivian ores.*

Chyandour Smelting Works, on the eastern outskirts of Penzance, owned by the Bolithos and later by the Consolidated Tin Smelting Co. Ltd. after 1891. This was the last to survive of the Cornish smelting-houses established before 1800.

and output from the Federated Malay States rose, for example, from 25911 tons in 1890, to 33102 tons in 1892 and to 46983 tons in 1894. The meteoric progress of the leading Malayan smelting works, owned by the Straits Trading Co., provided a parallel to this rise in Straits tin. The company, formed in 1885 with a monopoly on the purchase and export of all tin ore produced in Selangor and Perak, commenced smelting in 1887 in a works possessing only a single reverberatory furnace, sited on the island of Pulau Brani in Singapore harbour. Eight years later it had fourteen four-ton furnaces in operation plus a separate refinery, and with its capacity of 12000 tons per annum—an output greater than all Cornwall's works combined—was the largest smelter in the world.

Moreover, another of the world's great tin fields was about to enter the scene with the development of the Bolivian mines. These had remained virtually unchanged since the 1860's but after 1895 there came a swift rise. In 1898 4327 tons of barilla had been shipped but two years later 15088 tons, and this had risen to 28000 tons by 1906.[1] The Bolivian silver-tin lodes had hitherto been worked solely for their argentiferous content but the decline in silver value and the construction of a railway link with the Pacific coast made their large-scale working for tin an economic proposition. The completion of the Antofagasta & Bolivia Railway in 1892 as well as its various subsidiary mining branches and a network of telepherique lines in the years thereafter was the key which opened up the main tin producing districts of Oruro, La Paz, Potosi and Chorolque amid the peaks of the eastern Cordillera. What had hitherto been a prohibitively expensive thirty days' travel from the mines to the Pacific coast by mule or llama became a three days' journey by rail at a quarter of the cost and this encouraged the development of tin mining on a large scale. The stanniferous potentialities of the Bolivian Andes may be inferred from the occurrence of lumps of well-nigh pure cassiterite, sometimes weighing many pounds, in mountain gullies after the summer rains. In the mines themselves, anything except the richest ores was discarded as not worth working and most of the waste dumps and tailings from the old silver workings contained as much tin on average as the Dolcoath main lode in its later years. Indeed, had the extremely rigorous climate at an average elevation of 15000 feet above sea level,[2] the scarcity of labour, and consequent high working costs not militated against the Andean mines, Bolivia would have proved as great a factor in bringing about Cornwall's mining decline in the nineteenth century as had the Straits.[3]

[1] MJ 20.4.07.

[2] Water, fuel, and labour were all scarce and expensive, particularly as the lodes occurred in the most remote and highest regions. A few mines were worked as high as 18000 feet: 'as a general rule, one can say no snow, no tin.' (Captain John Penberthy, Tin in Bolivia, *RCPS*, 1909). Working conditions in the Andes mines were harsh in the extreme, with some of the greatest extremes of temperature ever encountered in the history of metalliferous mining. In one instance the temperature range in which the miners passed the day was minus 26° at surface and 131° in the deepest galleries of the mine (MJ 6.9.13), conditions exceeding even those of the notorious Comstock lode in Nevada.

[3] By 1913 Bolivian output was 44595 tons of barilla, and one mine alone (the Uncia, owned by

As a consequence of the expansion in world output, the price of black tin fell steadily and by the beginning of 1895 it stood at what was, for Cornwall, the crisis price of less than £38 per ton. This was far below the figure at which any Cornish mine could survive for long, and for the county the year was one of gloom throughout, with a more severe depression than had yet been known. What was worse, it seemed likely to be long lived and few enough of the existing mines were in a fit state to weather a brief, let alone a long, storm. In March, Botallack was abandoned, after an attempt earlier in the year to sell the mine as a going concern—not surprisingly, an unsuccessful attempt, for the loss was about £400 per month and no new ground of any promise had been opened up for years. Then, later in January, a dam penning back the water in the old Nineveh section collapsed—fortunately without loss of life for there were few men underground and the trouble was not unexpected—and as a result the lower levels at Wheal Cock were flooded.[1] The remaining 130 men were thereupon discharged, 50 having already been laid off late in 1894.[2] In April pumping was stopped at Tregurtha Downs at Marazion, last survivor of the brief upsurge in interest in 1881, and which had experienced a chequered career in the interim.[3] This was followed in May by the stoppage of the remaining western part of South Condurrow, where 30 underground and 70 surface hands were discharged,[4] whilst the same month the Cook's Kitchen adventurers opened negotiations towards a sale of their mine to Tincroft. Losses at Cook's Kitchen had been very heavy throughout 1894 but sooner than abandon the sett it was decided to retrieve what was financially possible by amalgamation. At Dolcoath there was the news that the mine was to be turned into a limited company, a turn of events which caused surprise in Cornwall for this had all too often in the past been the precursor of worse tidings to follow.[5] In Dolcoath's case, however, the conversion from cost-book to limited liability was essential, for no outside investor could be wooed into Cornwall under the former scheme. Past experience, not only of mining companies themselves,

Senor Simon I. Patino) had an output approaching 14000 tons—worth about £1½ million and a quantity almost as great as Cornwall had produced in a year in its prime. (See MJ 1.11.13). The adjacent Llallagua Mine, later also owned by Patino, subsequently surpassed even the Uncia in size, these two being far and away the largest tin mines the world has ever seen.

1 MJ 2.2.95. Like Levant, Botallack had desperately needed for years a new shaft to improve hauling capacity and to enable men to be raised mechanically. A meeting was held in August 1891 to decide upon this but from lack of finance it was not proceeded with. (RCG 27.8.91). Considerable sums were spent in 1892–3 straightening and enlarging Wheal Cock engine shaft to enable a skip to be used instead of the old-fashioned kibbles.

2 RCG 28.3.95. At the final meeting of the adventurers it was agreed that the engine and materials at Wheal Cock ' be offered to Lord Falmouth as a gift in consideration of the generous manner he has always treated the shareholders '—showing that the strife between mineral lords and mining companies was not entirely universal. Sales of ore from Botallack from 1836 to the closing in 1895 had been—black tin £829664, copper ore £220701, arsenic ores £6481. Total £1056846. (Bolitho Records, DDX 104, CRO, Truro.)

3 After three years of activity and an outlay exceeding £33000, the mine had been sold as a going concern for £1755 in 1885 and was worked spasmodically thereafter.

4 RCG 2.5.95. The loss on the four months to this date had been £1386.

5 To June 1895 (from 1799) Dolcoath had paid dividends totalling £913729 and made calls of £45252. (MW 12.9.25.)

but of the protracted proceedings of the old Stannary Court had brought this about. Dolcoath's registered capital was to be £350,000, with each holder of a cost-book share receiving forty new ones of £1 and the first decision of the re-constituted company was to sink a vertical shaft south of the main lode in new whole ground. This would by-pass the weak sections where the runs had taken place and should cut the southward dipping main lode in depth, giving a new opportunity for centralised working for the whole eastern part of the mine.[1] This was a major undertaking, showing massive faith not only in Dolcoath's future as a mine, but in Cornwall's future for tin at even greater depths than hitherto.

Unfortunately, under the prevailing conditions, this same optimism did not extend far beyond Dolcoath and the whole run of mines east of Tuckingmill was in a rapidly worsening state. All were now more or less interdependent as regards pumping and when Wheal Agar shut down early in 1895, followed by the stopping of its big 90″ engine in May, things boded ill for East Pool and, in the chain reaction that would thereafter follow, for South Crofty, New Cook's Kitchen, North Tincroft, Tincroft and even perhaps Carn Brea. The Agar adventurers, moreover, were in a mood for neither amalgamation nor arbitration. The losses they had made in the working of the sett had been heavy, about £100,000,[2] and the dispute with East Pool over pumping still rankled with them: only the fact that their lease of the sett ended in another eighteen months prevented them from holding the rest of the adjoining mines to ransom, for the materials and surface equipment would then revert to the mineral lord unless sold.

This state of affairs accelerated the acquisition of Cook's Kitchen by Tincroft, after a final call of £4000 to liquidate the debts of the former.[3] Carn Brea too was in extremely low water, with its shares—after being over £20 apiece only three years before—down to a few shillings. This, moreover, was a purely nominal price for they were in fact unsaleable, the market value of the entire mine being put at less than £1000 even though something like twenty times this had been spent in the previous few years in new machinery. In January 1896, with tin prices still receding, 300 men were laid off and with a loss of £6600 on four months' working then announced it was obvious that something must be done, or a new discovery made, if Carn Brea was not to be 'knacked.' To quote the manager, 'They had one of the finest setts in the county of Cornwall but

[1] The first turf of the new shaft, named after the Dolcoath chairman, M. H. Williams, was cut on October 1st 1896. In the interim, the company turned its attention to sinking and developing the Stray Park section of the mine.

[2] According to the *Mining Magazine* (April 1916) the calls from 1847–96 were £140,000 and the dividends £42000.

[3] Cook's Kitchen, when taken over by Tincroft, was in a fearful state as regards its machinery and equipment, having been worked with the most rigid economies for years. Typical was the pumping engine, which was described as '. . . simply a wreck, for the bob was broken, the nozzle patched in at least a dozen places, the cylinder was full of holes. Directly the engine was stopped and cooled down, the principal part of the engine almost dropped off.' (MJ 2.11.95.)

unfortunately they were in want of a discovery of tin.'[1] A further amalgamation between the newly extended Tincroft—where prospects were brighter than at Carn Brea—was therefore agreed upon without delay as a means of saving the mine, and the two mineral lords concerned, Basset of Tehidy and Lord Robartes, themselves took up large holdings in the new limited company that was formed (Carn Brea & Tincroft Mines Ltd.) when there was a poor response to the share issue.[2] Meantime, another major amalgamation had taken place amongst the mines south of Carn Brea. Here, the South Frances United shareholders, tired of repeated calls and of the heavy pumping charge, year in and year out, decided to throw their hand in, and their negotiations with Wheal Basset[3] received fresh impetus in December when the main 80″ pumping engine was completely wrecked by fire. This brought work in South Frances United entirely to a halt, the mine having sold 2809 tons of black tin in the previous three years.

The Basset Mines group was formed in January 1896 with a nominal capital of £100,000 under the guidance of Francis Oats, who was the largest shareholder, together with the Bolithos of Penzance. Oats, originally a working miner from St. Just who had been a star pupil of the Miners' Association 'migratory' or travelling schools in the 1850's, had emigrated to South Africa in 1874, where his mining skill raised him to the day-to-day control of the Kimberley diamond mines and to a directorship of the huge De Beers combine. He invested heavily in Cornwall, being one of its major financial supporters[4] in the dark days of the 1890's. At Basset Mines his faith in Cornish tin mining evidenced itself in the decision to enlarge and sink Marriott's Shaft at the old South Frances, and work Basset on a scale and to a depth to rival Dolcoath at their new Williams Shaft. Oats had brave, but what were later to be proved geologically unsound, ideas about tin holding down to depths as great as 1000 fathoms and a very considerable amount of money and thought was put into the re-building of South Frances as the long-term future centre of the Basset group. The main principle, as at Dolcoath, was to centralise all pumping and winding, using the most up-to-date appliances, and by-passing the bottle-neck imposed by crooked engine and hauling shafts, to enable profits to be made from a very much greater tonnage of ore-stuff milled each year.

During this time the deadlock over Wheal Agar had continued unbroken and the position of the surrounding mines deteriorated steadily. By the end of December 1895, despite the dry winter, East Pool's engine was overwhelmed by the increased water coming from Agar and all men except

[1] MJ 22.2.96.
[2] RCG 5.3.96. According to the prospectus, issued in July 1896, Carn Brea had paid dividends of £393250 to this date on an outlay of £162075; Tincroft, dividends of £350400 on an outlay of £96750. (Bolitho Records, DDX 104, CRO, Truro.)
[3] Commenced in April 1895. (RCG 25.4.95.)
[4] He was, again in company with the Bolithos, one of the largest shareholders in Levant.

those who could be found pitches above the 50 fathom level were discharged. The water then endangered South Crofty where it was useless to keep the pump going and where the men had already stopped work.[1] The mine was thereupon suspended, the sett having earlier been offered to East Pool but amalgamation terms could not be agreed upon.[2] The adjoining New Cook's Kitchen, dependent upon South Crofty for pumping, also stopped about this same time.[3] As a result, a mass meeting, orderly enough but determined, of the miners affected or about to be affected, was held in the Fair Meadow at Redruth, followed a week later by a similar meeting at Camborne.[4] A deputation from these men was appointed to ask Lord Robartes, the chief mineral lord concerned, to intervene in the dispute and set the Agar engine back at work. Continuing attempts were made by Strauss, then the M.P. for the Mining Division, to bring about a settlement and for a brief period later the engine was re-started and pumping commenced at his sole expense. Here, as never before, were brought together three of the main inherent mining difficulties to be faced in Cornwall; multiple mineral lords, each wealthy enough to be indifferent to a mine's stoppage; mutually dependent pumping, and a lack of any willingness to co-operate. There was no hope or even intention of the old Agar company re-working their sett but they would not accept arbitration as to a fair price for East Pool to buy the engines and machinery. Thus the matter dragged on through the spring of 1896, to the detriment of the miners and of the chances of reviving mining in the district.

In April, Wheal Friendly in St. Agnes was abandoned, whilst Wheal Kitty there stopped pumping and confined all operations to above adit level. Polberro had also been suspended[5] whilst West Kitty—one of the last great tin prizes in Cornwall and the only worthwhile mine to have been established in the 1870's—only paid costs or made profits too small to divide.[6] But the worst blow came with the precipitous abandoning of

[1] 'About a fortnight ago all the South Crofty miners left, as they were afraid of the water.' (MJ 15.2.96.)

[2] RCG 6.2.96, 20.2.96. South Crofty, although with a great career still ahead of it in the twentieth century, had made an overall loss of about £115000 to this date, with calls totalling £120000 and only £5000 in dividends. (MJ 9.5.96.)

[3] As a company New Cook's Kitchen was wound up in 1898. (RCG 12.5.98).

[4] RCG 6.2.96, 13.2.96.

[5] Reporting one of the last meetings of the Polberro company, the acid press comment was that 'the usual stock remarks were made by the chairman as to the wonderful property the shareholders possessed in this mine and excuses made for so small a quantity of tin being sold . . . the mine is certainly a wonderful one for the patience of the shareholder.' (RCG 9.5.95). Polberro had absorbed Trevaunance United in 1891, neither mine being brought to a cost-paying state. A 60″ was erected at the former in 1886 (bought from South Penstruthal) and a 50″ (from Penhalls) at the latter, £37000 having been expended on the 2 setts by 1896. (MJ 18.12.97.)

[6] West Kitty had continued rich enough to pay dividends up to this worst period of the tin slump. In sixteen weeks of 1890, for example, she sold £7000 of tin with working costs of only £3100 and with new leases then being negotiated there was some possibility of a fine for renewal being inflicted as at Dolcoath in 1883. (MJ 22.3.90). By the end of 1893 dividends of £108900 had been paid, as against an outlay in 1879–80 of £3600. (MJ 13.1.94). A steady decline set in thereafter, with the falling off in value of the original (Reynolds' or Wheal Rock) section only partly supplemented by the acquisition in 1894 of the 'South Mine' (Thomas's, or Polbreen).

West Frances, brought about by the relinquishing of Michael Williams' majority shareholding. As well as being chairman of Dolcoath, he was the largest individual shareholder remaining in Cornwall and his publicly declared decision[1] to withdraw all his holdings in Cornish mining coincided with a wave of pessimism as to the future of the whole industry. Doomed industries like doomed men may live long, but Cornishmen had come to realise that their mining industry was facing not retrenchment but the very real possibility of extinction. This was a more serious crisis than had been weathered before; not only a panic but a collapse was considered imminent from loss of confidence.

The reactions of Cornish mining men who had lived through the twenty-five years up to 1896 and who had seen their industry fall in that time from a premier position almost to extinction must have been sad indeed. The decline and its effect upon Cornwall had been gradual enough for its full impact to go unnoticed by those who lived amongst it day by day but to an emigrant miner of the 1870's returning home to west Cornwall to spend his last few years, the change in the scene must have seemed unbelievable; knacked bals on every hand, silent stamps and derelict dressing floors, empty houses and mining hamlets, whilst even those few mines that survived were in a threadbare and parlous state after years of the most dreadful economies.

The depression was sufficient even to hit the smelters, always the last to feel a mining slump, and during 1896 the Trereife works of the Consolidated Tin Smelting Company was closed, whilst the capital of the company was reduced drastically to £150,000. Consolidated's purchases had declined to a total of 4498 tons of black tin (3436 tons from Cornwall, 1062 from abroad), and the capacity of the Chyandour and Carvedras furnaces was now adequate to smelt this quantity. Even the perennially prosperous stream works too were suffering greatly from the unprecedented price of tin,[2] their returns having already been reduced by the continuing efforts of the mines to cut down dressing losses. The unemployed miners' stand-by of eking out a living by tributing on the dumps of the abandoned mines, selling small 'parcels' of tin-stone to bargain buyers who ran a few head of stamps at Bissoe or on the Red River, was also failing after a decade of such work.[3] Conditions among the labouring poor throughout the western mining districts were again causing grave concern,

[1] This was sparked off by the decision of the Tehidy estate to institute at this unpropitious time fresh claims for lands damaged by mining activity at Dolcoath, West Frances and South Frances. Originally it had been decided to stop the engines here in February and abandon the mine but they were re-started for a short period in order to let the Grenville adventurers put a protective dam in at the inter-connecting 174 fathom level. (MJ 25.7.96).

[2] 'The low price of tin has played terrible wreck with the tin stream companies.' (RCG 29.7.97.) Two years earlier the seventeen major surviving works in Cornwall were all on the Red River, comprising Gwithian, Trevorne, Tolvaddon, Brea, Magor, Entral, Reskadinnick (2), Bettyadit, Bridgemore, Roscroggan, Treskillard, North Basset, Tehidy Mill, Tolgarrick, Kieve Mill and Carnarthen. (See RCG 29.8.95).

[3] In 1890 alone over 5100 tons of tin-stone were sold in this way by little groups of tributers from Poldice and other Gwennap mines, and it was said that since Clifford Amalgamated Mines had closed there in 1870, over 120000 tons had been treated from its old burrows.

with conditions in some parishes described as 'positively appalling.' Emigration still provided an unwelcome outlet for the young and the healthy and 'every Friday sees dozens of stalwart young miners leaving Camborne and Redruth en route for Johannesburg,' with others bound for destinations as far apart as Cripple Creek in Colorado and Kalgoorlie in western Australia.[1] Even these men going abroad found there an echo of some of the troubles that Cornwall was suffering from, for reaction was reported against the employment of Cornishmen on the alleged basis that those who had learned their trade in Cornwall were too old-fashioned and set in their ways, and 'would not deviate from the time-honoured usage of their native land.'[2] This was an extension of the old adage that a good bal makes a good captain, for the reverse was now beginning to be applied to Cornishmen in some mining fields abroad. Certainly, through no fault of its own, the county had fallen behind other parts of the world and the insular nature of the Cornish—backed by a once justifiable conceit that no-one could work or manage mines like them—had brought about a reaction. Their over-confidence, and a disinclination to accept anything except the ways of their forefathers, were the penalties of the earlier prosperity when Cornwall had the easy-going mastery of the metalliferous mining world. Now times had changed and there was further comment by the *Mining Journal* of the benefits that might accrue from the importing of new blood and new ideas into Cornwall: 'let them feel that there is talent elsewhere than in the limits of their county; let them be willing to employ it and great advantages must result.'[3]

It was perhaps as a by-product of this outburst against out-dated Cornish mining methods and old-fashioned Cornish mining men that part of South Condurrow was fitted up as a model mine at this time for the up-to-date instruction of students. This was for the Camborne Mining School, and their example was followed by the Truro Mining School which took over Killifreth at the end of 1897 when it shut down,[4] whilst later the Penzance Mining School, founded as recently as 1890, took over Botallack for the purposes of practical instruction.

In the meantime there were signs of settlement in the Wheal Agar dispute, for Lord Robartes finally intervened in May 1896 when the water from South Crofty began to hinder the working of North Tincroft. Another month and the Agar and East Pool engines were again at work, tackling the enormous volume of water that had accumulated in the two mines and which constituted a growing threat to the northern section of the adjoining Carn Brea group. In January 1897 the South Crofty engine was also re-started by East Pool and Carn Brea jointly to help in forking,[5] whilst water was also hauled in skips by the winding engines of all three mines as a means of assisting the work. In December 1896 the Agar engines

[1] MJ 13.6.96. [2] MJ 3.10.96. [3] MJ 18.7.96.
[4] MJ 22.1.98. [5] RCG 18.2.97.

were for sale,[1] to be acquired by East Pool at long last for the sum of £4000. A new lease was granted, with the company re-constituted as East Pool & Agar United, still under the cost-book system. Other sales of the year were Wheal Peevor and West Peevor, both of which had been idle for several years, and the prices realised there by the machinery told its own tale of the mining depression.[2]

Although it might be said that, with the Agar and East Pool deadlock resolved, the worst of the crisis had been weathered, the following year (1897) saw little or no improvement in the price of tin. Moreover, two more mines went under, Blue Hills in January, and Killifreth in November. The former was one of the last of the old school of Cornish cost-book mines, a survivor from the days of gunpowder and hand boring, when tin prices were high and lodes were rich. And yet the mine was far from a small one, with one hundred employed there up to the last.[3] When the adjoining Penhalls—which had latterly been worked to all intents and purposes as one mine with Blue Hills—was abandoned in 1884, the latter had unsuccessfully applied for the sett, intending to work it above adit without steam power. It was, however, another nine years before the lords allowed Penhalls to be worked in this way. Meantime a useful subsidiary source of income to Blue Hills came from working the river for stream tin from 1884 onwards, to catch the waste from the greatly extended West Kitty dressing floors upstream. In 1894 East Blue Hills sett and machinery was acquired (for the sum of £107 12s. 0d.) including the lower set of water-wheel stamps in the valley which East Blue Hills had leased from them in 1881. The final meeting of the Blue Hills company was held at Carvedras at the end of 1896, its whole history since it had been re-started in April 1869 having been one of calls upon the shareholders' pockets, totalling £42661 in almost thirty years: 'a curious mine working quite on the old-fashioned principle; such tin as was discovered was taken away at once.'[4] The equipment was certainly old-fashioned to a degree, with no such new-fangled contraptions as air compressors, rock-drills, stone-breakers, Californian stamps or steel skips. The state of the 32 heads of Cornish stamps was typical of the whole mine, sixteen being driven by a ramshackle water-wheel which suffered from a chronic shortage of water in dry weather, the other sixteen by an old whim engine which was available for working them only at night when not in use for winding.[5]

[1] Comprising 90″ pumping engine, 34″ (88 heads) stamps, 24″ whim, double cylinder 12″/22″ horizontal whim, double 12″ capstan, 20″ and 18″ compressors. (RCG 24.12.96.)

[2] For Wheal Peevor, with 60″ pumping engine, 36″ stamps and 22″ whim, the best bid at the public auction was £435 against a reserve price of £650. For West Peevor, with a 50″ pumping engine, 20″ stamps and 22″ whim, the only bid was £220 as against a reserve of £600. (RCG 1.10.96, 15.10.96.)

[3] RCG 14.1.97.

[4] MJ 23.4.98. The Official Receiver was evidently suspicious or puzzled by the mine being continued for over a quarter of a century as a losing concern, for he was quoted as reporting that 'the whole business appears to have been a legitimate undertaking' (MJ 24.9.98).

[5] The miners' dry was in such an unsatisfactory and dilapidated state that a complaint was made by H.M. Inspector of Mines in March 1894. Another was thereupon erected, using old materials garnered from the old East Blue Hills section.

On such a basis, even producing tin of the quality for which Blue Hills was noted, it was obvious that it could not survive in the 1890's in competition with Straits or Bolivian tin. The wonder was that Blue Hills survived as long as it did—a perfect example of the tenacity of Cornish miners and mine adventurers against financial adversity.

The disappearance of Killifreth from the ranks of the remaining mines was in some ways a greater loss, for this was a promising mine which had paid dividends in previous years at times when most others could barely meet costs. Throughout the worst of the 1870's, prospects—and what was more important, sales of tin—had been sufficiently good to warrant the putting in of a 50″ engine in 1873 and a 32″ stamps engine in 1876. In 1891 a new lease had been granted, whereby it was stipulated that an engine of 70″ or larger should be erected on Hawke's Shaft in return for dues of 1/40 for three years and 1/24 thereafter, and an 80″ engine was thereupon acquired from the defunct North Treskerby.[1] In 1896 the mine was far from being exhausted and was still shallow; an attempt was then made to turn it into a limited company to secure capital for rock-drills and other essential development, but there were disputes between the old cost-book shareholders over this. Although the matter was reported to have been settled by arbitration in March, the company was wound up in July and the mine sold up in November, merely for the value of its machinery for scrap-iron and for re-working the levels above adit for arsenical mundic.[2]

By the end of 1897, it was possible to count on one's fingers the mines left at work in Cornwall. In place of the several hundred a quarter century before there were now only nine survivors. In Breage,[3] Wendron, Lelant and St. Ives, not a single working mine remained. In St. Just, only Levant survived and here as much on copper as on tin.[4] In this once premier tin mining district the roll of recently 'knacked' bals was a sad one—Botallack, the promising North Levant stopped in 1891, St. Just United, and the ill-fated Wheal Owles, where all work had ceased after the disaster. In the St. Blazey area, mining was all but a thing of the past, its future regarded as so hopeless by the chief mineral lord that shafts and dumps had been obliterated wholesale and the ground put down to forestry. Farther east, Drakewalls lay completely derelict and Phoenix United was being wound up in the hands of a liquidator. In St. Agnes there were only West Kitty and the suspended Wheal Kitty, whilst in Redruth not a mine remained. In Camborne and Illogan, the last citadel of Cornish tin, a mere four companies were in full operation—Basset Mines, Wheal Grenville, Dolcoath, and Carn Brea & Tincroft, whilst at East Pool activity

[1] RCG 30.4.91. This engine was started July 1893.

[2] RCG 22.7.97, 25.11.97.

[3] There had been great hopes for a re-working of Wheal Metal and Flow, commenced about 1881, but this closed down in or about 1896.

[4] In the first four months of 1896, for example, Levant sold 127 tons of black tin for £4727 and 1432 tons of copper ore for £6129. (RCG 28.5.96.)

was still confined to unwatering and to work at shallow levels. The diminution in the size of the industry in the previous few years was evident in the numbers employed; 5193 in 1898 (2749 underground and 2444 at surface) compared to 12049 in 1891 (6156 underground and 5893 at surface).

These few surviving big mines were relatively well-equipped and were held by shareholders who felt that if they hung on, the worst would pass and normal times would return. At East Pool & Agar both mines were drained to the 200 fathom level by early 1898 and something approaching normal working could be resumed, whilst at Basset Mines work was proceeding apace on the new Marriott's Shaft, cutting it down and installing new machinery. The outlay here was £18630 in 1897 alone, and after long deliberations and visits by Oats and the mines' engineer to South Wales and Germany to study the latest pumping practice, it was decided to put in a compound 40″/80″ Davey Cornish engine for pumping to 500 fathoms depth, with room in the engine-house for a second subsequent engine for the next 500 fathoms thereafter.[1] The only set-back at Basset had been the failure to acquire West Frances as part of the group. It had been agreed to purchase the sett in August 1896[2] but a dispute then arose over the question of dues. The mineral lord wanted all the West Frances ore stamped and dressed separately and not as part of the whole Basset output but this could not be agreed to by the mine company as it was obviously uneconomic, and in consequence the engines on the sett were sold off piece-meal early in 1898[3] after further fruitless negotiations.

At Dolcoath, trouble was brewing over the new shaft, with criticism over its size (18′ diameter) which was considered excessive, as well as the slow rate at which it was being sunk. After twelve months it was only down 85 fathoms from surface and had cost £14133 for less than one-sixth its proposed depth.[4] Second thoughts prevailed and its sinking was suspended early in 1898. All efforts were thereafter expended upon the western sections of the mine, newly extended by the acquisition of the old Carn Camborne and Camborne Vean the previous year. The latter, 326 fathoms deep, was thought to add greatly to the future possibilities of Dolcoath and was already partly drained by it.

Wheal Grenville, during 1897, had been turned into a limited company, the mine paying costs as it had been doing throughout the slump on the strength of good ore ground providentially opened up in 1890.[5] Even here

[1] A triple expansion Cornish engine was even considered at this juncture but was dismissed as too expensive.

[2] RCG 13.8.96.

[3] Comprising 60″ and 40″ pumping, 36″ (80 heads) stamps and 24″ whim. (RCG 10.3.98.)

[4] RCG 2.9.97.

[5] In that year the Grenville adventurers had decided to re-equip the mine throughout, acquiring four engines (including one with 90″ cylinder) from the ill-fated re-working of Tresavean, plus one other—an outlay necessitating a call of £15000: 'the mine being opened up with more regard to the future than for present profits.' (RCG 8.1.91). From 1872 to 1885 Grenville had made serious losses (£23375 from 1872 to 1876 when a new company took over; calls of £31650 and dividends of £12750 from 1877–1885) but thereafter to 1901 a profit of £65421 was earned. (Fortescue Records, DDF 388, CRO, Truro.)

there was difficulty in making ends meet, for pumping charges were particularly high and the total coal consumption on the mine was 10 cwt. for every ton of lode-stuff broken and raised. Wheal Grenville's becoming a limited company left only three cost-book concerns in the county[1]—East Pool & Agar, Levant and West Kitty. The latter had made a small profit in 1897 but the adjoining Wheal Kitty was on the brink of being totally abandoned, the pumps and all deep mining already stopped. In the east, Phoenix United finally disappeared in November 1898 when the engines and other machinery were disposed of. This had been the greatest tin mine anywhere east of Redruth and at one stage had been the county's third richest mine. From 1894, crippled by low tin prices and a lack of money with which to carry out the requisite development, Phoenix had dragged on month by month, paying the best part of costs each time by lowering wages, cutting out all dead work, and by rigid economy. In the three years of liquidation 1895–8 the mine sold 800 tons of black tin for £30518[2] and with tin at £10 per ton more there would have been no deficit. Alas, this was no compensation for the shareholders who had persevered here for so long and this was to be the end of tin in the Caradon area for a long time to come.

The diminished scale of mining had its inevitable repercussions amongst the smelters, and during 1898 the old established Carvedras works in Truro was closed down by the Consolidated company. This had been one of the premier smelting-houses in Cornwall and its closing meant that there were no longer any furnaces in or around Truro, once the focal centre of Cornish tin. Coosebean, the Truro works, Trethellan, Treyew, Calenick, and now Carvedras, had come and gone. The harsh facts of the case were that smelting at the latter was uneconomic and keeping the works open was incurring a loss. Accordingly it was decided to concentrate all Consolidated's activities at Chyandour, and the Carvedras lease was terminated at Ladyday 1898, with £1200 paid by the company to cancel the remaining fourteen years still to run.[3]

By the end of 1898, the price of tin had risen somewhat, and there were signs that the end of the depression was perhaps in sight. Demand for arsenic and wolfram (tungsten ore) had also sprung up and these two by-products were to prove useful sources of additional revenue to many mines in the ensuing years. Wolfram in particular brought £70 a ton, and its output together with the semi-metal proved especially valuable in

[1] 'Four years ago there was not a single mine worthy of mention which was not worked under the cost-book system.' (MJ 7.5.98.)

[2] MJ 11.6.98. From 1878 to 1897 approximately 800,000 tons of ore had been milled, producing 11615 tons of black tin. This, moreover, was produced without sinking the mine by so much as a foot. In fact, subsequent to 1867 there was no increase in the depth of the Phoenix workings, and only 10 fathoms in the seven years prior to this. (Bolitho Records, DDX 104, CRO, Truro.)

[3] At Carvedras 1511 tons of black tin had been smelted in 1894, 1668 tons in 1895, and 1585 tons in 1896, including a considerable proportion of Bolivian barilla. (Bolitho Records, DD RG 1/37, CRO, Truro.)

Like Treloweth, this later became a creamery, occupied by United Dairies in 1920. Today it is a warehouse and the external appearance is not greatly changed from its smelting days.

assisting Tincroft and East Pool & Agar from this time on. The rising price of tin brought the flotation of part of West Kitty in 1899 under the title of New Polbreen, a small enough venture but the first new mine that had been floated in Cornwall for many years. This was followed by the resumption of operations at Tregurtha Downs in August, at South Condurrow in October, as well as by the re-starting of Gooninnis Mine, St. Agnes.[1] Late in 1899 operations were also resumed at South Crofty,[2] with the pumping engine set to work in May 1900 and a start made on sinking a new vertical shaft, named Robinson's, on August 7.

A disquieting feature of the slight but marked revival had been the lack of capital forthcoming from outside sources, except in a very few cases. It was said that '. . . it is practically impossible to induce outside capitalists to consider, much less to embark in, even the most promising and sound mine undertakings in Cornwall or Devon'[3] and there was little doubt that the investing public had lost confidence in Cornish mining, to the extent that no market existed whatever for shares in 'progressive' (i.e. developing) mines. Some of the reason for this lay in the greater attraction to investors of foreign mining fields, in particular those producing gold, yet beyond this there was no doubt that Cornwall had gained a bad name with investors generally. In the opinion of many up-country, Cornwall as a mining field was finished[4] and to combat this state of affairs, as well as to protect and promote the general interests of local mining, the London and West Country Chamber of Mines was formed during the year. Although its formation could do little other than good, it received virtually no support from the mineral lords and it was apparent that even the traumatic experience of the mid-90's had not closed this division in the ranks. At Dolcoath the unfair dues borne by the mine were evident enough when the results for the first half of 1901 were announced. For the shareholders there was a five per cent dividend only, whilst the Basset estate as mineral lords received almost twenty-five per cent of the gross profit Dolcoath had made—a levy 'that is probably not endured by any Limited Liability mine outside England.'[5] Work on the new Williams Shaft had been re-started, with a new 85′ high lattice steel head-gear erected in 1900 and the sump down 192 fathoms from surface by the summer of 1902. The second half of the year at Dolcoath brought to the shareholders (then

[1] Here, on the old Wheal Turfrink sett, a new shaft was started from surface. (RCG 13.4.99.) A 50″ engine from Penhalls was later erected, together with a new double cylinder horizontal whim from Holman Bros.

[2] RCG 23.11.99. New Cook's Kitchen had also then been acquired, for £650 plus an exchange of shares.

The 80″ engine which restarted pumping at Tregurtha Downs at this same time was destined to be acquired by South Crofty for Robinson's Shaft three years later when the former mine closed in August 1902.

[3] Report of H.M. Inspector of Mines for the South Western District, 1900 (quoted MJ 16.6.00).

[4] By 1900 output was a very small percentage of the world total, 4268 tons out of 82400. By 1904, even after a slight measure of recovery, it was still only 5 per cent of world output, to which the Straits then contributed approximately 64 per cent, Banca and Billiton 16 per cent, Bolivia 10 per cent, and Australia 5 per cent.

[5] MJ 1.3.02.

1650 in number) the shock of no dividend, a piece of news that augured badly for Cornwall as so much attention centred on the mine. The falling off in value of lode-stuff raised had brought production cost per ton to over £56 and the question was bound to arise sooner or later as to what grade of ore it would pay the company to treat.

Other unwelcome news from Dolcoath was of the severe ankylostomiosis (hook-worm) epidemic there. This was a mining disease hitherto new to Britain but known abroad—particularly in the Sicilian sulphur mines—giving rise to intestinal worms bred by the insanitary habits of the men working underground. At Dolcoath it arose from the lack of anything like adequate underground sanitation for the considerable numbers employed, allied to high temperatures and restricted ventilation. Medical evidence showed that it had first occurred in 1894–95,[1] probably carried by a miner returning from a spell of working abroad, and that it had been at its worst from about 1897–1900. Known in Cornwall thereafter as 'Dolcoath anaemia,' its main symptoms were pallor of the complexion and shortness of breath, neither of which were likely to arouse suspicions amongst managers with no experience of the disease. Thus it occurred unsuspected for a long time and its presence was chiefly detected at Dolcoath by the inability of the managers to get any sustained work done underground. In 1903 it was said to have 'obtained a firm hold in Cornish mines,'[2] whilst at Dolcoath practically every underground worker was infested.[3] Steps were, however, taken to bring the disease under control. No deaths occurred but the disease had caused fatalities abroad in mines on the Rand and elsewhere, as well as amongst men working in the Mont Cenis and St. Gotthard tunnels.

The more deadly phthisis was also prevalent in Cornish mining at this time, the increase in the death-rate from lung diseases since 1890 being most marked, especially amongst men under fifty. Ladder climbing was now virtually a thing of the past and medical investigation showed that rock dust was the cause—aided by bad ventilation and the poisonous fumes of nitro explosives—and this was confirmed by the prevalence of 'miner's disease' amongst those who had spent time working in the gold mines of the Rand.[4] There, in dry mines with enormous numbers of rock-drills in use, the mines were akin to flour mills, with a dust problem far worse than anything Cornwall experienced. The use of respirators was dismissed as impracticable and a water-spray system when drilling, as well

[1] According to the *Mining Journal* it had existed in Cornwall in the early 1850's 'but it was not recognised.' (MJ 2.4.04.)

[2] MJ 25.4.03.

[3] MJ 17.2.12.

[4] The damp climate of their native county played havoc amongst these returned miners. It was not difficult to trace the trouble back to its root cause in the rock-drill, for those who had worked them had, at this date, a poor chance of survival to old age. In one medical survey carried out in the Camborne area it was shown that the death-rate before the age of forty of those who had worked rock-drills was double that of other underground workers in the county. (See *Trans. Inst. Mining Metallurgy*, Vol. XIII, 1903–4.)

as after blasting, was introduced to control the problem not only in the Rand and in Cornwall but everywhere that rock-drills were used in metal mining.

Although the price of tin had risen by 1903 to £130 per ton, Cornish mining was still far from recovered from the ordeal of the 1890's. The Cornishman's cry then had been that all would be well given only a good price for the metal, yet now it was becoming apparent that something more, in the form of new capital, was needed. The opening, neglected by capitalists in this country,[1] was seen by French and other continental interests who at this time expressed interest in the financial possibility of Cornish tin. On this score and the rising price of metal, East Pool & Agar shares rose in a mere four months of 1903 from 15s. to 60s. and South Crofty from 9s. to 40s., although such rises were by no means general. At Dolcoath considerable changes had been made; a big new air compressor was put in to enable more drills to be used, a locomotive was bought for hauling ore over a 22″ narrow-gauge railway from Stray Park to the stamps, and widespread changes were made in mechanizing the dressing floors, both to cut down on labour costs and to reduce the loss of tin slimes into the Red River. No less than 526 rag frames, 38 revolving frames and 80 frue vanners[2] were put in during 1902–3, reducing the numbers employed in tin dressing from 252 in 1897 to 148 by 1903 without any decrease in output. In fact, from 1894 to 1904, to keep sales of black tin constant, it had been necessary to double the output from underground, for Dolcoath's average produce fell in this period from about 80 lbs. to 36 lbs. per ton.[3]

A major blow to the mine's future came in January 1903 with another underground fire near New East Shaft.[4] At first this was thought to have done little permanent damage but by March it was evident that the fire had, in fact, done grave injury. New East Shaft was rendered unusable

[1] Despite the lack of capital for Cornish mining, two Redruth-based companies were floated in 1905 alone to open mines in the Federated Malay States.

[2] Frue vanners and Wilfley tables—high-speed shaking mechanical concentrators—were two of the great improvements made at this time in the technique of profitably dressing low-grade tin ores. Apart from being simple, easily maintained and of large capacity, they could handle coarse sands equally as well as slimes.

[3] 'With a falling produce, Dolcoath, as the bell-wether of Cornish mining, has exercised a depressing influence upon the industry.' (MJ 23.4.04.)

As a result of this necessity to reduce working costs, considerable thought was put into revision of Dolcoath's system of dressing etc. By 1908 the old Cornish stamps had been entirely replaced by 60 head of Californian, plus two of Husband's stamps—the latter equivalent to 20 head of Californian. Before going to these, the ore had been screened, selected and crushed to about 1½″ size in the stone-breaking and sorting sheds. After stamping, the tin pulp went to the Frue vanners, and thence the concentrate went to be calcined and re-washed before final preparation for the smelter.

[4] Unlike the earlier fires at Dolcoath, this one was a serious conflagration and it was considered fortunate that the 400 men underground could be warned by the system of bells installed throughout the mine. All were safely at grass within two hours of the discovery of the fire at 8 p.m. The fire had to be left to burn itself out for there was nothing that could be done except to leave the wood to smoulder until it was consumed or until it ceased to burn from lack of oxygen, and it was five days before the workings were clear of smoke to enable the first party to go underground. (MJ 31.1.03). Apart from the previous fire in 1897, it was reported at this time that 'during the past six years four small outbreaks have taken place.'

and the remaining section of ground around it—at this date the most productive in the sett—where the great run of 1893 had destroyed all below the 302 fathom level, was now rendered quite unworkable from the 290 fathom level upwards. Rumours of the fire's effects were worse even than was to be the case and Dolcoath £1 shares (15s. paid) could be bought in March at 3d. each.[1] In April a reduction of one half in the lord's dues was granted to help the mine financially and the company turned gallantly towards finding what new ore ground it could in the other parts

ANNUAL OUTPUT OF TIN IN CORNWALL, AND AVERAGE ANNUAL PRICE 1895–1905

	Black Tin (tons)	Metallic Tin (tons)	Price per ton (black tin)	Price per ton (metallic tin)
1895	10582	6627	£36	£ 97
1896	7657	4833	£34	£ 64
1897	6214	4039	£36	£ 65
1898	6586	4281	£39	£ 74
1899	5659	3678	£69	£127
1900	5991	3894	£76	£138
1901	6533	4246	£63	£121
1902	6690	4014	£68	£121
1903	6471	3882	£72	£129
1904	5991	3778	£71	£128
1905	7174	4450	£80	£144

of the sett. Certainly the old mine was in low water at this time and the £53421 outlay on the new shaft must have seemed money ill-spent in view of the falling-off in value of the bottom of the mine.[2]

Moreover, if Dolcoath closed, the outlook for men wishing to emigrate was becoming bleak. The Rand had provided a ready outlet for the surplus mining population of Cornwall for some years but no longer could it be depended on. As the *Mining Journal* put it, 'since 1884 the average Cornish miner has regarded the goldfields of South Africa as a sort of Eldorado or Tom Tiddler's Ground to which, when tired of cold pasties, an eight-hour shift, and 18s. per week, he might retire for a few years, and where his voluntary exile might bring him anything from £20 to £30 per month.'[3]

Dolcoath persevered however, and elsewhere in Cornwall there was evidence of a new spirit in mine working, within the confines of locally provided capital or with money taken from current revenue. West Kitty, still a cost-book company but with a set of shareholders from whom the faint-hearted had long since been weeded out, was putting a considerable sum of money into the new Wheal Friendly shaft, north of their own sett. Friendly had been abandoned in 1896 and was taken up by West Kitty in

[1] MJ 12.3.03.

[2] The half-finished Williams Shaft was described as 'an expensive white elephant' in the *Mining Journal*, (16.4.04.)

[3] 13.2.04.

1900 primarily because of the pending exhaustion of the Reynolds and Thomas's sections of the old mine, and it was decided to develop Wheal Friendly as the main future of the mine. There was a wide divergence of opinion amongst the West Kitty management as to the wisdom of this but matters went ahead, the shaft itself being 12′ by 7′ and everything in connection with it ' laid out as for a permanent mine.'[1] By 1904 it was down to 76 fathoms below adit, and a compressor and rock-drills were put in during the year.

The prices of tin and wolfram were now high enough to warrant attention being paid to re-working old burrows and heaps of tailings from former mining operations. The leading example of this was Gunnislake Clitters where a considerable sum was made from crushing and dressing the old Greenhill Burrow—an immense heap of burnt leavings from which only arsenic had been extracted.[2] A new and up-to-date mill was put in with 25 heads of Californian stamps and with Wetherill electromagnetic separators used for the extraction of the wolfram. This modern plant at Clitters contrasted strongly with other mines in the west where dressing plant was still as simple as ever, without any such refinements as frue vanners or Wilfley tables, merely allowing the pulp coming from the stamps to settle in long launders from where it was dug out for final treatment in buddles and kieves. Under such circumstances an enormous waste of slime tin occurred and the whole arrangement was quite unsuited to the mass treatment of low grade ores. Evidence of this was the amount of tin still running to waste in the Red River and Basset River—sales from the stream works on these realising over £37000 in 1904 alone.

The winter of 1903–4 proved to be an exceptionally wet one, adding the expense of extra pumping to working costs and even drowning out the bottom levels of some mines. At Basset Mines, for example, no tin ground was available below the 190 fathom level for almost nine months and although attention was turned to side lodes during this time a loss was made in 1904 of £2194 despite high prices for tin. Here as elsewhere the steadily decreasing lode values were causing much anxiety and the mine needed either to make a rich strike or to find other means of increasing output. East Pool made steady progress at this time, December 1903 seeing the first use of electric trams to carry ore over the public street tramway to the stamps at Tolvaddon. Prior to this 8d. a ton had been incurred hauling the whole output, something like a thousand tons a week, by horse and cart and the saving was said to be over £500 a year. The earlier telpherage scheme had been unsuccessful, partly from construction difficulties and partly from the difficulty of its crossing the turnpike road.

[1] MJ 11.2.05.

[2] The success at Clitters United was not repeated by a company at the old United Mines in Gwennap. In 1900 an 80″ engine was put up here on Poldory, together with a 34″ engine and 120 heads of stamps. The burrows were processed and a certain amount of ore raised from the shallower levels but by 1905 all the equipment was for sale and a heavy loss reputedly made. The same company had also started similar work at West Seton, in the Tuckingmill valley.

The giant stack of the Truro Smelting Works, as seen from the Malpas Road about 1895. This is virtually the only surviving visual record of this works which was established in 1816 and after a chequered career under various owners, was finally closed down in 1871.

Ingots of tin from Trereife smelting works stacked on Penzance quay awaiting shipment. Barrels of grain tin at right.

Calenick village, near Truro, with a partial view of its famous smelting house, owned by the Lemon and Daniell families, and then by the Michells and Bolithos. The road over the bridge was once the main coaching road to Falmouth. The works was closed in 1891 and has largely disappeared with the exception of its clock tower.

Towanroath Shaft at Wheal Coates, St. Agnes, about 1910. In this working a horizontal engine for pumping was installed in the corrugated iron structure behind the old engine-house. The nose of the balance-bob and the rising main beside it mark the mouth of the shaft from which the skip-road can be seen curving up and out to the dressing floors higher up the cliffs. The men are engaged in ' stocking ' one of the pump rods.

A ' core,' or shift, of miners about 1890. Note the numerous candles and water-kegs and coils of fuse. The wooden structure houses the landing brace of a drawing or whim shaft. The mine is probably Dolcoath (Valley Shaft) or South Condurrow.

In the Agar section little was being done other than pumping, the inclusion of the sett with East Pool being more by necessity and *force majeure* than anything else. At Wheal Grenville, one of the very few mines which had paid any dividend in 1904, a new lease was approved in 1905. Its terms were of widespread interest, for mineral lords and their dues were ever to the fore as a topic, and the whole subject was one in which renewed interest was being taken. Under the new terms at Grenville, no dues were to be paid when the mine was making a loss, whilst 1/5th was due when there were profits. This was a tolerably equitable arrangement which established the principle long sought for that dues should not be based on output alone, irrespective of whether losses were or were not being made. With this new lease in hand, arrangements were made to extend the Grenville sett and in July 1906 Grenville United Mines Ltd. was formed,[1] to include South Grenville, East Grenville, Polgine and the part of South Condurrow that had not passed to the Camborne Mining School.[2] Of these four, South Condurrow alone was a going concern, having been re-started in 1900 but making a loss of £100 a week and with its engine too overloaded to sink the mine deeper.[3] At Grenville itself at this time, Fortescue's Shaft was down to the 350, and Goold's to the 300. Here, as elsewhere, there was a severe shortage of skilled miners and it was said that 'the vigorous miner of forty, training his youngster to follow his calling, is almost as rare as the dodo.'[4]

These were sentiments to be expressed by more than one mine manager in the ensuing few years as the pendulum of Cornish mining began its swing back from slump to boom. For the first time for years there was an inflow of capital into the county for mining purposes, ending the period during which hardly a penny piece had been offered or made available on any terms to the industry. In almost every one of the old mining districts new mining ventures were started, by syndicates and companies, by individuals and speculators. To quote one Cornishman—who knew tin as well as anyone—writing of this period thirty years later, 'setts were acquired which landlords had given up hope of ever again seeing again in demand,'[5] whilst by May 1906 there were repeated articles headed 'The Tin Boom' in more than one local newspaper.

This optimistic burst of activity, brought about by the high price of tin, bore a strong resemblance to the boom of the early 1860's although fewer

[1] To this date, Wheal Grenville, from its commencement in 1845, had paid dividends totalling £127359 on an outlay (calls) amounting to £111750. (MJ 9.5.08.)

[2] This latter had been given the name of King Edward Mine in the year of his Coronation (1901) and was described in 1903 as 'about as pretty a working model of what a Cornish mine should be as could well be desired.' (MJ 14.3.03). In 1909 the Redruth and Penzance Mining Schools were amalgamated with the one at Camborne (that at Truro having ceased to exist) to form the School of Metalliferous Mining (Cornwall), now world-famous as the Camborne School of Mines.

[3] This 60″ engine (Marshall's) had come to South Condurrow from Wheal Jane in 1888, erected at the latter second-hand in 1881 and of great age. According to the *Mining Journal* (27.2.04) 'during the past ten weeks Marshall's engine alone has pumped over 120,000 tons of water.'

[4] MJ 2.12.05.

[5] Horton Bolitho, letter dated May 1936, (Bolitho Records, DDX 104, CRO, Truro.)

but far larger companies were to be involved this time. The new spirit of optimism was also engendered by glowing forecasts of the possibilities of cheap pumping by electricity.[1] Far more dangerously, it was also presumed that by doing things in a big way and on a large scale, success was assured—an expensive fallacy for metal mining still remained, as it does today, one of the very few business enterprises where success could not be assured even by the most abundant capital, the finest skills and the utmost determination. Amongst the larger ventures in or shortly after 1906 were Condurrow United, at Camborne; Botallack Mines Ltd., floated in November with a capital of £150,000 to re-work Botallack and Carnyorth; Boscaswell United, to re-work the old Boscaswell Downs and Pendeen Consols; Falmouth Consolidated Mines Ltd.,[2] comprising the old setts of Wheal Jane, West Jane, Nangiles, Wheal Falmouth & Sperries, covering in all no less than 948 acres; Tresavean—despite an earlier re-working for copper and tin from 1881–87 which had resulted in a reputed loss of £100,000;[3] Phoenix United, at Linkinhorne; Great Dowgas, near St. Austell; and Wheal Vor, using electric pumps in Borlase's Shaft where the great 100″ had worked half a century before. On a smaller scale the following were also opened or prospected: Wheal Sisters, South Phoenix, North Crofty,[4] Ventonwyn, Bunny Mine, Wheal Coates, New Wheal Eliza, Tregrehan Consols, Polbreen Consols, Gooninnis, Boswin, South Parbola, Polmear, Commerce, Providence, Gwallon, and others. During the year North Levant & Geevor Ltd. was registered, with the first sales of black tin made in August, whilst South Crofty was formed as a limited company also. Amongst the older mines, West Kitty persevered at the new Friendly Shaft, the loss being about £4000 a year and the company sorely in need of new capital. One decision was to cut down cartage costs[5] by installing two heads of pneumatic stamps at Friendly in lieu of the existing twenty heads of Californian stamps that the company worked on its Jericho dressing floors over a mile away at the head of Trevellas Coombe. The old Reynolds Section of the mine was practically

[1] Although the first use of electric pumps in Cornwall was made in 1905, one of the Dartmoor tin mines—Hexworthy—had considered them sixteen years earlier. Working was entirely by water power but electricity was suggested in 1891 'in the event of further motive power being required.' (RCG 19.11.91.)

[2] As 'Falcons' these became favourite market shares for a brief time.

[3] *Trans. Inst. of Mining & Metallurgy*, Vol. XXI, 1911–12.

[4] Here an old shaft was re-opened to 200 feet, enlarged to 8′ x 5′ and made vertical, equipped with steel headgear, cage and cars. An electrically driven hoist and air compressor were installed, supplied by a suction-gas generating set which also ran the mill. Arsenic was intended to be one of the main staples and flues and chambers built entirely of firebrick and strapped with steel beams were put in —'the most elaborate ever built in the county.' London & West Country Chamber of Mines, Records 1904–8.

[5] In 1906 the company was paying 1s. 4d. per ton cartage from Friendly Shaft and 11d. per ton from Thomas's and Reynolds' section, to the dressing floors. The use of a Foden steam wagon was proposed at an estimated cost of 8d. per ton.

Another of West Kitty's problems was that of coping with the numerous mineral owners under whose separate grants the mine was worked. There were no less than thirty of these involved in all and this was said to be a situation unparalleled in any mining field of equal size in the world. (MJ 11.2.05). In mining circles in St. Agnes, lord's dues were said to be a topic for discussion almost equal to the weather.

worked out and let solely to tributers, although working in West Kitty as a whole was still on a considerable scale, with a cage started in 1905 at Thomas's Shaft to raise and lower the men mechanically for the first time. 358 were then employed in all, but from the gradually declining state of the mine this was down to less than 300 by December 1907.[1] The other surviving St. Agnes mine was transformed in 1906 into Wheal Kitty & Penhalls United Ltd., to acquire the old cost-book Wheal Kitty (previously sold into new hands in 1904) plus the Penhalls sett to the north and the briefly re-tried Gooninnis sett to the south. A new central shaft, Sara's, was commenced in October 1906, 15′ x 7′ within timbers, and on this was erected a 50″ engine.

A late starter in this boom[2] was St. Ives Consolidated Ltd., formed in 1908 to re-work St. Ives Consols, Wheal Trenwith, Rosewall Hill & Ransom, Giew and other setts. Chief activity was concentrated on the Giew section where a small dressing plant was erected. Here pumping was commenced using, as an interim measure, an electric motor to drive the old Cornish pumps which still remained in the shaft from the previous working—a ingenious arrangement which nevertheless worked tolerably well. At the main Consols section there was difficulty over the supply of water for dressing purposes and of its disposal thereafter, and the company decided later to erect the mill ' about 1½ miles away on the sea coast and to utilise sea-water for milling and dressing purposes '[3] with transport via an aerial ropeway from the various sections of the mines.

Despite the activity and the considerable capital outlay in these many mines, all did not go well. At Condurrow United, where the choice had been the traditional Cornish engine for draining the mine[4] and after a fine all-granite ' cathedral ' of an engine-house had been erected for it, a long series of chokes and blockages in the 280 fathom deep shaft delayed matters, consuming time and money alike. At Wheal Vor, where there had been a pale echo of the re-working of 1852 by the lighting of bonfires to celebrate the re-starting of the mine in May 1906,[5] the electric turbine pump gave constant trouble and was eventually burnt out.[6] This resulted

[1] The same month the old Wheal Coit was added to West Kitty's sett to extend it westward. To this date West Kitty had paid dividends of £155700 on calls amounting to £25800—the latter sum including the £22200 that had been expended there before the West Kitty company commenced in 1879. (MJ 23.5.08.)

[2] Something of the enthusiasm of this period for a resurgence of Cornish mining is evident in E. W. Meyerstein's *A Key to Cornish Mining* (published by the *Mining Journal*, 1907), the author being a director of many of the new flotations. Comparison with South African mines was made, optimistically summed up in the words that ' as compared with Cornish mining, the Rand is a field of yesterday, with a life that may end tomorrow, while Cornish mining was born in a prehistoric age and its life limit still remains to be defined.'

[3] Report dated 6 March 1913, Acc. No. 547, CRO, Truro.

[4] This came from the short-lived re-working of the Gwennap United Mines, an 80″ engine originally built by Harvey & Co. for West Chiverton Mine in 1867.

[5] WB 24.5.06.

[6] In a reminiscence, written in 1936, Horton Bolitho likened this attempt to un-water old Wheal Vor to ' trying to bale out the ocean with a teaspoon.' He considered the re-working a hopeless proposition ' . . . on the logical grounds that if in the second working 1853–60 with £200,000 capital, the mine was proved hard and poor, now after nearly 40 years under water it could hardly have got rich . . .' (Bolitho Records, DDX 104, CRO, Truro.)

in the flooding of the mine—which, after three years' work, had been drained to the 144 by December 1909—as there was no stand-by. At Providence Mines by June 1908, £25000 had been expended in putting in a new mill and in clearing chokes and caves in the old workings and the adit to the sea, without any production whatsoever;[1] whilst at Botallack the task, estimated to take three years, of sinking the new vertical (Allen's) shaft required under the terms of the lease was consuming a vast amount of capital. The intention was to get below the old Wheal Cock and work this section in depth. The shaft itself was 20′ x 6′ with five compartments, sunk entirely by hand for the first 500′ or 600′ using sixteen men on each of three daily shifts.[2] 180 men were employed on the mine as a whole, causing a local labour shortage that would have affected Levant most seriously had the two companies agreed not to employ each other's labour. At surface four expensive gas engines were installed to work the electric pumps[3] as well as 40 heads of Californian stamps, whilst a traction engine was bought for the transport of materials to and from Penzance quay. With the new shaft down 370′ from surface, the men refused to sink further on account of the danger of water-filled workings nearby.[4] This necessitated un-watering the old Botallack and Crowns sections in 1909 before sinking at Allen's could be resumed, a year's work which involved the company in additional expense which it could ill afford.[5]

At Tresavean, electric turbine pumps were put in and initial pumping commenced in April 1909 but the chokes in the shaft (Harvey's) were so bad that four months later they had only un-watered 27 fathoms. Although Tresavean was destined to have the longest life of any of these re-workings (to 1928) it was to be almost two and a half years before the bottom of the mine (then 286 fathoms below adit) was seen.[6] In the east of the county, at Clitters United, where the company had done so well in the opening years of the century, attention had been turned about 1904 to the tin left

[1] Report dated 1909, Bolitho Records, DDX 104, CRO, Truro. Herein it was stated that in this and the earlier working of Providence, exploration eastward had always been prevented because of 'the fear and risk of breaking into the Old Providence Mine where the surface workings were bathed by the waves at Carbis Bay.' This re-working of Providence was hampered also by the complaints of local residents at what was considered 'the eyesore at Carbis Bay' from the 'stream of red water which poured onto the beach from the Providence battery . . .' *Western Morning News* 29.6.08.

[2] Oats Collection, RIC, Truro. Normal rate of sinking was 10′–12′ per day. Waste rock from the shaft went to fill in the surface depressions where the old Grylls' Bunny tin floors had been excavated in times past.

[3] Until the establishment of the generating station of the Cornwall Electric Power Company at Hayle in 1910, mines using electric pumps and hoists had to install their own generators, except in the Carn Brea area where a smaller generating station had been built in 1902 for mine use. After 1910 and the extension of supply to Botallack in the west and Truro in the east, the use of mains-generated electricity spread swiftly in the larger mines.

[4] One side of the shaft became rather wet and the shaft-men, not unwisely, 'rather nervous.' A drill hole was cautiously put out, tapping water in less than 6 feet. No record of workings existed at this spot. (Oats Collection, RIC, Truro). This incident illustrated once again the ever-present dangers in re-opening setts in an ancient mining area such as St. Just where the ground from Pendeen Watch to south of Priest's Cove was honey-combed by previous centuries of mining activity.

[5] There was also a fear that the sea had broken into Crowns and so into Botallack, as the pumped water at one stage became salty. This cleared but recurred again later and it was then found that during gales waves and heavy spray at Crowns Engine Shaft were responsible.

[6] The un-watering of Tresavean, *Trans. Inst. of Mining and Metallurgy*, Vol. XXI, 1911–12.

standing in the old levels above adit in this and other old Gunnislake mines. Thereafter a more grandiose scheme to re-open the old copper workings of Hingston Down Consols for tin was tried but, as at Condurrow, the shaft was a bad and crooked one, giving endless trouble. To keep the mill fed, cargoes of foreign ores were imported via Calstock, although it was obvious that this could only be a stop-gap measure to keep the company's expensive plant in full work, and in May 1908 Clitters United was stopped. Thus, despite a good price for tin, the hoped for revival had not arrived. The deep mines of Camborne were obviously faring better, with tin at three times its former price, but there were still doubts about the future. No worthwhile discoveries had been made by all the work from the Tamar to St. Just, nor in the deep mining around Carn Brea: 'over twenty years have elapsed since the last important rich bunch of tin was cut in the county.'[1] At Dolcoath, admittedly, the bottom of the mine was looking encouraging enough to bring about a resumption of work on the partly finished Williams Shaft, although to some extent perhaps as an alternative to the Stray Park section where the outlook was poor. As before, the shortage of miners militated against the all important necessity to keep the stamps fully occupied with the enormous tonnages of low-grade tinstuff that had now to be raised: 'if we had more men, we could get more tin'.[2] At Basset Mines, results for all the money and effort put into Marriott's were disappointing indeed. In 1909 a profit of £86 12s. 6d. was made—a disappointing figure in view of the good price of tin and the enormous capital outlay—but better by far than the £4358 loss in 1908.[3] Carn Brea & Tincroft was perhaps in the greatest difficulties, with an overall loss from its reconstitution as a joint concern in April 1900 to December 1909 of £4830. This was not in itself a deficit heavy enough to endanger the continuing existence of the mine but it showed only too clearly how ill-equipped and behind the times it was. The cost of production was £97 per ton of black tin and even with an output of 80 tons a month and metallic tin at £150 a ton, costs were still not being paid. At surface 8d. a ton was incurred for 'carriage and horse & cart work' taking ore the short distance to the stamps, whilst the dressing plant was primitive in the extreme by twentieth century standards.[4] Carn Brea & Tincroft

[1] MJ 19.6.09.

[2] Manager's report, MJ 16.3.07. In an endeavour to attract more men, Dolcoath erected 58 new cottages for its employees in 1907.

[3] Basset Mines' chief problem lay in the cost of pumping. In 1902 the coal consumption of the group was no less than 15773 tons (Lyle's 80″ pumping engine 4143 tons; Lyle's whim 569 tons; Marriott's 40″/80″ compound and compound whim engines 4581 tons; Basset stamps 2680 tons; West Basset stamps 1406 tons; Carnkie pumping engine 802 tons; Thomas's whim (primarily used for raising and lowering the men) 271 tons; Pascoe's compressor and whim 646 tons; calciner 387 tons; Lyle's capstan 48 tons; smith's shop etc. 240 tons). With 49007 tons of tin stone stamped by Basset in 1902 this meant that a ton of coal was burnt for every three tons of ore raised. Nor was this year exceptional, for 16024 tons were consumed, for example, in 1903. (Basset Mines Ltd. Annual Reports, Fortescue Records, DDF 388, CRO, Truro.)

[4] MJ 5.3.1910. This raised an echo of William Teague's words in 1872 (see page 138) and it is doubtful whether much had been done to the dressing floors in the interim to prevent them from once again becoming 'a disgrace to civilisation.'

it should be remembered, was one of the biggest mines in the county, employing 963 persons,[1] yet despite its size and the scale of operations it was becoming time to either ' mend or end ' it, with re-equipment of the dressing department most desperately needed.

Another old mine which had met with mixed fortunes was Levant, which even with improved prices for tin and copper after 1900 had not done much more than meet costs overall. For a time the mine would make profits sufficient to divide, whilst in other years calls were necessary to make up the working deficit. The company was still a private one, headed by the Bolithos, and all tin was sold privately to Chyandour and not by ticketings. Ore from the deep levels beneath the sea had been the mainstay of the company during most of this period and with well over 600 employed, the mine was far and away the largest outside the Camborne area. It was, however, difficult and expensive to work, with problems found nowhere else in Cornwall. The best ore ground in the deep western levels was yearly becoming farther and farther away from the main hauling shaft and output had to be handled no less than seven times before it reached the stamps. Ventilation and heat were also major problems—air temperatures of 90 degrees being usual throughout the submarine section—all made worse by the men having to spend hours out of their working shift in getting to and from the stopes 1½ miles out under the sea, even with the man-engine in use. In the 1890's there had been attempts to effect economies and bring the working of the mine more up to date. In 1891–2 a small steam locomotive was tried for hauling underground[2] but it was found unsatisfactory and ponies were brought into use instead—one of the only two instances of their use in any deep Cornish mine. In 1895 a set of Californian stamps was put in and at much the same time telephone communication was installed underground, whilst in 1897 a winding engine with coupled 15″ cylinders was set up in the 260 fathom level for hauling ore up the New Submarine Shaft. This was started in April 1898 and as it was worked by compressed air, necessitated the installation shortly thereafter of a big new air compressor[3]—the cost of which, with its new buildings, amounted to £8000.[4] No less than 511 men

[1] 730 men, 166 boys, 67 girls. The great decline in the employment of female labour, not only here, but in all the mines, was striking. In 1861 an estimated one-third of the 40000 employed in mining operations in Cornwall and Devon were women and children yet by 1906, out of a total labour force of 6380, scarcely any adult females were employed, only a few girls, and not more than 600 boys at most. (See MJ 11.11.1911). It is also interesting to note that, according to evidence given to the Committee formed to report on the Stannaries Law Amendment Bill of 1887, it was rare for a married woman to be employed on Cornish mines, and had been for years.

[2] See *The Cornish Beam Engine*. This was the only instance of locomotive haulage underground in Cornwall until very much later. Geevor introduced an electric locomotive for hauling in 1939, followed by one at South Crofty in or about 1943.

[3] This was started in December 1901, the largest machine of its kind in Cornwall. Built by Holman Bros., it was 60 feet long overall, and was of the four cylinder cross-triple-expansion type, with 17″ high pressure, 27″ intermediate and two 30″ low-pressure cylinders. The existing compressor was sadly inadequate and the Levant men maintained that they could do one third more work if they only had air at a higher pressure for their drills. (MJ 14.10.99.)

[4] MJ 17.7.01. There had previously been complaints by some shareholders at the cost of the buildings

and 165 boys were employed in the mine at this period—their ranks swollen by a number who had returned from South Africa on account of the war—and in 1903 it was decided that the New Submarine Shaft, which had been sunk from the 178 to the 338 and was to go to 350, should be made capable of hoisting men. But there were inescapable difficulties in the mine's situation seven miles away from the nearest shipping port, for this added the extra burden of cartage of all ore and coal to the mine costs. In 1898 a traction engine was acquired and proved such an economy that a second was later added.[1] In the period from 1896–1902 Penzance, Hayle and Newlyn were all tried as ports for the shipment of Levant copper ore and the import of its coal but finally space on Penzance quay was acquired for coal and ore storage.[2]

In March 1906 Levant received considerable publicity in the local and national press with reports that the men were refusing to work underground because the mine had been broken into by the sea. This was in fact true, but only to the extent that sea-water was entering the workings at each high tide through a fissure opening into a cross-cut which connected with the old Phillips' Shaft sunk to 232 fathoms depth at the base of the cliffs below the dressing floors. Under such conditions the men were not unwise in their refusal to go underground at any time near high water, even though the company maintained that it was merely a matter of 'only a few gallons of water per minute splashing into the shaft.' Remembering what had happened at Wheal Owles thirteen years before and realising what chance they had of escaping if a further plug of soft ground was washed out by the sea, the men waited until the shaft had been cleared of debris and sealed off from the mine entirely by being filled with concrete to a depth of twelve feet below the fissure.[3] Some lingering fears remained, however, and men left as and when they could find jobs elsewhere—'choosing between a safe and an unsafe, and between a comfortable and an uncomfortable mine.'[4] At Christmas 1905 there had been 640 on the payroll but by June 1907 this was down to 479. Even a wages bonus scheme, amounting to 10 per cent of all dividends, failed to

housing the engine and their 'elaborativeness'—'the designers should have their names placed on the cement tablets outside so that their names might be handed down to posterity.' (MJ 27.4.01.)

[1] Another less praiseworthy attempt at economy was made in 1891 when a cheap German nitro-glycerine compound was tried. This was so dangerous in use and to store that the Levant miners came out on strike sooner than work the mine with it. Their case was amply proven when the importing company were subsequently fined £60. (MJ 27.6.91.)

[2] The transport cost in 1908, even using tractions with two big trailers, was 3s. od. per ton from Penzance (or Hayle) to the mine compared to 3s. 9d. per ton sea freight from South Wales. This added almost 50 per cent to the costs of the coal being used, then amounting to 400 tons a month. The law that a man had to precede the traction engine carrying a red flag—in Levant's case on every journey from Trewellard to Penzance or Hayle—infuriated the mine manager, and matters were not made any better by the fact that the local authorities would not permit the engine to cross the Ross swing bridge by the harbour, or allow it to traverse the main (Market Jew) street. Thus it had to detour round the back of Penzance on every journey it made. (See MJ 24.6.99, also Levant Cost Books, RIC, Truro.)

The old Trereife works at Newlyn were considered as a storage depot when these closed in 1896 but they were not close enough to the harbour. Levant Cost Book, RIC, Truro.

[3] MJ 10.3.06, 24.3.06, 7.4.06.

[4] MJ 15.6.07.

halt this drift away and was abandoned at the end of 1907. By then the mine desperately needed re-equipping throughout, the state of the skip shaft which 'used to wind 25 skips an hour and now did three'[1] being typical of much of the mine. As an interim measure towards cutting costs and helping to retain the men, a cage was introduced in 1909 in New Submarine Shaft but the main long-term requirement was a completely new shaft from surface. This was a provision which the Crown, as the chief mineral lords, insisted upon under the terms of lease renewal, although annual extensions of the existing lease were granted whilst negotiations continued. In 1907 the dividends had been £10434, followed in 1908–9 by a loss of £12820—'an object lesson in cost-book finance, the comparison of which is not without significance'[2]—and with this loss, plus the growing difficulties of working the mine, the fate of Levant hung in the balance.

The years 1911 and 1912, despite the as yet unrewarding results of all the new activity in the county,[3] continued to be ones of prospective boom. Work was re-started at Wheal Peevor, at Silver Valley near Liskeard, at Wheal Droskyn & Ramoth near Perranporth, and at Garlidna and Boswin, near the old Wendron United. In Breage part of the Grylls mines was resuscitated under the name of Wheal Boxer, whilst a London syndicate acquired Wheal Hampton at Marazion. Killifreth was also re-started on a considerable scale, having been worked above adit in the interim by tributers, whilst at Wheal Coates work was progressing steadily, with ten heads of Californian stamps and a horizontal engine and pumping plant erected.[4] At Balleswidden and Ding Dong, the burrows were being examined and stamped, with 16 heads of Cornish stamps put up at the latter for this purpose.[5] Nor were the possibilities of tin alluvials neglected, for a steam navvy (owned by the St. Erth Tin Dredging Co.) started work on the upper reaches of the Hayle river in 1911, whilst in 1913 a floating bucket dredge was ordered by the Molinnis Moor Dredging Co. for a stretch of ground near Bugle.[6] The possibility of dredging the sea-bed in St. Ives Bay for its tin content was also explored[7] whilst sampling of the Loe Pool alluvium below Helston also took place.[8] Elsewhere in this period

[1] Levant Cost Books, RIC, Truro.

[2] MJ 16.5.08.

[3] By 1910 output had only increased to 4797 tons, compared to 4268 tons in 1900. World output in this same period had risen from 82400 to 110100 tons.

[4] This came from Wheal Merth, Lelant, closed shortly before this date. As in the former working at Wheal Coates, the double skip-road came from Towanroath shaft into the open and ran up the cliff side to the mill, hoisted by a double cylinder 12" horizontal engine.

[5] From September 1912 to March 1915, 13649 tons of ore were picked from the Ding Dong burrows, which yielded 51 tons of black tin. (Report dated 1924, Acc. 317, CRO, Truro).

[6] To be built by Werf Conrad Ltd., of Haarlem, Holland, who supplied many of the Malayan tin dredges at and after this date. (MJ 12.7.13.) So far as is known this dredge never entered service.

[7] Official permission for this was granted, despite the opposition of St. Ives fishermen, who feared disturbance of their fishing grounds. A parallel exists today (1966) in the present similar proposals to dredge St. Ives Bay.

[8] Loe Pool and the Cober river which fills it are today hardly thought of as a tin streaming locality. A century ago these waters rivalled in colour the well known Red River on the north coast and

of boom which reached its peak early in 1913 there was also activity at Kingsdown, near St. Austell;[1] Wheal Florence and Wheal Neptune, near Marazion; North Wheal Vor; at Roscrow, near Ponsanooth (the old South Tresavean); Devon United, near Tavistock; and at Bedford United and Dimson Mine at Gunnislake. Near the latter, Prince of Wales Mine had also returned to life, having been acquired from its French owners—who had done little except maintain pumping there—by a London company in 1911. By 1913 the mine was un-watered to the 145 and stoping recommenced.

Amongst the old established mines during this time, fortunes were mixed. Grenville United paid £51244 in dividends in 1912–13 and the possibility of taking over the struggling Condurrow United was considered by the company. There, after delays and difficulties, the shaft had been drained and secured to the 250 fathom level by March 1911 but results thereafter had been all too depressing. At East Pool & Agar, steady progress was being made, with the company reconstituted under limited liability in 1912,[2] whilst at Basset Mines 1912 saw a profit of £5405. The previous year, Geevor Tin Mines Ltd. had been formed to take over North Levant & Geevor Ltd., whose sales of black tin in 1910 had been 98 tons. At Dolcoath, work on the Williams Shaft had been completed by May 1910, 3000 feet deep and the culmination of fifteen years of spasmodic work. The enormous sum of £90362 had been expended on the shaft and its equipment and although Dolcoath paid a 20 per cent dividend in 1913, there was no hiding the disappointment felt at the marked falling off in the lode in the bottom (510 fathom) level of the mine. In 1911 the 490 had been rich, whereas from the appearance of the 510 it now seemed possible that even the legendary Dolcoath was bottomed at last. At West Kitty, where the main (Reynolds) section had been finally abandoned in January 1909, the old cost-book company had gone into liquidation in 1910 and was later acquired for £1000 by West Kitty Mines Ltd., before being merged shortly thereafter into the much more grandiose St. Agnes Consolidated Mines Ltd.,[3] which owned virtually every unworked sett in

according to one writer (A. Esquiros, *Cornwall and its Coasts*, 1865), on the occasions when the Bar was cut and the Loe Pool waters released, the sea even as far away as the Scilly Isles was, in due course, tinged red.

[1] Part of the old Hewas sett. This represented the eighth recorded occasion on which Hewas had received attention—up to 1774, 1794–96, 1813–22, 1846–48, 1853–60, 1871, 1881, 1912 onwards.

[2] East Pool had been one of the great tin prizes of Cornwall, second only to Dolcoath. She was started in January 1835 for copper, divided into 128 shares on which £9 per share was called up (£1152). By December of the same year the mine was in the dividend list and up to 1860 the adventurers had divided £38492, almost entirely from copper. During 1860–61 there was no dividend but these recommenced in April 1862 and by May 1869 a further £19712 had been paid. The shares at this date were subdivided by fifty, to make 6400, and throughout the worst of the '70's, '80's and '90's East Pool made handsome profits and not a single call. The first came in 1897 and then only to cover the cost of acquiring Wheal Agar engines etc. To the end of 1907 from May 1869 the total calls had been £27200 (£4 5s. 0d. per share) and the dividends £495520 (£77 8s. 6d. per share). With such a fine dividend record it was small wonder the East Pool committee could boast that during seventy years ' not a single share had been either forfeited or relinquished.' (MJ 16.5.08.)

[3] This company, registered in December 1909, was subsequently described as ' a concern with large ideas and little capital.'(MM, July 1920.)

the area between St. Agnes Beacon and Wheal Prudence, east of Trevellas Coombe. The intention was to build a central power station on a site above Trevaunance Quay, using diesel engines to generate electricity for supplying the whole of the group. Work was proceeding at East Blue Hills but most activity was still concentrated at Wheal Friendly Shaft, with the two other pumping engines (Thomas's 40″ and Reynolds's 50″) idle but retained for emergency use. The pneumatic stamps and other dressing equipment erected at Friendly in 1907 had only proved a limited success, for apart from a failure of their concrete foundations and local objections to the calcining of tin there, it had been found—somewhat belatedly—that there was a shortage of water during part of the year.[1] Thus it had been necessary to revert to the old Jericho stamps which were to some extent brought up to date. Here there was at least an ample water supply and also room for further expansion when St. Agnes Consolidated got into its hoped-for stride. As it was, even treating a mere sixty tons of ore-stuff a day—at an exorbitant cost of 5s. 0d. per ton exclusive of cartage—there was almost an acre of dressing floors which the restricted Wheal Friendly site could never accommodate.

From evidence given to a Royal Commission in 1912, it was apparent that the problem of hook-worm was still prevalent despite the passing of ten years since the recognition of the Dolcoath epidemic. Statistics for the year showed the serious extent of the infection amongst men working underground: South Crofty 11 per cent, Botallack 12 per cent, Basset 51 per cent, Dolcoath 69 per cent, whilst earlier official figures for East Pool were 47 per cent in 1904 and for Tincroft, 25 per cent in 1908.[2] Stamping out the disease was inevitably a difficult problem, not only from the inability to supervise sanitation amongst so many workers but from the very extent of the underground workings where the disease bred. Dolcoath alone was said to have one hundred miles of levels etc. by 1905[3] and the task before the manager was one that was probably never mastered throughout the rest of the mine's history. Another unhappy feature of the previous few years had been the rise in the accident and death rate in Cornish mining as a whole.[4] By 1910 this was a matter of growing concern,

[1] In dry spells, the West Kitty manager was said to be frequently seen 'searching the heavens with his telescope for the faintest sign of a cloud.' (MW 28.9.07.)

[2] MJ 8.6.12.

[3] MJ 8.7.05. This was an under-statement rather than the reverse. Forty years before, in 1869, there were said to be about fifty miles of levels and cross-cuts, plus twelve miles of shafts and winzes, whilst the rate of sinking and driving (later greatly increased) was about one mile per annum. (Journal RIC, Truro, Vol. XI, 1869).

[4] Fatalities, per 1000 men employed underground—

	in Cornwall	all U.K. mines
1902	1.15	1.37
1903	1.45	1.35
1904	2.00	1.35
1905	3.70	1.52
1906	3.63	1.42
1907	2.39	1.46
1908	2.78	1.47

being twice the average United Kingdom mining rate. In 1906, for example, fifteen were killed underground, compared to thirteen in 1905, with three killed in the worst accident of the year—in May 1906, as a result of a fall of ground in the 200 fathom level at Basset Mines. This was the commonest cause of all fatalities underground at this time, although the falling away of a hanging stage between the 160 and 170 fathom levels in March 1912 at East Pool & Agar also resulted in the deaths of three men. In extensive stopes, as were now common, it was difficult to see the state of the hanging wall even with the acetylene lamps which, by 1912, had been introduced in lieu of the time-honoured candles.[1]

As regards the smelting of Cornish tin, the period saw important and far-reaching changes. The decline in Cornish mining in and after 1896, allied to the rise of the Bolivian mines—as well as those of Nigeria, which became a producer of tin on a scale of world importance from 1905 onwards[2]—meant that furnaces in the vicinity of Liverpool, the main port through which these foreign ores were handled, had a commercial advantage over those in Cornwall. The Penpoll Company was the first to see the gain to be made by siting a smelter near Liverpool and in 1901 they leased the Canada Works at Birkenhead, run as a subsidiary establishment to their main works at Point. In 1905 this was given up in favour of another works on the opposite side of the Mersey at Bootle. Three years later Williams Harvey decided to follow suit and erected a new works close by that of the Penpoll Company. This bore the same name, Mellanear, as the Hayle works which was closed down some years thereafter when that at Bootle was in full commission.

By this date too there were other tin smelters operating in Britain, with small works at Neath and Swansea (T. S. Sutton & Sons, and James Stephens) and at Birmingham (R. Jones & Co.) active by 1901 at least.[3] The focal centre of tin smelting had shifted from Cornwall and the effects of this were apparent with the disappearance of the Consolidated Tin Smelting Co. Ltd. in 1912. This company had suffered mixed fortunes since the late 1890's and even the closing down of the Carvedras works in 1898 and the concentration of all smelting at Chyandour had not prevented losses being made in some years. Considerable purchases of foreign ores had to be made to supplement local black tin, for all Cornwall could not keep a big smelter going week in and week out, and Cornish works were

[1] Acetylene lamps, once they could be made to burn through a whole shift, were superior to candles not only on account of their greater illumination but because they did not melt in hot levels. They would, moreover, burn in air too foul to give more than a faint blue flame from a candle and in a particularly bad end it might be necessary to light a bunch of six to give the usual light of one in good air. Thus they were also cheaper in use under such conditions.

[2] In 1912 Nigerian output was 2886 tons, a figure that rose to 6143 tons by 1914 and 8187 tons by 1916.

[3] Suttons commenced tin smelting in or about 1883 and continued until 1914, whilst Capper Pass & Son of Bristol—a firm still engaged in smelting today—were also listed amongst the buyers of barilla at the Liverpool ticketings in the 1880's. (Foreign Ores Tin Ticketing Book 1873–93, Bolitho Records, RG 1/112, CRO, Truro). These smelters do not appear to have bought Cornish ores at the Redruth ticketings, however, confining themselves entirely to foreign ores.

badly sited for the import of these ores which had to be trans-shipped into coasting vessels for Penzance from the larger U.K. port where they were consigned. In one particularly bad year, 1907, Chyandour works made a loss of £14976, smelting in this twelve months 1806 tons of Cornish black tin and 1530 tons from abroad. In 1909 this deficit was turned into a modest profit of £2347 but with losses of several thousand pounds in the next few years and little evident prospect of much improvement the decision was thereupon taken to close the works in June and voluntarily to wind up the company in October 1912.[1]

This left only three works active in Cornwall, those of the Penpoll, Cornish and Redruth companies, and the former's works at Point had, by this time, become only a subsidiary of the main one at Bootle. In 1913 came yet another change, with the acquisition of the Cornish Tin Smelting Co. by Messrs. Vivian Younger & Bond Ltd., one of the leading brokers of tin in London. The Seleggan works was the largest in the county at this date and had been extended soon after 1900 by the addition of four new furnaces. In 1912 forty per cent of all Cornish tin had been smelted there, with the works carried on under the sole ownership of Alfred Lanyon, one of the founding partners in 1887.[2]

In terms of mining, the autumn of 1913 saw the first signs of impending disaster for Cornwall for the price of tin on the London market fell steadily from £288 in January to £184 by July. The selling up of Great Dowgas in September—regarded as one of the best equipped of all the new mines in the county[3]—was followed by the dismissal in November of 270 out of the 360 employed at Wheal Jane (Falmouth Consolidated Mines).[4] This, like Dowgas, was regarded as one of the most modern, well managed and secure new mines in the county. A few weeks later the three pumping engines at Carn Brea were stopped[5] followed by the official closing of this once great mine in January 1914, when the company decided to confine working solely to the North Tincroft section. In view of the danger to adjoining setts, however, pumping was re-commenced at the instigation of the mineral lords. The drop in the price of black tin

[1] Bolitho Records, DD/BL, CRO, Truro.

According to a report in the *Mining Journal*, an inability to secure any renewal of the Chyandour works was the primary cause of smelting ending there, although this seems unlikely. (MJ 30.6.12).

In fact, when Consolidated Tin Smelting Co. Ltd. ceased business in 1912, all hope was not given up by Messrs. Bolitho of continuing smelting and an effort was made to form a new public company, using recently patented electric furnaces. In an undated draft prospectus, it was proposed to take over the goodwill and brand-names of the Consolidated Tin Smelting Co. for £10000 and it was stated that a trial electric furnace had been tried at St. Ives with most satisfactory results; 'although the production was comparatively small, it was carried on at a profit . . ' (Bolitho Records, DDX 104/21, CRO, Truro.)

[2] *A Century of Metal Broking*, Vivian Younger & Bond Ltd., n.d.

[3] The equipment included a 400 h.p. horizontal cross compound condensing pumping engine by Holman Bros.—originally built and finished for an exhibition—six heads of Holman pneumatic stamps, four gas engines, a lattice steel headgear which cost £700 and was described as the second largest in the county, and a mass of other machinery. (See MJ 13.8.13).

[4] MJ 29.11.13.

[5] MJ 13.12.13.

was now being felt in earnest and later that same month, 100 hands were dismissed at Botallack.[1] The company had met with indifferent success and in March 1911, with the new central shaft down to 145 fathoms and £70,000 expended it had been necessary to reconstruct matters financially to secure additional working capital. The inclusion of the Wheal Owles sett came at the same time, although this was a manoeuvre that added nothing to the immediate prospects of the mine.

The seal of this winter of bad news was set by Dolcoath when the disappointing state of the bottom of the mine, and the decreasing values in the principal stopes, was announced at the February meeting. By June 1914 tin was down to £139 per ton and it became evident that 1914 was to become a year to be remembered along with 1874 and 1896 in the annals of Cornish mining. All the hopeful prospecting and the flotation of new companies that had taken place since 1906 came to an end and the only survivor of any size was Tresavean. Dowgas, Botallack, Wheal Jane, Wheal Vor, Phoenix United, Great Condurrow, Boscaswell, Wheal Coates, St. Agnes Consolidated, Gwennap United, St. Ives Consols, and a dozen others had come and gone since 1906 and in them at least a million pounds of invested money was to be lost.[2]

In July ninety men were dismissed at Levant, where a loss of £2540 was reported on sixteen weeks' working.[3] The same month the Condurrow 80″ went out of steam and work started on drawing up the pumps, whilst in August the managers at Phoenix United considered developments 'so very unpromising' that they advised an immediate closing down.[4] In Devon the last of the Dartmoor mines, Birch Tor & Vitifer, was also suspended, partly on account of the falling metal market and partly because of the shortage of labour. With some intermission, work had been carried on here since 1854, with the mine sunk to 70 fathoms from surface and as many as 120 persons employed at one time.

The outbreak of war itself brought a further blow, for the London Metal Exchange closed temporarily, bringing the buying of concentrate by the smelters to a complete stand. For nine long weeks no tin ticketings were held, a state of affairs that was severely felt in Cornwall and in many cases it was only the public-spirited action of the Bolithos, whose bank advanced almost £30000 against stocks of black tin, that serious unemployment was avoided. During the rest of the year the news grew steadily worse of the depression; Dolcoath curtailing all development work from September; Botallack and Wheal Jane completely shut down

[1] Followed in March by the complete closing of the mine.

[2] In the years from 1906–20, an estimated £1,250,000 was lost in attempted re-working of water-logged mines in Cornwall. (See MW 14.8.20).

[3] Here, in April, all development work had been stopped, as well as all stopes 'bordering on the unpayable.' In October underground prices were reduced by 20 per cent and all other wages by 15 per cent. The adventurers had been unable to secure a new lease at better terms and the lease was thereafter renewed annually on the same terms. Cost Book, RIC, Truro.

[4] £140,000 had been expended here in this re-working, equipping and draining the mine, but little development work was done underground.

and their plant for sale; Boswin, Mulberry, Wheal Hampton and other new mines suspended. Out of the thirteen ticketing mines only East Pool & Agar and Wheal Kitty reversed the trend of a noticeable decline in the tonnages produced. The only bright spot was Tresavean's first sale of tin, in September, and this mine alone was to survive out of the many ventures of the 1906–13 boom.

The biggest defection of 1915 was that of St. Ives Consolidated which was compulsorily wound up in April, leaving a deficit estimated to be £140,000.[1] But it was now the wartime conditions of working rather than the rise and fall of the metal market that governed the state of Cornish mining, the first effects being felt by Dolcoath where 128 of the best underground men had left by February to join His Majesty's Forces.[2] The company at this time were still looking ahead and expressed their readiness to lay out £5000 in sinking for tin in the apparently barren ground below the 550 fathom level if the mineral lord would do the same. This, however, was never to be proceeded with. By September 1915, 223 men from Dolcoath's labour force, 191 of them from underground, had left for the armed forces. The demands of the army became even more pressing for miners for the tunnelling companies in the land mining campaign[3] that was undertaken as a way to end the deadlock of trench warfare. This bore particularly hard upon Dolcoath and Tincroft (as well as upon Basset Mines and Grenville) for they were not given the exemption accorded East Pool and South Crofty from where no more recruits had been accepted since August. This was on account of the valuable wolfram, essential for armaments production, produced by these two mines. Output of wolfram—at one time an ore constituent that had been regarded by the miners as an enemy because of the difficulties of separating it from the cassiterite[4]—had been stepped up from a total of 180 tons for the county in 1913 to 347 tons by 1916, this bringing high prices and contributing considerably to the income of the two Pool mines.

A minor casualty of the year was the Redruth and Chasewater Railway, running from the mines south of Carn Brea to Devoran on the Fal, which closed in 1915. Through ninety years of mining bonanza and borrasca this little line had retained its independence, its final closing being precipitated by Basset Mines' decision to purchase coal which was imported

[1] Operations were continued for some time in the hands of a receiver and in 1917 the setts were acquired by the Thermo-Electric Corporation Ltd. Mining continued thereafter at Giew alone and the entire remaining plant was shipped to the same company's Wolfram Camp mines in Queensland. (See MJ 17.2.17).

[2] The effect of the loss of the best of the underground labour force, and of wartime working generally, was evidenced by the development distances driven at Dolcoath in the war years. In 1914 this was 723 fathoms, in 1916 263 fathoms, in 1918 only 67 fathoms. (MW 22.3.19.)

[3] For details of the work of the miners (including the Cornish Tunnelling Corps) on the Western Front from 1915–18, see MW 26.7.19 onwards.

[4] Due to their having more or less similar specific gravities. Wolfram was also very similar in appearance to cassiterite, resembling the latter 'so nearly in appearance when in the stone, that none but experienced tinners can tell the difference. This has often been the cause of disappointment to parties, by supposing they had rich tin until it came to the test of fire.' G. Henwood, *Four Lectures on Geology and Mining*, 1855.

via Portreath and not Devoran as hitherto. By this date both Basset and Grenville were losing concerns, the latter incurring a deficit of over £5000 in the second half of 1915. As at Dolcoath shortage of labour was making itself felt, with no opening up of new ground going on and the reserves in all sections being steadily depleted. The increasing cost of coal also hit Basset and Grenville particularly hard, and the latter, then 395 fathoms deep, suffered a series of pumping and other breakages of machinery in 1916 which brought the loss for July—December to almost £9000. This, moreover, was despite a greater tonnage of ore being milled than ever before, underlining the falling ore values in the mine.

THE POST-WAR COLLAPSE: 1917–1922

'*Cornish mining is a dying industry unless adequate capital can be secured.*'

Mining Journal, 3 March 1917

IN THE TURMOIL of war, the news of Dolcoath's loss of £4990 in the latter half of 1915 was not the matter of moment to the county that it would otherwise have been. This was the first loss made since the company's formation in 1895 and by economies and with an improved metal market the succeeding six months' working showed a welcome profit of £6900. Almost the whole of the central and eastern parts of the mine were now exhausted and activity was concentrated on the Wheal Harriet section as the future hope. The intention was to drive the 450 westward into Stray Park, but to improve the indifferent ventilation and the great heat in this part of the mine[1] it was necessary to sink to the 490 to give connection with New Sump Shaft and the deeper easterly levels. Unfortunately by the end of 1916 results showed that there was little payable tin ground in the sections westward as had been hoped, and that the main lode was completely broken up. Dolcoath's average produce had declined inexorably year by year, from 80 lbs. per ton in 1895 to about 25 lbs. by 1918 and it seemed that nothing could be done to reverse or even halt this decline. Miles of levels were driven but no further ore-bodies of sufficient value had been found to break the depressing pattern of the annual decline in lode value. With a loss again of £2910 in the latter half of the year Dolcoath's fortunes were at their lowest ebb.

As if to compensate for this, however, the last few days of 1915 saw the discovery at East Pool of a lode of extraordinary size and value, in a cross-cut north from the 240 fathom level.[2] Named Rogers Lode, this

[1] The temperatures and working conditions of the Hot Lode at United Mines in Gwennap fifty years before are well known: less well known, but equally deserving to be on record, are those in Dolcoath at this time. The temperature of the water in the shaft was over 110° and electrically driven fans had to be introduced to enable work to proceed at all. (See MJ 16.2.16). Earlier, in 1903, the 375 at Wheal Harriet had had to be abandoned on account of the excessive heat. (MJ 31.1.03.)

[2] Struck on Boxing Day 1915. (MJ 9.9.16.)

was the best discovery that had been made in Cornwall for very many years, and at this juncture it gave much needed encouragement to West of England mining, particularly as its existence had been predicted by geologists.[1] Even though 230 East Pool & Agar employees had joined the armed forces by March 1916, before the new compulsory Act prevented their recruitment, the effect upon the company of the new discovery was dramatic. In 1915 a loss of £972 had been made, in 1916 a profit of £22485, and in the next two years its dividend was to double and re-double. In the latter year sales of black tin rose to 914 tons, the highest output in Cornwall, realising £94495, whilst 526 tons of crude arsenic brought £12434 and 100 tons of wolfram £17543. To South Crofty this new lode was also a discovery of importance and in March work was started driving a cross-cut towards it into the North Crofty sett on which an option was already held. Inspired, perhaps, by this evidence that all was not perhaps hopeless, the Dolcoath management took heart and, using a diamond drill borrowed from the flourishing East Pool, commenced an underground drilling programme in Stray Park.[2] Dolcoath was hampered by having no arsenic with which to supplement sales and in 1917 precious spare labour was diverted to the construction of the necessary calciners and flues for burning arsenical ores. Arsenic soot brought £100 a ton (or £130 refined)—ten times the average of pre-war years—and it was a by-product that Dolcoath could no longer afford to ignore.[3]

Tin was now £300 per ton, a price which might well have put the industry back on its feet had times been normal. As it was, the rising costs and the loss of good miners militated against anything except hand-to-mouth working, whilst what was worse the government was soon to control the price of tin. In St. Agnes, after a long period of vicissitudes, West Kitty (St. Agnes Consolidated) finally closed down in October 1916. Over 150 hands had been employed and the abandonment meant that less than a hundred working miners remained in this one-time stronghold of Cornish tin. In the main Redruth-Camborne area the demise of Basset Mines could only be a matter of time whilst at Grenville the loss for January to June 1917 rose to £10047. In both cases they produced a particularly pure and clean cassiterite and this meant that there was neither arsenic nor wolfram to bolster up falling sales of tin.

Wolfram was now fetching £178 per ton of concentrate and was of extreme national importance in the production of tungsten steels and the government itself, in the shape of the Duchy of Cornwall, had commenced

[1] After the formation of the new East Pool company in 1913 (as East Pool & Agar Ltd. with a nominal capital of £100,000) a considerable outlay had been incurred. A new electric air compressor was put in and all hand work underground replaced by the use of rock-drills except at a few isolated pitches worked on tribute. East Pool & Agar Ltd. Reports, 1913–39, *per* F. Michell, Redruth.

[2] MJ 25.8.1917.

[3] The importance of the value of wolfram and arsenic sales to a mine during this period was evidenced by the case of South Crofty. Here, in 1917, these two by-products realised 18s. 9d. per ton of ore crushed as against 21s. 0d. per ton for the black tin, the mine receiving considerable benefit from the arsenic refinery which had been completed in 1909. In addition, in order to increase wolfram sales, the company acquired the little Castle-an-Dinas Mine, north of Goss Moor, in 1918.

Basset Mines: above: *18″ gauge four-coupled tank locomotive of German manufacture acquired in 1906 for hauling ore from the South Frances section of the New Stamps at Carnkie* below: *ragging frames on the dressing floors at the head of Church Coombe. Note the derelict engine-houses of Wheal Uny on the skyline.*

The recommencement of mining operations in 1906 at Great Condurrow United, Camborne, prior to the erection of the 80″ engine-house.

The Wheal Friendly section of West Kitty Mine, St. Agnes, in or about 1905, with 60″ engine. This was the final section of West Kitty to be worked, after Reynolds and Thomas's sections had been abandoned.

A group of bal-maidens and others, locality unknown, about 1900. Note the distinctive Cornish shovels and the traditional head-dress.

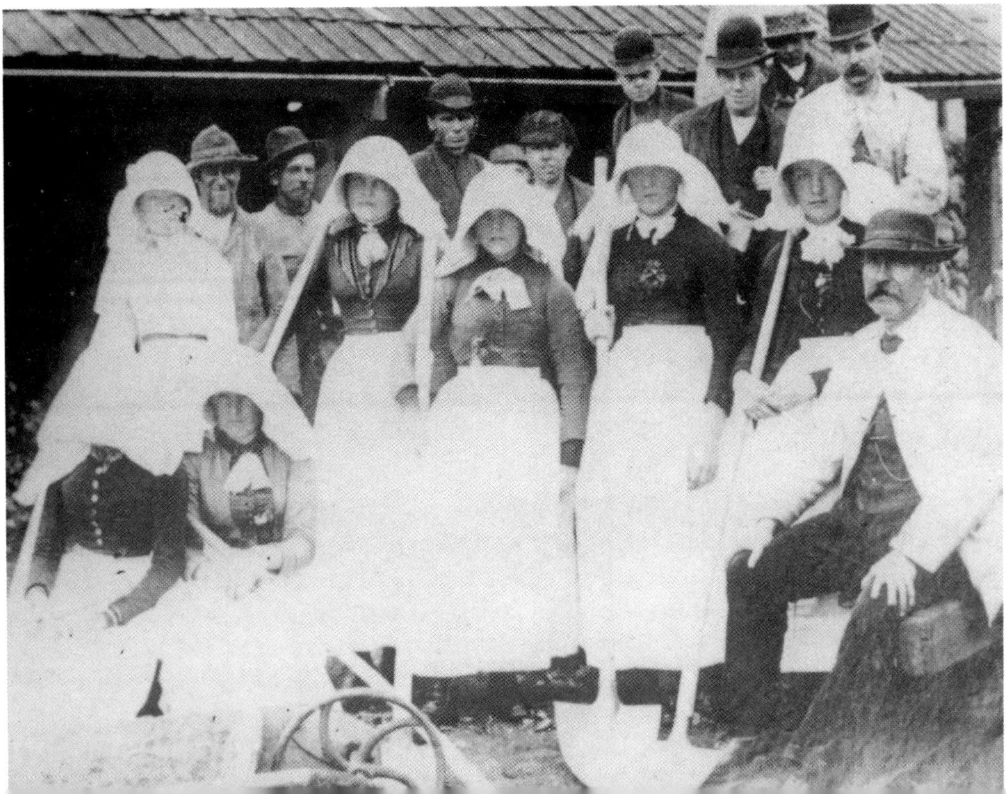

West Kitty stamps (30″ engine) and dressing floors at Jericho, at the upper end of Trevellas Coombe, St. Agnes. Ore was carted here from the mine, about two miles westward.

Sailing vessels discharging coal at Trevaunance Quay, St. Agnes, shortly before the first World War. Almost to the last days of mining in St. Agnes parish, coal was imported in this way, revealing some of the legacy of obsolescence which played its part in crippling latter-day Cornish mining.

The miners' dry at Levant, about 1914. This was erected in 1889 at a cost of £1000 and was regarded as a model of its kind. Note the numerous chests, and the heated tubes (centre) *for drying underground clothes. The railings in the foreground guard the passage leading to the top step of the ill-fated man-engine.*

A hand-bill appealing for funds and donations, issued by the Mining Division Unemployment Relief Committee, 1921–2.

Cornish Tin Miners' S.O.S.

There are 2,000 unemployed, mostly **Tin Miners,** in the area we represent, making a total of nearly **7,000** men, women, and children for whom we have to provide.

The present situation is exceedingly critical, as these men have been out continuously for nearly three years.

The need was never greater than at present, and if we are going to be in a position to assist them during next winter it is **imperative** that the public should immediately rally to our assistance.

We appeal for help on the following grounds:

1. These men are not on strike, and have never been on strike.
2. They consented to a deduction of 20% from their wages to keep the mines going.
3. They received no subsidy nor bonus during the war.
4. Their conduct has been most exemplary—indeed wonderful, and has earned the praise and sympathy of all.
5. These mines are not closed because of a trade dispute, or because the men refused to work for a certain wage, but solely on economic grounds.
6. Owing to being out so long all their savings are spent and our funds are nearly exhausted.

Don't let Cornish women and children starve!

N.B.—Parcels marked "OLD CLOTHES—OWNER'S RISK" will be carried by the Railway Co.s at a cheaper rate.

Donations and gifts of clothing are urgently required, and may be sent to Rev. W. A. Bryant, Deacon Hill, Camborne, Hon. Secretary.

work on and around Kit Hill to supplement the production from the Camborne mines. Activity, started in January 1916, was confined to the sinking of costeaning pits across the hill plus the re-opening and draining of a few minor adits. The Clitters mill was re-opened, connected to Kit Hill by an aerial ropeway along the length of Hingston Down, and some work was done at Clitters itself as well as at the abandoned Hingston Down Consols. Disgruntled Cornish comment was made as to the way Kit Hill was costeaned and quarried in a fashion that the Duchy would never have allowed any of its mineral lessees[1] and it is doubtful whether results were economically justified even under conditions of wartime necessity.

At this bleak juncture, in 1915, G. F. Basset, the foremost mineral lord left in Cornwall, sold out his entire mineral rights to a London syndicate.[2] This was at once an ill omen and good tidings for Cornwall. The pulling out of Basset could only be construed as a lack of faith in mining's future, whilst on the other hand, the passing of control to interests more closely allied with mining would give greater unity and co-operation. The deal would have made headline news in the Cornish press in peacetime but under war conditions it received little publicity, some of the sting having also been removed by earlier abortive negotiations to the same end in 1906. The outcome, subsequent to this transfer of 1915, was the joint acquisition of the mineral rights over some 8000 acres of the Tehidy estate by Dolcoath and East Pool & Agar for the purchase prices of £60,000 and £30,000 respectively.[3] The latter could afford this from its funds in hand but at Dolcoath it was necessary to raise debentures to find the money. As an investment it had its attractions, for Dolcoath had paid on average £8558 in dues every year since 1895, although against this it could be argued that Dolcoath's future was not perhaps as bright as its past. Thereafter, however, these mineral rights of both East Pool & Agar and Dolcoath were disposed of to a new company, Tehidy Minerals Ltd., formed in February 1919. A subsequent deal in August of the same year transferred the extensive mineral rights over about 25000 acres of Lord Clifden (the Agar-Robartes family) to the same company.[4] This estate was more extensive, and, in view of its china-clay producing ground, of more value than that of Basset and the purchase price was £10,000 in cash, plus the allotment of 190,000 fully paid shares in the new company.[5]

East Pool & Agar was, by now, far the richest of the surviving mines, the profit of £59651 in 1917 having almost doubled (to £103784) in 1918.[6] The riches of Rogers Lode and the ground to the north were responsible for the setting up of an off-shoot concern in August 1918, known as Tolgus Mines Ltd., formed to explore and develop the tin that should, according to geological reckoning, lie beneath the old copper workings of the Tolgus group. A tunnel (10′ x 8′) was commenced from

[1] MJ 5.1.18. [2] MJ 16.10.15. [3] MJ 2.2.18.
[4] MJ 30.8.19. [5] MW 23.8.19.
[6] To this date, the largest annual dividend ever paid by a Cornish mine.

the 255 fathom level of Agar towards these, with high hopes.[1] Output from the parent mine was greater than it had ever been, displacing, for the first time for forty years, Dolcoath from her premier position as Cornwall's largest producer. The output of the nine surviving mines in 1918 was—

			employed		
	tons milled	black tin	underground	surface	total
East Pool & Agar	75400	1280	220	300	520
Dolcoath	61500	827	254	340	594
South Crofty	67588	581	249	215	464
Grenville Utd.	44772	451	178	199	377
Geevor	24956	418	205	98	303
Tincroft[2]	54356	379	264	345	609
Lanner Syndicate[3]	26986	311	179	113	292
Levant	12136	278	176	136	312
Whl. Kitty		62	63	32	95

Missing from this table was the once important Basset group which shut down in the latter part of the year and went into liquidation on 31 December 1918. In 1917 only 26486 tons of ore stuff had been milled compared to 32419 tons the previous year and by December 1917 the accumulated loss at Basset Mines Ltd. was £23422. The closing was inevitable by the end of 1918: 'want of capital, heavy pumping charges, and thirty years of adversity have combined to bring about this decision.'[4] Notice was given to all the men early in December and the last engine to stop was Lyle's 80″ which ceased pumping on December 21.[5] The previous month there had been a likelihood that East Pool & Agar might take over the working of Basset, and surveys of the latter were carried out with this end in view. It was said that East Pool 'has in the bank £150,000 of undistributed profits. It can, therefore, like Wellington's army, go anywhere and do anything'[6] but despite this the purchase of Basset was not proceeded with.

[1] MJ 17.8.18. The financial advantage of a tunnel over the development of a new mine from surface was obvious, for the latter scheme was anticipated to cost £100,000 at least. Very rapid progress was made with this tunnel, a rate of 55 feet a week being attained in May 1920, although it was in extremely hard ground. Four Holman drills were used side by side, worked continuously in six-hour shifts. A double tram-road was installed and all the broken ground dumped in old stopes to save any hauling to surface. To obviate any delay, the face of the tunnel was blasted each time onto a steel net and this was then hauled back by winch so that the next shift could re-start work without the usual delay. A smith's shop and drill repair shop were also fitted up underground to save further time. (See MM May 1920).

[2] Carn Brea & Tincroft re-named in 1917.

[3] Viz, Tresavean.

[4] MW 28.12.18. 'Basset possesses the finest pumping plant in the whole county, and needs it, for it is probably the wettest mine. It consumes as much coal as the whole city of Truro.'

[5] MM January 1919. Pumping charges then constituted 40 per cent of Basset's total operating costs. During 1919 the tin remaining in the stopes was cleaned out, and the shafts and levels cleared, using 200 men working three shifts a day against the rising water-level which on one occasion rose 60′ in 36 hours. The balance bob at the 240 was the only thing that could not be raised in time and by mid-April all underground work had been completed, with surface operations and the fencing of shafts completed by Christmas 1919. The non-ferrous and iron scrap, the oak pump rods, the machinery and other materials realised £21946 (see MW 7.2.20).

[6] MW 9.11.18, 23.11.18. By April 1919, in fact, East Pool had liquid assets in hand of £187859. (MW 26.4.19).

The demise of Basset Mines ushered in a year which was obviously bound to be one of extreme difficulty in making the transition from war to peace conditions.[1] As matters stood, Cornwall was defenceless in the event of any sudden difficulties in the notoriously fickle tin market. To make immediate matters worse the first winter of the peace was the wettest for fifteen years and at Grenville United there was the added burden of water from the abandoned Basset group.[2] By March it was more than the Grenville engines could deal with and water began to rise slowly throughout the mine. By the end of April many of the levels from the 262 downwards had to be abandoned and during the month a number of men were put on a week's notice.[3] By August Grenville was drowned to the 190, with about 180 fathoms of levels submerged, although there the water was held, helped by a providentially dry summer. Desperate attempts had been made to secure further capital to re-equip the mines, for as one commentator said—' Only a thick poultice of bank notes can restore the company to prosperity '[4]—and the company was financially re-constituted in August. An application for government aid was made to the Non-Ferrous Mines Committee set up by the Board of Trade and in October this was reported to have been granted.[5] Hope was then widespread that as a result of the deputation to the Board of Trade in January 1919 by the mine adventurers, official assistance would be forthcoming for the other mines to bridge over the crucial reconstruction period.

During 1919 Wheal Kitty at St. Agnes had gone under, operations being entirely suspended early in the year after the mine had been two years in the hands of a receiver. But it was in the main Camborne district that the worst blows were still to come, with disastrous losses being recorded. At Tincroft, where war-time profits had been high,[6] a loss of £12245 was incurred in the first half of the year. Output of arsenic and wolfram had brought brief prosperity to the mine but the decline in the value of the former and the complete collapse of any demand for the latter—now unsaleable after being £178 per ton in 1918—left Tincroft dependent solely on her failing tin reserves for survival. Dolcoath was in similar dire

[1] For a long resumé of Cornish mining at this date, its immediate past and its future prospects, see MJ 5.1.18.

[2] Later it was discovered that most of Grenville's trouble was top (or surface) water as a result of the incessant rain, and that a cross-course was keeping back almost all Basset's water.

[3] MJ 12.4.19.

[4] MW 6.9.19. A portion of the £50000 or so dividends that had been paid out in 1912–13 would have proved invaluable to the company at this time had they been retained as a reserve fund. As had been the case with Basset Mines, pumping was the burden that Grenville could not sustain. The coal consumption was 14–15000 tons a year and the pumping charges, which had been £12570 in 1912, had risen to £22661 by 1918.

[5] M.W. 4.10.19. ' The Grenville directors are certainly to be congratulated upon being the first to " draw blood " from the Government.' (See also MW 4.11.19). The sum involved was, however, small, amounting to only £2,000.

[6] From July 1917—July 1918, for example, Tincroft had made £26344 profit.

straits, with working costs per ton of ore crushed rising between January and June 1919 from 41s. 11d. to 47s. 9d. and this on a falling metal market.[1] The loss here, in this same period, was £19510 and what was worse, completion of the programme of diamond drilling throughout the western part of the mine had discovered nothing of importance. By September it was said that Dolcoath's 'continued existence under present conditions is a miracle which the management themselves hardly know how to explain.'[2] Even East Pool was having a troubled time, due to the excessively wet winter which flooded the mine to the 240 for a long period. This prevented the richest stopes from being worked and on this score the company's profits in 1919 were decimated to a mere £5521.[3]

For a time, however, the next month, these commercial trials and tribulations of the industry paled into insignificance at the turn of events at Levant. Here attention was rivetted on the men upon whom the industry turned rather than upon its finances. On Monday 20 October part of the rod of the man-engine broke, smashing this and its human load down into the 266 fathom deep shaft. The death roll was not known for some time, access to the shaft and rescue conditions in it being difficult in the extreme. One injured miner was trapped for 53 hours, only to die at surface, whilst the last body was not recovered until four days after the smash. Thirty-one died and twelve were seriously injured in this disaster, and even though this number was far fewer than had at first been thought,[4] it was still the second worst in the long accident-filled history of Cornish mining. The widows and orphans now added their grief to the dereliction and departures that had saddened St. Just for fifty years past.

Levant, at the time when she received this unwelcome notoriety in 1919, was something of a phenomenon: an old-fashioned mine, with crooked shafts, primitive dressing floors, and—most antiquated feature of all—still worked under the old cost-book system by a private company. Even during the early war years the mine had made little or no profit and in August 1915 'it was resolved to instruct the manager to discharge all hands except those necessary men to stamp the broken stuff, reserving about 10—15 men to stope any place which will pay well to work.' Some months later, however, this and the decision to give notice to the lords about drawing up the machinery, was annulled and the mine continued.[5] Arsenic was of increasing importance to the company, being chiefly sold to buyers at Bissoe, and in June 1918, somewhat belatedly, the company decided to build their own arsenic works and refinery on the mine.

[1] Coal costs per ton of ore crushed had risen at Dolcoath from 3s. 10d. in 1914 to 10s. 4d. in 1918.

[2] MJ 20.9.19.

[3] During the year no less than £36586 was expended by East Pool on pumping and repairs to pumping plant—as much as on the whole cost of ore production.

[4] MW 25.11.19.

[5] Levant Cost Books 1912–20, RIC, Truro. From 1916–19 the calls had been £3627, the dividends £4134.

Unfortunately this was completed only a few months before the post-war slump in arsenic prices. Labour troubles had plagued the mine throughout the later war years, on account of the low wages paid, and it was singled out for strike action by the Dockers' Union (then representing Cornish miners) in 1918. In August the men struck and intimidation was used to prevent the pumping engine from being worked.[1] Extra police were drafted to St. Just during the several weeks that the strike lasted and work was resumed in the middle of September only when a bonus as well as a general increase in wages were granted by the company. The mine, however, was in a failing state—although not from exhaustion of its mineral reserves, for the only limits westward as regards good ore ground were those of practical mining—and in May 1919 the manager was instructed to discontinue the working of this submarine section and to concentrate all the men[2] in the upper levels. At the time of the disaster there were hopes of a major reconstruction of the Levant company including the sinking of a new vertical shaft (for which a horizontal winding engine had then been recently bought from Basset Mines), and an amalgamation with the adjoining Geevor had also provisionally been agreed upon.[3] These schemes for Levant, however, all fell through and after reconstruction of the company under limited liability, work continued in the shallower landward parts of the sett whilst the man-engine shaft was cleared out and made into a ladder-road for the men to climb through.[4]

Wages at this stage of the tin depression were indeed low enough in view of the hard labour involved, let alone to compensate for the bad conditions underground and the dangers inherent in the work. Average wages for miners were 41s. 3d. to 45s. 9d. per week; for surface workers, 37s. 6d. to 43s. 0d., whilst pumping engine men received only 32s. 6d. for their not unimportant task. Discontent with these rates and growing union activity resulted in a strike at South Crofty during late December 1919, extending into the New Year.[5] This was confined to surface workers and stemmed largely from the refusal of a number of them to join any union. One by one all the South Crofty men were coerced into becoming members, with the company expressing its disapproval by discontinuing forthwith the monthly bonus system that it had paid to all wage-earners. In February

[1] *Ibid.* Some of the ponies underground (then seven in number) were 'safe and looked after, but the one at the 350 would soon be drowned by the rising water.'

[2] About 120 men and boys worked underground at this time. The submarine section had not in fact been abandoned entirely at the time of the disaster. It is interesting to note that from August onwards, as a result of the coal strike, the compressor and stamps had both been stopped to conserve coal stocks, with only the pumping engine, a whim and the man-engine in use when the disaster occurred. *Ibid.*

[3] MJ 3.1.20. In the previous two or three years Geevor had been engaged in opening up and unwatering the old Wheal Carne section, with encouraging results.

[4] To this date (1920) Levant had sold, in a century of working—

136562 tons of copper ore	£859303
21959 tons of black tin	£1531977
3800 tons of arsenic soot	£32657
MM March 1920.	£2423937

[5] MJ 3.1.20,

there was a strike at Giew, and elsewhere during the first three months of the year there were growing and widespread demands for a complete stoppage by all Cornish miners unless all wage rates were raised.[1] But further consideration of this was to be forgotten by both sides thereafter in view of the imminent collapse of the industry. A long-awaited government report on the future of the mines appeared in March[2] but it contained little that was new about the past and even less that was concrete for the future. For Cornwall, the time for platitudes and official sympathy was past and in June a deputation went to the Board of Trade asking for a specific £200,000 as a grant, or failing that, as a loan, to be divided amongst five mines for development purposes; Dolcoath £80000, Grenville United £50000, Tresavean £30000, Tincroft £20000 and Wheal Kitty £20000.[3] In July, however, the county was informed flatly that the government could not see its way to ask the House to vote financial assistance in this way. Setting aside the question of whether the government showed wisdom at this juncture in refusing to bolster up what may or may not have been dying concerns which no private enterprise capital was prepared to back, it is true to say that the authorities pandered to the coal miners and colliery owners at this time and that Cornwall's political and business lobby was simply not strong enough to secure the same treatment.[4] In spite of the refusal, hope was not abandoned of some measure at least of official support and despite a growing realization that this was now an age of government without principle or conscience, another deputation proceeded to London later in July.

Meanwhile the position in the county had worsened rapidly, following the appallingly swift drop in the price of tin during 1920. In February, when the crash started, it was £396 per ton and by the end of the year £213 and even this was not the worst for it was to go down to £156 per ton by the succeeding March. In May 1920, after announcing a loss of £18063 for the second half of 1919, Dolcoath had decided to let the mine fill to the 340, to discharge 200 men and to give provisional discharge notices to all the others. By holding the water at the 340 the possibility still existed of proceeding—if government aid was given—with the

[1] MJ 20.3.20, 27.3.20. The employers made a partial offer, upon which the men held a ballot. Results of this in the Camborne district showed that 1542 men were prepared to accept, 524 were against its acceptance. In the St. Just district, the two sides were evenly divided. (MJ 3.4.20.)

[2] Even at this grave juncture, with Cornish mining in the most desperate straits, there was still the occasional foolish venture. The African & Australian Co. Ltd., a company with a capital of £200,000, had acquired Wheal Hampton and also Parka Mine (the old Park of Mine, at St. Columb, see page 164) in 1920 and the latter was described in terms almost incredibly sanguine: lodes varying in width from 1′ to 5′, with values 'ranging from 40lbs of tin per ton of ore to 1 cwt. 3 cwt. and 7 cwt. per ton of ore . . . so rich indeed is some of the ore that it was not considered safe to bring it to the surface until a proper place of security could be provided: it has therefore been stored underground.' Hundreds of tons of 'low-grade ore' assaying about 1 cwt. per ton' were also reported there, all above adit level. (MW 20.3.20).

[3] MJ 5.6.20. The same month the 682 workers at Dolcoath sent their own petitions to the President of the Board of Trade in an endeavour to save the mine. (MW 26.6.20.)

[4] 'It is the deep rooted opinion in Cornwall that . . . Cornish miners would have had perhaps more than justice instead of gross injustice if they could have boasted of a million votes or if they had shown themselves an intolerable and vindictive body of malcontents.' (*The Cornishman* 7.9.21.)

scheme for a tunnel northward from the 338 fathom level at Williams Shaft into the Roskears.[1] This would cross-cut the presumed lodes of North and South Roskear throughout a distance of over 4000 feet. Even though this was far cheaper than sinking a new shaft from surface it was obviously still going to be a long and costly business, at a projected driving rate of 200 feet per month, requiring about £120,000 in outlay.

At Grenville United a further 150 were discharged in June, followed the next month by the laying off of all except seventy remaining men. Tincroft also discharged 120 in July, added to which there were fears for the future of Holman Bros., major Camborne employers, who were on short time and had also been obliged to discharge men. The crisis was at its worst, with the black news already mentioned that there was 'positively no hope of any state subvention or grant of any sort.'[2] Dolcoath decided to curtail operations even further[3] and in December a loss of £11275 in the six months to June 1920 was revealed, and this despite the radically reduced scale of working. The loss since the Armistice had been more than £50000[4] and it was decided to discharge all the remaining men and discontinue all operations on the sett. This, in December 1920, marked the end of Dolcoath, the premier Cornish mine for half a century past. From the recommencement of working in 1799 to the end of June 1920, the mine had sold copper ore value £2,328,435 and black tin value £6,842,351, a total of £9,170,786.[5] No bigger or richer mine had been worked for tin in Cornwall, and the closing of Dolcoath marked the end of an era in Cornish history.

The next mine to be abandoned was Grenville United, where all pumping was stopped in December and where, with virtually no cash in hand, there were debts of £9000 to meet. In January 1921 the decision was taken to wind up the company, the chairman having some hard things to say about the old mine, notably that 'nothing too strong could be said about the plant . . . It was absolutely worthless; it was no better than scrap-iron.'[6] The same week, a receiver was appointed to wind up Tincroft's affairs[7] and at Tresavean, as a preliminary measure to stopping

[1] This idea, following the principle of the Tolgus tunnel started by East Pool, was first mooted in August 1918 but printed plans in connection with it were not submitted to shareholders until June 1920.

[2] RCG 28.7.20. Adding once again to the sadness of this period was news of a further serious accident at Levant in August 1920. A series of explosions of gelignite, from some unexplained cause, occurred near men tramming ore in the 190 fathom level, resulting in the death of four and serious injury to others. (MW 21.8.20.)

[3] Underground mining (stoping) ceased in July 1920, and all work in the mine in April 1921.

[4] From January 1919 to June 1920, £48848.

[5] RCG 10.11.20. In another account (MW 25.12.20) reporting a statement at the Dolcoath meeting, the total value of the minerals sold from 1746 to June 1920 was stated to be £10,596,471 and the dividends £1,356,430 in this time.

[6] RCG 12.1.21. Despite these comments, it is interesting to note that the 90″ engine from Grenville was moved to South Crofty some months after this and worked there, successfully and trouble free, until it was destroyed by an accident in December 1950.

[7] No hope existed of finding further capital to keep Tincroft going, for in the twenty years that the company had been in existence (i.e. since its financial reconstruction in 1900) the shareholders had not received 'a farthing interest on their capital.' (MW 11.9.20.)

the mine, advance notice of discharge was given to the 350 men, although the intention here was to keep the pumping going ready for an eventual resumption of work. A ten per cent reduction in wages at East Pool had already been accepted by the men in the hope of helping keep the mine afloat whilst even at South Crofty—regarded as the best placed of all the remaining mines[1]—operations were curtailed from February onwards to await an improvement in the position. This was followed later the same month by the discharge of all labour at both these mines, with the exception of those few men sufficient to maintain the pumping engines.[2] This threw almost another thousand mine workers onto the unemployment lists, with between four and five thousand jobless in west Cornwall as a whole.

In the St. Just district the same sad story had unfolded itself in the opening weeks of the year. In January both Levant and Geevor had been driven to giving a fortnight's notice to all hands, neither mine being able to stand the heavy losses that a continuation of working entailed. At a friendly meeting between the men and the managers, the former offered to take lowered wages in the hopes of keeping the mines open but the disparity between working costs and what the output of black tin would bring was too great for this even partly to bridge the gulf. By February both mines were shut down,[3] with 624 unemployed miners in St. Just parish as a result. In the east of the county, activity in the several Duchy of Cornwall mines had also come to an end. These were closed down in October and the whole plant sold up the following month.

Thus, by March 1921, thirty months after the Armistice, with tin at less than half its value months before, only one Cornish mine remained at work—Giew, which had formed part of the re-working of St. Ives Consolidated.[4] Basset, Grenville, Tincroft, Dolcoath and Wheal Kitty had gone under, with South Crofty, East Pool, Geevor, Levant and Tresavean suspended. Cornish tin mining, from the days of Elizabeth, had experienced ups and downs in full measure, but never before had it come to a virtual standstill throughout the county. Even large-scale tin-streaming had also come to a halt, for the last big works at Reskadinnick closed down on the Red River.[5] Not only its staple industry but Cornwall itself had come to a watershed in its history. Throughout the mining districts, the abiding roar and rattle of stamps came to an end, a sound that—distant or near at hand—had been a ceaseless accompaniment to life

[1] In 1920 the company made a profit of £16946 compared to a loss at East Pool of £28008 in the year.

[2] RCG 16.2.21.

[3] At Levant, where underground work had been suspended on 8 January, operations were continued re-equipping the mine whilst funds were still available. The man-engine shaft was opened, stripped out, and re-timbered to the 100, levels cleared to work the western ground over the 190, new electrically driven air compressors and mill (with two heads Nissen stamps and five heads of Californian) put in, etc.

[4] Working ended here in November 1922 when the 115 employees were discharged and the mine shut down. (RCG 15.11.22.)

[5] In 1920 87 tons of black tin had been sold. Later, about 1923, this works was re-started.

in the Mining Division of west Cornwall for many decades. Disturbing this unaccustomed silence was only the thin whistle of wind through rusting head-gear and the broken window-panes of silent engine-houses.

The output figures showed only too plainly how the industry had been decimated. In 1920 3065 tons of black tin had been produced, in 1921 679 tons and in 1922 only 370 tons. This might be seen as perhaps the nadir of the decline that Cornwall had experienced for fifty years past. Some, a majority perhaps, of the big mines were exhausted—in the commercial sense of the word—yet even so the causes of this final collapse were primarily external: the war-time conditions with an incentive to maximum output but a shortage of skilled labour, and the great rise in deep mining costs, particularly in the price of coal which affected Cornish mining so radically. These factors occurred at a time of drastically lowered values of tin, due to the cessation of war-time prices and the world-wide change from a war to a peace economy. Allied to this there was the development of the industry in Malaya, Siam and elsewhere, the number of tin dredges growing rapidly after the war-time interruption following their successful introduction about 1912. Cornwall, with dear coal and no cheap native labour, could not raise low-grade ore—every ton of which was hard rock requiring to be bored, blasted, trammed, hoisted, hauled, stamped, and expensively dressed to yield its 30 pounds or so of black tin[1]—in competition with the soft alluvials of Malaya. The vulnerability of lode mining was only too apparent when comparison was made with the latter, for companies there—many of them Cornish owned and directed—merely stepped up output to meet low metal prices and were in some cases able to go on paying dividends throughout this period that had crippled Cornwall.[2]

HOPE DEFERRED: FROM 1923 ONWARDS

> '*Past disappointments have made stoics of Cornish folk. Generations have swept through sloughs of despond buoyed up with hope of brighter times.*'
>
> *Mining World*, 3 March 1928

A PARALLEL CRISIS, with its origins in the war years, also affected Cornish tin smelting at this time. The conflict had inevitably caused a radical alteration in the pattern not only of British but of European smelting, brought about by the blockade of Germany where a considerable proportion of Bolivian ores had formerly been treated. In 1913 German smelters had produced 11800 tons of tin, but in 1915 only 1080 tons, and

[1] At Dolcoath, for example, in 1919—the year before it closed—71 tons of rock had to be worked to yield one ton of black tin. (MW 29.3.19.)

[2] Malayan producers were helped by the action of the Federated Malay States government which bought tin in 1919–20 to help support its price. In 1921 the Netherlands East Indies authorities joined in this scheme and some 20000 tons of tin was stocked in this way, until the return of better prices in 1924. In Bolivia, where the mines were in very few and enormously wealthy hands, output was cut back and also stock-piled until the market improved.

it was obvious soon after the outbreak of war that a considerable expansion of tin smelting capacity was possible either in Britain—traditional home of tin smelting—or, more logically, in neutral America where a large proportion of the finished metal was consigned. In November 1914 there was a rumour that the Bolivian government and mines were considering setting up their own smelting works in America, but this was scotched the following month by the announcement that Williams Harvey had secured the contract with Patino—owner of fully one half of Bolivian output—to smelt all his ores for the duration of hostilities.[1] Early in 1915, however, the lead producing American Smelting and Refining Co. decided to build a tin smelting works and enter the Bolivian trade, a significant turn of events which made Williams Harvey decide to erect a branch works in America, or alternatively to join forces in a venture with the new entrant's chief competitor, the National Lead Co.—the latter itself being a company with a fifty million dollar capital and the second largest consumer of tin in the world. The intensified blockade of Britain by U-boat meant the loss of commercially and strategically valuable cargoes by torpedo in the Atlantic[2] and it became more and more apparent that smelting in the United States by Williams Harvey, either alone or in conjunction with others, was essential if the company was to maintain its leading position in the trade. The company, still owned and run entirely by Cornishmen, had come a long way since the first beginnings of its forebear at Portreath in 1814. In 1915 tin ore for smelting was bought to the value of £2,349,732, 15600 tons of metal were produced and a profit of £110600 15s. 7d. returned. Moreover, apart from the fact that 'events have more and more tipped the scale . . . towards the American side of the fish-pond'[3] extra smelting capacity was needed as the Bootle works was so congested that all 'first slag' was having to be left for treatment at some future date instead of being dealt with constantly as was usual.[4] Output had risen from 11497 tons in 1912 to 15600 tons in 1915 and early in the latter year, to increase capacity, the old Mellanear works at Hayle—of which the company had wisely decided not to dispose shortly before hostilities started—was re-started and brought into full production. It was also planned to expand the works at Bootle greatly, to allow the production of alloys and to produce 100 per cent pure tin by the new electrolytic process. Meantime negotiations with the National Lead Co. had been continued and in the summer of 1916 the latter acquired a half-share in Williams Harvey for £225000,[5] payable partly in cash and partly in shares

[1] Prior to this date Patino's ores had been smelted by Goldschmidt in Essen.

[2] Barilla was also diverted across the Pacific for smelting in the Straits where the gigantic works of the Eastern Smelting Co. had been built at Penang in 1911.

[3] For this and other detail on Williams Harvey & Co. Ltd. at this period, see Harvey Records, WH Correspondence Files 1914–16, CRO, Truro.

[4] In July 1915 an estimated £45000 was tied up in the value of this slag.

[5] Williams Harvey's existing capital was £160,000, having been increased from £80,000 in 1908. The company was a private one, registered under limited liability in 1901, having as its directors the same persons who had acquired the old partnership in 1888.

in the new company to be formed to run the American smelter. This was to be situated at Jamaica Bay, nine miles from New York, close to the American Smelting & Refining Co's works which had been built at Perth Amboy, whilst elsewhere on the east coast of America eight smaller plants for the treatment of Bolivian ores were also set up by other companies.

The works of the new Williams Harvey Corporation—appropriately bearing the same Cornish name as the two British works—was not brought into production until early in 1918 and by this time Senor Patino had also acquired a one-third share in the company as well as in the original English branch. From Williams Harvey's viewpoint an alliance with the biggest single producer of tin in the world was of paramount importance for the future, even though this put the Cornish shareholders, 35 in number,[1] into a fragmented minority.

Meanwhile, in Britain, to take advantage of the greatly changed tin smelting position, another new but very much smaller smelting company had been started. This, the London Tin Smelting Co. Ltd., was owned in one-third shares by Henry Merton & Co., of London (who also acted as sales agents for the company), the Bolithos of Penzance and Lord Clifden. At first all ores were smelted at Merton's works alongside the Thames at Northfleet but in 1916 smelting there was discontinued 'owing to a bad title discovered by Mr. Gilbert [Lord Clifden's agent], and unsatisfactory working demonstrated by Horton Bolitho.'[2] In fact, Merton & Co. was partly controlled by German interests and, on account of this enemy participation, was wound up in 1918, at which date the Bolithos acquired their share. Meantime, in July 1916, it had been decided to build new smelting works at Penryn,[3] near Falmouth, and in the January following the first of the two furnaces was completed and smelting commenced. Local as well as foreign ores were thereafter smelted in what was to be the last works of its kind set up in the Duchy. By 1918 £13000 had been expended out of the working capital of £20000,[4] with the balance consumed by trading losses on the sale of metal in 1916–17. Bank loans of £10000 and more had to be negotiated to enable the company to continue but in 1919 a profit was made 'totalling 15 per cent on all capital.' This, moreover, despite the fact that 'for a long period, owing to manoeuvres by other smelters, no high grade ores were obtainable and the company continued to turn out good common tin from Bolivian ores of very low grade quality.'[5]

[1] Primarily the Harvey and Pearce families.

[2] Historical Notes on the London Tin Smelting Co., compiled by Horton Bolitho, May 1921, (Bolitho Records, DDX 104/21, CRO, Truro.)

[3] In Church Road, Penryn. At this same date, Chyandour was also listed (*Post Office Directory*, 1917) as an address in Cornwall of the London Tin Smelting Co. Ltd., but so far as is known this was solely for office purposes.

[4] This was pitifully inadequate for tin smelting on any scale. By comparison, Williams Harvey had bank balances (at London, Liverpool, New York and Truro) of £140715 in December 1914, and even this was not exceptional, for this total had risen to no less than £245000 by March 1915.

[5] Horton Bolitho, *ibid.*

When peace-time conditions returned and the first brief post-war boom had subsided, the world's tin smelting capacity, standing at about 176000 tons per annum, was more than a third in excess of world consumption. The big American works were the first casualties and by 1923 all had been shut down, together with the small smelters which had also been erected at La Paz in Bolivia and at Arica in Chile. Once again most of Bolivia's output was smelted either in Germany or at Bootle, where Williams Harvey's works was the third largest in the world.

The latter company had experienced a reversal of fortune which was typical of results everywhere in the trade; from 1914–19 profits had been £482008, but in 1920 the English branch alone lost £55021.[1] The following year was only marginally better, due to calciner troubles, the coal strike which closed the entire works from 16 April to 11 July, and to the necessity to spend £31000 on a new fume extraction plant.[2]

In Cornwall the effect of excess tin smelting capacity had been quickly felt and in January 1920 the London Tin Smelting Co. ceased to buy ore. Most of the men were discharged and the remainder occupied solely in 'cleaning up slags, remnants, residues, furnace bottoms, etc.' whilst attempts were made to secure fresh capital for the company. By September, however, the whole works was standing idle, although there were hopes of contracts with various Nigerian interests, and a year later the Penpoll works was shut down. At Hayle, Mellanear was closed entirely in the year and all the men dismissed.[3] In 1923 the same fate befell Bain's smelting works at Redruth, leaving that at Seleggan the only one active in the whole county. Thus, in the five years after the war, the already thin ranks of the Cornish smelters had been reduced from five to one, and even in this London-owned survivor, ninety per cent of all ore emanated from either Africa or South America.[4]

Fortunately this drastic contraction in the smelting industry had but little effect in making unemployment worse, for the numbers concerned therein were minute compared to those in the bigger mines, where any stoppage threw hundreds out of work.[5]

[1] Harvey Records, WH Correspondence File 1921, CRO, Truro. In 1919, 18625 tons of metal were produced; in 1920, 19134 tons.

[2] When the Mellanear works at Bootle was originally built, a fume extraction system had been put in but from 1916 onwards, due to the deterioration of the fans and the wartime difficulty of obtaining renewal parts, this had gradually fallen out of use and completely so by 1919.

[3] In 1919 the output of the four Cornish works had been—Redruth 915 tons, London 816 tons, Cornish 718 tons, Penpoll 395 tons. The small works of the Bissoe Tin Smelting & Arsenic Co. Ltd., which had been established in 1890 (MJ 5.12.91), had closed earlier in that year. (MJ 3.12.19). For several years this works had been concerned primarily, if not exclusively, with the production of arsenic and even in its pre-war career what tin was smelted there was probably only small parcels of concentrate bought from owners of a few head of stamps etc., where free-setting tributers and others brought their output. Most of this Bissoe works still stands (1966), whereas the big square stack of the main Bissoe Tin Smelting Works (closed in 1878) was thrown down in October 1932.

[4] During this same period, the Cornish tin ticketings died a natural death—the *Mining Magazine's* comment being that 'this ancient comedy appears to be nearing its end' (June 1920). Their demise was largely brought about by the decision of East Pool, South Crofty and Dolcoath to revert to sales by annually renewed private contract after November 1919.

[5] Nonetheless, unemployed tin smelting hands were classed equally with tin miners by the local Unemployment Relief Committee, the only other workers to be so,

By July 1921 Camborne had 1215 miners unemployed and Redruth 1148, and even these figures were swollen to a combined total of 2749 by October. Something like one man for every household was out of work, dependent on fifteen shillings a week unemployment benefit plus a shilling for every child. This, moreover, was for a limited number of weeks and all benefits ceased thereafter, the unemployed then becoming a charge on the rates alone for relief. In 1922 these stood at 20s in the £ in Camborne, necessitating the abandoning of all such services as street lighting. Distress was rife and even though soup kitchens were set up, semi-starvation faced some families. In 1920 the Mining Division Unemployment Relief Committee was formed to alleviate the worst cases and later to supervise what few municipal measures were possible to give employment. Food and clothing depots were set up at various points and in the winter of 1921 bedding and coal were distributed to many families. Donations were received from Cornish people in other parts of Britain and from many places abroad, from California to New South Wales. Old mine relief funds, including the balance of the Levant disaster fund, were also made available but there was insufficient money to institute any major schemes which could give lasting work. Tehidy Minerals gave permission on liberal terms for the dumps and outcrop of any lodes to be worked and an approach was made to the Great Western Railway to employ men on doubling the railway track on the section west of Scorrier. In July 1921 it was decided to form a choir 'along the lines of the Welsh Miners' Choir'[1] and during 1922 this brought in a useful sum approaching £2000. In September 1921 a house-to-house collection for funds had been made in Falmouth by a group of twelve miners 'in costume dress' but this had an unfortunate sequel, for it was promptly copied by other non-official groups in various towns in the south-west collecting for their own pockets. Initially there were objections to giving any financial grants towards the cost of emigration—although as one small measure of assistance Messrs. Bain offered free passage on their steamers from Portreath to any man seeking work in South Wales—but from March 1923 onwards, with the long-term outlook for Cornish mining bleaker than ever, funds were set aside to help emigrants, including a number of men who went out to the Hollinger gold mines in Canada. During these months a few minor schemes for public works such as road-building, afforestation, and land reclamation were instituted by the local authorities,

[1] For this, and all other incidents here given relating to the relief measures taken in this depression period of the twenties, see the Mining Division Relief Committee Minute Books (Nos. 1 and 2), RIC, Truro.

This Tin Miners' Choir, clad in their underground working clothes, and in hard hats with clay and candle attached or with the white skull caps customarily worn beneath the miner's hat itself, made a great impression wherever they went throughout Britain. Unfortunately its success had repercussions, for in December 1923 its members resigned *en bloc* and were officially disowned by the Committee when they announced their intention of going on tour on their own account. A number of other 'unofficial' miners' choirs were also formed and in July 1923 it even became necessary to have five hundred posters printed 'to warn inhabitants and visitors against men going round begging in our name.'

aided by a sixty per cent grant for such schemes from the government, but these made only slight inroads into the basic unemployment problem.[1]

Of more immediate concern to the surviving mining interest was a disastrous underground collapse in May 1921 around the main East Pool engine shaft. This was the worst occurrence of its kind for many years, even more extensive than the runs that had crippled Dolcoath [2]and might have had disastrous consequences had men been working underground in the normal way. A large area of workings below the 130 in proximity to the shaft began to collapse in April, with a further succession of falls in these old gunnises[3] taking place during the following weeks, culminating in a massive cave-in on 18 May when an area of ground about 70′ wide and 300′ high ran together. Up to this time there had been hopes of preventing the collapse of the main engine shaft and the total ruin of the mine but all efforts were unavailing. Earlier, the ground had been secured by reinforced concrete pillars but the subsidence was far too extensive for these to contain it. The bottom set of electric pumps (at the 240) was lost and when it was obvious that the damage to the mine was irreparable the remaining pumps were raised, the Agar engine stopped, and dams put in to seal off all this old part of East Pool.

Despite this setback at such a crucial time and a loss of £42053 for 1921 as a whole, the company resolved in August to sink a new shaft north of these old workings, to open up anew the northern lodes where the mine's future prospects lay. This would be virtually a new mine and would give the opportunity of more economical, large-scale working and a centralised plant. Even so, the estimated cost of the new shaft, to a proposed depth of 1000 feet, was £110,000[4] and to some extent this sum might be said to be that lost by East Pool in this disastrous underground 'run' in 1921. On 12 January 1922 work on the new shaft (known as Taylor's) was started, about 720 feet north of the old Wheal Agar,[5] where the company decided to erect the 90″ engine which they had bought from Carn Brea, as a reserve, in 1918 for £3250.

Meantime the effect of the East Pool collapse upon South Crofty had been nothing short of calamitous, for the water from the abandoned workings now augmented the normal pumping load handled by Robinson's 80″, this being already a matter of some concern in view of the stoppage of pumping at Tincroft. The company was thus driven to the purchase of another big engine and Fortescue's 90″ was acquired from the

[1] A typical scheme was the clearing and reclamation of some rough land above Troon where 100 men were employed to clear nine acres at a cost of £2000. This was then let for grazing pasture at £10 p.a. and it was obvious that it would have made better economic sense to expend the same money in a subsidy towards mining than to schemes of such doubtful value. (W. H. Thomas, *op. cit.*).

[2] 'One of the most serious catastrophes which ever happened to a Cornish mine.' (RCG 26.4.22.)

[3] Some illustrations of these enormous excavations where the richest part of the lode had been removed at East Pool are to be seen in J. C. Burrow and W. Thomas, *'Mongst Mines and Miners* (1893, reprinted 1965).

[4] MW 1.4.22, 22.4.22.

[5] The possibility of making use of the Agar main shaft instead of a new one was considered but dismissed, as it was in too bad a condition to be worth renovation.

liquidator of Grenville United in December 1921 for erection on New Cook's Kitchen shaft, this fortunately having a large reserve compartment and could thus be rapidly adapted for pumping. To assist the mine with this, a government loan of £30,000 was made in May 1922 to South Crofty under the Trade Facilities Act, the first official assistance of this kind that Cornwall had received.[1]

Levant, Tresavean and Dolcoath had also made similar applications for government help.[2] The first named had been reconstituted as a limited company after the old cost-book concern had been wound up at a meeting held in October 1921—the last such to be held in Cornwall under this system.[3] Geevor now held over half the shares in Levant, and £10,000 was applied for to help put the mine back into efficient working order. At Tresavean, a government loan of £30,000 was made available in September 1922 to re-start the mine[4] but Dolcoath's resuscitation was less simple, for the plan was to sink an entirely new shaft in the old North and South Roskear setts, which had been worked down to 265 and 160 fathoms below adit respectively. This would open up the area in the same way as had earlier been proposed from the 388 fathom level of the now water-filled old mine. Earlier, in 1921, the company had issued additional shares to preserve its assets in the form of surface machinery and equipment, and the decision to sink the new shaft, subject to the acquisition of sufficient new capital or of government assistance, was taken in February 1922. Over a year was to pass before this scheme came to fruition, for there were difficulties in raising the £70,000 of private capital as stipulated by the Treasury before they would advance the £50,000 loan. By January 1923 only £35,000 capital had been forthcoming and appeals were made for local support for the scheme.[5] This raised the figure, eventually, to £55,000 and the Treasury was prevailed upon to increase their loan to £65,000. On this basis it was announced in July that the reconstituted company was to sink the new Roskear Shaft to 1680 feet and to complete it within two years. The company's boast that they intended to ' create a new record in shaft sinking ' was somewhat marred in Cornish eyes, particularly amongst those of unemployed Camborne miners, by the contract going to Messrs. Pigott & Sons, the Welsh pit-sinkers. The shaft itself was to be 16′ diameter, brick-lined throughout, and use was to be made of the headgear and pumping equipment from the abandoned Williams Shaft. Results at the new East Pool shaft were encouraging and in May 1923, at approximately 1215 feet depth, another rich lode (later known as Moreing's)

[1] An attempt was also made in April 1922 to secure a similar grant to re-start Phoenix United where the Duchy had maintained the 80″ engine since working ended in 1914. (Bolitho Records, 31M/BPR 16/22, DDX 104, CRO, Truro.)

[2] Crushing re-started at Levant in February 1922 and the whole mine a few months later (with ten rock-drills working the morning shift, five in the afternoon).

[3] Levant Tin Mines Ltd. took over from the old cost-book company as from New Year's Day 1920.

[4] RCG 4.10.22. One hundred miners were re-started that month. The mine had been idle since February 1921, with pumping maintained.

[5] MW 27.1.23.

was struck. This hastened the revival of the Tolgus project, in this case by the projected sinking of a new shaft in lieu of the earlier and now unusable tunnel which had been started from the old Agar section and had been driven about 1000 feet before the collapse of the mine. As Tolgus Mines Ltd., the first turf of this was cut on 7 May 1923: 'a most unique event, as the cutting of the sod of a new shaft on new ground had not occurred in the district for a hundred years.'[1] The shaft was on the old Great South Tolgus copper sett, abandoned in the late 1860's, and its position was chosen to open up in depth what it was hoped would be the junction at the eastern end of Rogers, Moreing's, Great and Engine Lodes in East Pool.

But neither this scheme, nor the sinking of Taylor's Shaft at East Pool & Agar provided much employment and the problem continued to be a major one in west Cornwall. The small amount of road-building and land reclamation that had been instituted by the local authorities had done little to right the position. An enterprising attempt to get Tresavean re-started on the basis of a government bonus of 7s. 6d. per week being paid to the company for every man employed had been proposed in February 1922[2] but the chief public venture of the period was the driving of a sewage drainage tunnel at sea level from the North Cliffs to Camborne. This scheme, apart from giving work to men whose skills in tunnel driving stood idle, laid open to shallow depth a promising area of mining ground north of Camborne, although good care was taken to keep any mention of this aspect of this municipal and government financed scheme well in the background.

By the end of 1923 a recovery in the price of tin[3] brought a gradual resumption of more normal working in the suspended mines and by the end of 1924 Levant, Geevor, Tresavean, South Crofty and East Pool were back in normal production. South Crofty in particular had a good year and in the twelve months ending April 1925 returned a profit of £33129.[4]

The worst of the unemployment crisis was now past, with the figures for Camborne and Redruth—still, at this date, separate entities in local government affairs—down to 733 and 818 respectively by 1925 and by 1927 it was possible to disband the Unemployment Relief Committee, its seven year task considered more or less complete.[5] The approximate numbers employed in the working mines were South Crofty 420, East

[1] MW 19.5.23.

[2] Pumping had been continued at Tresavean throughout, at very considerable cost, and the mine was ready for an immediate resumption of work. £170,000 had been laid out here by this date. (MW 11.2.22.)

[3] To £235 a ton (metal) by December 1923. Throughout 1924 it ranged between £219 and £278 per ton.

[4] RCG 23.5.28. From 1909–18 the company had also paid £165625.

[5] During this time, about £80000 had been gathered into the Fund, some £25000 from the Chief Constable's (Lt. Col. H. B. Prothero Smith) relief fund and £30000 from various other donations, plus £24000 raised by the various official choirs, etc. Although nominally serving only the Mining Division (Redruth and Camborne), the area of activities had been extended to cope with the relief work necessary in Gwennap, Chacewater and Helston, where the respective Boards of Guardians were unable to cope. It had also been decided about 1922 to allocate one sixth of all available funds to the equally badly hit St. Just district.

Wheal Kitty, St. Agnes; a view of the surface buildings, headgear and engine-houses at Sara's Shaft (commenced in 1900), about 1930.

Tolgus Mine (or shaft) near Redruth, about 1924; the site of one of the most expensive failures in twentieth century Cornish mining.

Polhigey Mine, Wendron, about 1929: an extremely well-equipped mine, using electric power throughout and complete with aerial ropeway to the mill and pony transport underground.

Engine-houses at Wheal Peevor, near Scorrier, about 1912 when the mine was being re-opened but before the 70″ pumping engine had been erected in the centre house.

The original wooden headgear and ore-bins of Robinson's Shaft, South Crofty, about 1910. Today this is the principal of Cornwall's two surviving mines.

Tresavean Mine, Lanner. In the final deep working for tin of this famous old copper mine, electric turbine pumps were used, powered by three compound Bellis & Morcom high speed engines coupled to alternators. Note the stamps engine-house, a relic of former days, and the extremely tall 150′ stack (all now demolished).

Pool & Agar 400, Tresavean 340, Levant and Geevor 300, Tolgus 150, Dolcoath 120, and Tincroft 100.[1] Activity had recommenced at the latter, in May 1923, primarily for arsenic for which there was a short-lived rise in demand. Mining had also recommenced at Wheal Kitty and at Killifreth. The demand for arsenic was such that the moribund Wheal Busy sett was also bought by the Killifreth company and something like a hundred men were employed there removing arsenical ores from the shallower levels. The revived demand for arsenic also helped East Pool, where 1001 tons were sold in 1927, swelling the company's profits for that year to £24076.

More normal conditions had thus returned to Cornwall, although the havoc of the 1921–2 crash, on top of a decade or more of inexorable decline, had brought the county's output down to a mere 1.24 per cent of world production in 1924. Cornwall's output by world standards was now negligible. The Federated Malay States by this date yielded 45000 tons per annum, the Dutch East Indies 28000 tons, Bolivia 33000, Siam, China and Nigeria together a further 23000 tons and it was against the outputs of these giants that Cornwall was now competing in world terms. The county's output reached 3270 tons in 1929—only two per cent of a still rising world output—but even this modest figure was never again attained thereafter. At opposite ends of the world, Llallagua and Uncia high in the snow peaks of Bolivia, Gopeng and Tronoh amid the mud paddocks of Malaya, had now replaced Dolcoath and Basset, set amidst the stone-hedged fields of west Cornwall.

Some conception of the competition that had brought Cornwall to her knees can be gleaned from a brief consideration of the two first named mines. During and after the first world war, these had gone from strength to strength until by the early 1920's they were producing at their peak about 2200 tons of black tin a month. In 1919 electric locomotives worked into the Uncia Mine via a massive stone-arched adit, whilst in 1925 the similar main level into Llallagua extended a full six miles. A total of 3750 were employed and a private line fifty-seven miles long gave connection with the main Bolivian railway system. They were, moreover, largely owned by one man, and even though one *Mining Magazine* correspondent might disparagingly comment that 'a grocer with a tin opener and a cork-screw could make profits out of either mine '[2] they were still tin producers on a scale to make Cornishmen fall silent. They were in fact quite unparalleled in the history of lode tin and from 1906 to 1926 produced the incredible quantity of more than 193470 tons of 'fine tin' to a value of about £212,000,000. Their annual dividends throughout the latter part of this period were over £1,250,000 for several years in succession—more

[1] RCG 3.3.26. A total of 1830 persons was then employed in these mines, less than half the figure of a few years before.
[2] May 1920.

in fifty-two weeks than Dolcoath in its entire career of one hundred and seventy years.[1]

Ignorant or uncaring about giant competitors such as these, massive enough to make some of the activity in Cornwall seem almost trifling, Cornishmen still lived in hope, pinning their faith on the expected results

ANNUAL OUTPUT OF TIN IN CORNWALL 1906–64

	tons	London annual price per ton		tons	London annual price per ton
1906	4522	£181	1936	2099	£205
1907	4407	£173	1937	1986	£242
1908	5047	£133	1938	2010	£189
1909	5193	£135	1939	1632	£226
1910	4797	£155	1940	1619	£256
1911	4872	£192	1941	1509	£261
1912	5254	£210	1942	1363	£276
1913	5288	£202	1943	1359	£276
1914	5056	£151	1944	1289	£301
1915	4968	£164	1945	1152	£301
1916	4697	£182	1946	793	£328
1917	3936	£238	1947	898	£427
1918	3954	£330	1948	862	£551
1919	3272	£257	1949	898	£605
1920	3065	£296	1950	890	£747
1921	679	£165	1951	841	£1077
1922	370	£159	1952	903	£964
1923	1021	£202	1953	1103	£732
1924	1986	£249	1954	940	£719
1925	2339	£261	1955	1034	£740
1926	2327	£291	1956	1044	£787
1927	2593	£289	1957	1028	£755
1928	2761	£227	1958	1087	£735
1929	3271	£204	1959	1252	£785
1930	2488	£142	1960	1199	£797
1931	598	£118	1961	1210	£889
1932	1337	£136	1962	1181	£896
1933	1542	£195	1963	1226	£909
1934	1999	£230	1964	1226	£1239
1935	2050	£225			

from East Pool, Tolgus and Roskear new workings. At the former, the ex-Carn Brea 90″ engine was started in January, assisted by two sets of electric pumps, and by April five levels (totalling 1400 feet) had been opened. East Pool's profits in 1924 had been £7898 and this even though stoping had commenced only in the second half of the year. For what was in effect a new mine this was creditable progress and once again the company's mill at Tolvaddon (which in the intervening two years had perforce stood idle but maintained in good repair) was brought back into use. At Tolgus

[1] See MW 20.6.25, 8.8.25, 14.4.28.

the new shaft—unlike the Dolcoath one, rectangular in shape and the same size as Taylor's at East Pool, 17′ by 8′—was down to 1450 feet by June and at 1525 feet was expected to strike the Great South Tolgus Lode, with hopes running very high.[1] The previous year there had been a heavy inflow of water at about 580 feet depth but this had been successfully checked by use of the new cementation process to seal the ground. At the Roskear shaft of the Dolcoath company 1470 feet from surface had been reached by October, despite having to sink through some extremely hard greenstone. By March 1926, after sixteen months' work, the projected depth of 1860 feet had been attained and it was then decided to sink a further 140 feet (completed by 29 April), but even this increased depth failed to bring the shaft into granite as had been expected.

By 1926 the price of tin had recovered to almost its wartime peak of £300 per ton and as a result there came the inevitable resurgence of interest in the industry. Activity recommenced at Polhigey Mine, and at Wheal Reeth (part of the old Great Work sett) in 1927, whilst prospecting or exploration of sorts was once again proceeding at Wheal Hampton, Carzise, Tregembo, Tindene, Wheal Squire (St. Hilary), East Lovell, Lushington, Wheal Vlow, and Wheal Buller. Wheal Kitty at St. Agnes was also resumed in July 1925 with the mine drained by the October following.[2]

At the Tolgus Shaft, where cross-cutting south had commenced in July 1926, confidence was still unabated, even though £150,000 had been laid out and nothing of significance had yet been found. In December the project was stated to be ' nearing success,' although this was not perhaps unconnected with the contemporary news that further capital was required by the company.[3] Then, in February 1928, after further long trials, it was announced that work there was to be discontinued as no ore bodies of commercial significance had been found.[4] This was serious news indeed for the 2000 feet deep shaft and its cross-cut had opened up what was virtually barren ground in this geologically most favoured area. The same month there was the sale of all the plant and machinery of Killifreth and Wheal Busy[5] and this was followed in August by the shutting down of Tresavean, the last and biggest survivor of the pre-1914 boom. Work there had been

[1] MW 27.6.25, '. . . the property will one of these days be the most famous mine in Cornwall.'

[2] Another venture of 1925 was the putting into service of a floating bucket dredge to re-work the alluvial flats of Goss Moor, near Roche. A company entitled the British Alluvial Tin Syndicate Ltd. obtained a lease of 1000 acres from Lord Falmouth and their dredge—built by Werf Conrad, 80′ long and 35′ beam, at a cost of £15500—started work in April 1925. It was then described as the first ever to work in this country—' an event in the history of British mining.' (MW 25.4.25). This same sett had been worked by gravel pumps before the war from 1910–12 by two companies in succession but without any success.

For an account of the various companies that ' lost about £120,000 capital here ' in this century, see H. M. Creswell Payne, *The Story of the Parish of Roche*, n.d.

[3] MW 4.12.26, 11.12.26.

[4] MW 18.2.28. It was stated that Tolgus Mines would continue to be developed from East Pool & Agar by driving east from the latter's workings. Later, in December 1929, Tolgus Mines Ltd. was taken over completely by East Pool, who already held a large share in it.

[5] MW 11.2.28.

resumed in 1923 and three years later annual output had risen to almost 500 tons of black tin. In this re-working after 1909, Harvey's Shaft had been carried down a further 100 fathoms or more (to a depth exceeding 400 fathoms below adit) but with a marked falling off in lode values everywhere in the bottom of the mine and disappointing trials of the eastern workings in the old Comford copper section in 1926–7, it was decided to cease operations.

Once again the hoped-for revival of the industry that had seemed so assured was slipping away. Many brave and bold words had been spoken and written in 1926–7 of the new era that was said to be dawning for Cornish tin, but on the basis of measuring mining results 'in terms of coin of the realm' all was far from well. South Crofty and Geevor were prospering, it was true, but East Pool—after a brief period of boom from 1925–27[1]—had a bad year in 1928, returning a loss of £4385, even though 940 tons of black tin were sold for £118644. The only bright feature of its year was the successful re-entry back into the Rogers Lode old workings—a perilous task which was safely achieved in July.[2] But the worst news was the progress of Dolcoath, where the sad story of Tolgus was to be more or less repeated. In September 1926 it had been necessary to issue further debentures amounting to £50,000 and expenditure had risen to £160,000. Throughout 1927 reassuring progress reports were issued but at the end of the year, from sheer shortage of money, there was a brief cessation of working. A further £50,000 debenture issue was negotiated and mining resumed. Unfortunately results were not up to the high expectations that had been held out and the ore that had been raised—rising to a maximum of about 100 tons a day—was proving refractory compared to that from the old sett.[3] By August, on a falling tin market, it was becoming apparent that there was a distinct possibility of the failure of the company. Despite a direct appeal, 'for the first time in the long history of Dolcoath, the shareholders failed to support the board' and of the further £40,000 in debentures asked for to keep the mine going, only £17000 was forthcoming. In November work underground was stopped, although pumping was continued for another nine months whilst fresh negotiations continued in the hope of finding more capital. In August 1929, however, there was insufficient money even for this, and all activity at the mine ceased: 'a great blow to the community as well as to the prestige of Cornish mining.'[4] The same month meetings were held in Camborne to apply once again to the government for funds to keep the company solvent. It was estimated that a further £100,000 would be

[1] Profits in these three years amounted to £110932, but to offset this the sum of £115514 had been expended on sinking and equipping the new shaft.

[2] Earlier, in 1923, a cross-cut had been driven back to the old Agar shaft to provide ventilation and also a second emergency exit to the new mine.

[3] MW 12.1.28, 26.1.28. On this score it became necessary for Dolcoath to rent four additional calciners early in 1928. During the year about £30000 of black tin was sold.

[4] MW 10.8.29.

needed to re-start the mine, now completely under water, whilst the Trade Facilities loan (amounting, with interest, to £77379) also stood in the way as a prior charge. Expenditure at the Roskear shaft then stood at £297517 and there appeared little hope of any further official loan. All the schemes for obtaining more private capital also came to nought and in March 1930 a receiver was appointed by the Treasury solicitor to wind up the company's affairs. The Roskear workings had failed in their purpose of establishing a second Dolcoath, and the time-honoured name of Cornwall's greatest tin mine passed into history.

Later in 1930 the world trade depression, and the inevitable corollary of a slump in tin prices, brought another hiatus, comparable almost to 1921, in the operations of the few remaining Cornish producers. Wheal Kitty was the first to go under, shut down in August after five years of resumed work.[1] The next month Polhigey was closed, the end of the only spirited attempt to work a Wendron mine on a large scale in modern times. It was said here to be ' impossible to work at a profit under the existing conditions '[2] and in October the great Levant also ceased all work. Over two hundred had been employed throughout the late 1920's, although the working had made steady losses.[3] By March 1929 the deficit was over £29000 and Levant's 10s. 0d. shares stood at a nominal sixpence each when the Treasury (whose 1923 loan was still outstanding) appointed a receiver.[4] In the same month South Crofty announced that all operations except for pumping and essential maintenance were to be stopped and 400 men were laid off. At East Pool some were discharged and a further 200 given notice, the company deciding to keep the mine at work on a limited scale by working one shift a day instead of the normal three.[5] The loss on the year for the company had been £10838, despite the sale of 900 tons of black tin and 330 tons of refined arsenic.

This, by November 1930, left East Pool and Geevor as the only two producing mines. In the decade since the disastrous collapse immediately after the war, the industry had gained no firmer footing and indeed had lost ground. Tresavean, Levant and Polhigey had gone under, whilst all the prospecting for which such confident hopes had been expressed only months before was at a standstill once again. The year 1931 also saw the

[1] RCG 20.8.30. 220 men were discharged at this time.

[2] RCG 17.9.30. A minor but noteworthy feature at Polhigey in this working was the use of ponies for haulage underground, apart from East Pool and Levant a practice unknown elsewhere in Cornwall. At the latter mine they were stabled underground but at Polhigey they were hoisted to grass at the end of the day's work.

[3] The only profit in this period was in 1924 (£1648) as against annual losses totalling £30226. The worst year was 1926 which left a deficit of £7065, whilst working in 1928 was interrupted by ' a breakaway (which) occurred underground at a point where most of the mine water at the shallow level comes in and it was thought prudent to stop underground work until the exact nature of the damage had been ascertained and the necessary repairs executed.' This happened on New Year's Eve 1927 and for ten months of the following year it was necessary to draw on the old mine dumps to keep the mill occupied. (Oats Collection, RIC, Truro).

[4] Sales of tin since 1920 (see page 259) had been £385000, even with operations confined largely to the shallow levels which had hitherto been regarded as exhausted.

[5] RCG 8.10.30

end of tin smelting in Cornwall, with the closing of the sole surviving works at Seleggan. In the brief 1926–7 boom, when world output of tin concentrate rose in a four-year period by 50,000 tons, the works had been considerably extended, completing a programme of expansion which had been commenced in 1919 but had thereafter had to be suspended. Four new furnaces, each of two tons capacity, were installed and further additions put in hand, giving a smelting capacity of 350 tons of black tin per week.[1] In 1930, however, the works passed into the ownership of the Consolidated Tin Smelters Ltd., a group formed in December 1929 to acquire the Cornish Tin Smelting Co. Ltd., the Penpoll Tin Smelting Co. Ltd.,[2] Williams Harvey & Co. Ltd.[3] and the Eastern Smelting Co. Ltd., Penang. As the first step in a process of rationalisation within this large smelting combine—hastened by the world depression which had again brought about excess smelting capacity—it was decided to close the Cornish works in 1931, and on 10 August of that year the long history of tin smelting in Cornwall came to an end. When the furnace-men at Seleggan drew the fires for the last time on that summer day it was indeed a historic occasion in tin smelting. For a thousand, perhaps two thousand years this had been a unique Cornish industry, from the 'Jews'-houses' of antiquity, to the numerous blowing-houses of the Middle Ages, the melting-houses of later centuries and the big smelting works of modern times. Smelting had been more than a mere thread running through the history of Cornish tin mining during this time; it had formed a major part of its fabric.

Yet although smelting in the county had come to an end,[4] perhaps for all time, mining in Cornwall was far from finished. The period of suspension ended in October 1931 when work was recommenced at South Crofty, employing a hundred men on a one shift a day basis and by

[1] *A Century of Metal Broking*, Vivian Younger & Bond Ltd. This was about seven times the works original capacity in 1887.

[2] Although the Penpoll (Point) works had been shut down in 1921, the company continued smelting at the works bearing the same name at Bootle. In August 1928 a new Penpoll tin smelting works was opened there, close beside the old works, with oil-fired furnaces and a capacity of 12000 tons of black tin p.a. This brought United Kingdom smelting capacity up to 57000 tons (metal) by 1929—the high water mark of the industry—the capacity in 1925 having been only 34000 tons.

[3] Earlier in the year Williams Harvey had been acquired by Patino Mines & Enterprises for a reputed £1,332,000. (MM April 1930.)

[4] Some of the names associated with Cornish smelting, however, still survive. Williams Harvey's works at Bootle (a subsidiary of Consolidated Tin Smelters Ltd., London) is the largest of Britain's two remaining tin smelters (the other being Capper Pass & Son Ltd., North Ferriby, Yorkshire and Bedminster, Bristol) and their brands ' Penpoll,' ' Cornish ' and ' Mellanear ' perpetuate names made famous in the history of tin.

During the 1939–45 war there was to some extent a repetition of the events that affected tin smelting during that of 1914–18. In 1942 a big government-owned smelter was started in the United States, in Texas City, to treat Bolivian concentrates and obviate the loss of valuable cargoes in the Atlantic. After the war, in 1957, this closed down but it has again been re-opened by a private company smelting concentrates from a variety of world sources. From the English tin smelters' point of view the most serious development of recent years has been the setting up of two new works in Nigeria. These are capable of reducing all the ores produced there and this has destroyed what was a British monopoly of Nigerian tin smelting. From 26374 tons of tin smelted in the U.K. in 1960, output had fallen to 16849 tons by 1964.

September 1932 about one half of the normal complement of 400 men had been re-started. Unemployment during this period had again been severe, rivalling the conditions in the worst days of the '20's, with 1998 men out of work in Redruth and 1343 in Camborne in the summer of 1932. Despite the decline that had taken place, mining was still not only the largest employer of labour in the Redruth—Camborne area, it was almost the only one. Although high enough, it should be remembered that similar unemployment rates prevailed in various other areas of Britain where basic industries were suffering similar recession, on Tees- and

OUTPUT OF BLACK TIN IN CORNWALL AND MAJOR PRODUCING COUNTRIES AS AN APPROXIMATE PERCENTAGE OF WORLD PRODUCTION 1850–1930

	Cornwall	*Malaysia*	*Indonesia*	*Australia*	*Bolivia*	*Nigeria*	*Siam*	*Total output*
1851–1860	35%	35%	30%					18,000 tons
1861–1870	37%	36%	26%					24,000 tons
1871–1880	25%	18%	22%	23%			8%	37,000 tons
1881–1890	17%	37%	18%	18%			8%	53,000 tons
1891–1900	9%	54%	20%	7%	5%		5%	77,000 tons
1901–1910	5%	50%	17%	6%	17%		5%	97,000 tons
1911–1920	3%	38%	18%	5%	20%	4%	6%	117,000 tons
1921–1930	1½%	33%	20%	2%	23%	6%	6%	150,000 tons

Tyne-side and in the valleys of South Wales. Again, in Cornwall, a small measure of public works was instituted jointly by the government and local authorities to prevent humanity rusting too obviously on the street corners, and a variety of proposals was made for the permanent resuscitation of mining. These ranged from specious suggestions to nationalise all the west country mines to grandiose schemes for forming all those around Camborne—Redruth into one gigantic combine with a capital of £750,000, with centralized pumping and dressing operations—both of these proposals having been mooted as possible solutions a decade before. Another smaller scheme put forward in August 1932 was for a mining tunnel to be driven from Roscroggan in the Tuckingmill valley through the old Dolcoath adit southward to Grenville and beyond, but it soon became apparent that there was little government enthusiasm for such ventures, especially when a grant of £50,000 was suggested.[1]

After a measure of revival after the worst of the slump, although now with only three mines surviving in the county, the industry drifted on throughout the rest of the '30's, neither gaining nor losing ground. Cornish mining was now suffering from the legacy of what had been done, and said, in the late 1920's, and in particular the disastrous failures of the new Dolcoath and Tolgus Mines on which so much attention had been focussed. It was, moreover, a period of dear money and of a restrictionist

[1] RCG 10.8.32.

policy in world tin production[1] and Cornwall received bare mention in the mining press. Not surprisingly the enthusiastic articles recalling the glories and past dividends—never the calls—that had been such a feature a few years before, had dried up.

In 1935 South Crofty acquired the Dolcoath and Roskear setts as a long term extension of her reserves of ore ground, whilst the more settled economic conditions after this date again brought a slight but welcome mining revival. Work was re-started at Porkellis, Wheal Reeth[2] and Wheal Breage in 1934, whilst in 1935 Great Work Tin Mines was formed to re-open to shallow depth part of the old sett of that name. These, however, were small prospects, working on a scale that would have been considered insignificant in earlier days. A major proposal in 1937 by the British Non-ferrous Mining Corporation Ltd. was to drive a large adit from sea level near Portreath up into the old copper workings of the North Downs area but this scheme, which would have been enormously costly but might have had interesting results, was not proceeded with. The same company was also behind a major attempt, started in 1935, to work for tin the eastern end of the old United Mines sett in Gwennap, under the name of Mount Wellington Ltd.[3] A great deal of exploration work was done here, including drilling in the West Wheal Jane sett, whilst nearby there was also activity south of Carharrack at Whiteworks. In St. Agnes, the moribund Wheal Kitty Tin Ltd., which had abandoned the mine in 1930, resumed activities in 1937 at the Turnavore Shaft of the old Polberro where some work had already been done by them in 1928–9. The shaft, 108 fathoms deep, was re-equipped and sunk a further 80 fathoms, with a borehole below that for another 20 fathoms or more, in search of the Kitty lodes going west.

Apart from Geevor, which continued quietly prosperous throughout the 1930's, the progress of the other two remaining Cornish mines of any

[1] In 1929 a Tin Producers' Association had been formed, a combination between the big smelters and many of the larger world mines, in an endeavour to regulate output. This was only partly successful but in 1931 it was given sanction by the governments of Bolivia, Malaya, Netherlands East Indies and Nigeria. As the International Tin Committee (fore-runner of the International Tin Council of today) this operated a quota scheme regulating world output and price of tin, going a long way towards achieving its object of '. . . preventing rapid and severe oscillation of price' (see E. S. Hedges, *Tin in Social and Economic History*, 1964). Had a similar body been in existence twenty years earlier, to stabilise prices and to negotiate for the tin industry as a whole, the war-time and subsequent careers of several big Cornish mines might have been very different. As it was, however, Cornwall's output was so low and her prospects deemed so poor by world standards that she was exempt from the Tin Restriction Scheme of the 1930's.

[2] At Wheal Reeth in January 1937 there was another case of old workings flooding out the mine which had remarkable similarities to the earlier occurrence there in 1891 (see page 209). A collapse of ground separating the workings from the old Wheal Boys section let a flood of water into the mine, trapping three men underground, who fortunately managed to scramble to the safety of an air-pocket in an old stope. This was late on a Saturday afternoon and earlier in the day forty men at least had been underground. Eventually the three trapped miners were rescued and brought to surface unharmed on the Monday following. (RCG 27.1.37.)

[3] The company leased not only United but the old Consols, Wheal Clifford, Wheal Andrew, Nangiles, West Jane, Wheal Jane and other setts—an area of ground over three miles in length along the strike of what had been Cornwall's richest copper lodes. (See RCG 2.11.38).

size or importance provided a striking contrast—South Crofty rich, and East Pool & Agar poor. The former, on a capital of about £117,000 had paid dividends amounting to £290,400 from 1907 to 1933[1], although in the next three years a loss of about £40,000 was incurred. In 1937 the mine made a profit of £29250, and £23800 in 1938, but in the following year its career was interrupted by a strike which was to prove the longest in the county's mining history. There had been rumblings of discontent amongst Union men at the mine for some months and early in January 1939 all the members of the Transport and General Workers' Union there came out on strike.[2] The company promptly gave extra wages to the remaining men and by February pitched battles between militant pickets and police with drawn truncheons had occurred in Pool village and at the entrance to the mine.[3] The company made no secret of their refusal either to recognise the Union or to agree to the demand for a guaranteed minimum wage of 45s. per week.[4] Feelings ran high, particularly as there was still a sufficient number of non-Union men to keep operations going. The mine was three years ahead of the mill and, according to the manager, 'could face a complete stoppage for two years.' With only 234 men on strike out of a total labour force at South Crofty of 435, it was plain that different tactics would have to be used if the strike was to achieve anything, and accordingly all cargoes of tin concentrate from the mine were declared 'black' by Union labour at all Cornish ports and at Liverpool.[5] To circumvent this, the company decided in April to ship 100 tons of tin to smelters in Holland and the old, deserted port of Gweek on the Helford River was chosen as the loading place for a chartered Dutch coaster. A fleet of twenty-eight lorries was hired to carry the tin from the mine but, inevitably, local Union officials got wind of the plan and another minor riot occurred at the mine, with windscreens smashed by strikers' stones and heads broken by police batons in the ensuing *fracas*.[6] The tin, however, was got away, bringing in about £15000 to the beleaguered coffers of South Crofty, although this was to be the only ore sold during the strike. In August it was still in full swing, with both sides as determined as they had been nine months before[7] and it was only the outbreak of war and the intervention of the government that ended the dispute sometime during September.

At East Pool & Agar the problems were of a very different kind during this time. Initially, after a full resumption of work on a three shifts a day

[1] RCG 1.2.33. Output in this period had been 13864 tons black tin (sold for £1,759,389), 12504 tons arsenic (£397,167) and 1414 tons wolfram (£150,688).

[2] RCG 11.1.39.

[3] RCG 8.2.39.

[4] RCG 22.3.39. The existing average weekly pay at South Crofty was then said to be 32s. 6d. The bitterness of the strikers was particularly marked at this time as the mine's profit for 1938 (£23800) was just then made public: 'Men at work at South Crofty, which paid 25 per cent dividend, had to go to the public authorities for boots.'

[5] South Crofty's output was then normally sold by private contract to Williams Harvey at Bootle.

[6] RCG 26.4.39.

[7] RCG 2.8.39.

basis in 1933, results were good, with profits of £12458 and £21805 in that and the following year[1] but in 1935 a loss was made. During this time numerous exploratory bore-holes were put out from the various ends and cross-cuts in the hope of discovering new lodes to supplement Rogers and Moreing's Lodes, which were the sole mainstay of the mine and were failing in depth. Unhappily, nothing of any great economic value was discovered by this drilling programme or by trials in the old North Pool and West Tolgus setts in 1933–4. The loss in 1936 was £4872 and in 1937 £1112, and early the following year it was announced that ' in accordance with recommendations made by the general managers it had been decided to close down the mine and that the ore now available should be extracted as cheaply as possible and after this has been done that the pumps be withdrawn and the mine closed.'[2] This was a blow indeed, for East Pool had been a most important mine and one for which particularly bright hopes had been held in the 1920's, even though it now only employed 225, less than half either Geevor or South Crofty. Application for government aid to continue the mine was made to the Minister for the Co-ordination for Defence, in conjunction with the local authorities, and a month later the decision to close was modified and instead the company decided to continue mining selectively on a small scale.[3] On this basis the mine was kept open, although the all important footage driven amounted to only 1040 feet in 1938 as opposed to 3545 feet in 1937.[4] The loss in the latter year was £906 but it was then becoming apparent from the worsening international situation that a rise in the prices of tin and wolfram—of which East Pool & Agar had been an important producer in the past—might restore something of the mine's old prosperity.

At or soon after the outbreak of war, however, all the attempts at a mining revival in Breage, in Wendron, in Gwennap, in St. Agnes[5], and elsewhere, came to an end and once again no new mine had been established by this activity in the second half of the 1930's. At least it could be said that the war was of benefit in saving East Pool & Agar, for with the loss of Malaya in 1941, home supplies of tin and wolfram assumed great strategic importance to the British war effort and direct government assistance was given to the mine from June 1942. This enabled operations to continue even though the grade of ore being milled was completely uneconomic. Elsewhere in Cornwall, a war-time ' crash ' programme of trials and prospecting was put in hand from 1942–44 under the guidance of the Cornish Tin Mining Advisory Committee—a body formed to give

1 1934 saw the last use made of the electric street tramway which hauled East Pool ore to Tolvaddon mill, an aerial ropeway coming into operation in August.

2 RCG 23.2.38.

3 RCG 25.3.38. Since the starting of the new mine consequent upon the destruction of the old East Pool in 1921, the gains up to 1938 of £160498 had largely been annulled by the outlay and working losses of £150398. East Pool & Agar Ltd. Reports, 1913–39, *per* F. Michell, Redruth.

4 RCG 24.5.39.

5 Polberro Tin Ltd., working at Turnavore Shaft, sought government assistance to keep going in March 1941, but this was not forthcoming and the mine closed. (MW 29.3.41.)

advice and submit propositions to the Non-Ferrous Minerals Development Control, a section of the Ministry of Supply.[1] Yet little additional output was achieved by this means. During centuries of mining activity Cornwall had been too thoroughly combed by skilful prospectors at times when tin was high and too thoroughly ransacked by starving tributers at times when tin was low for there to be much left that could be mined at a few months' notice.

On a very much smaller scale, the condition of the industry at the end of hostilities in 1945 had unhappy similarities to the state of affairs ruling after the first World War, for war-time conditions of working, rising costs and the labour shortage—as well as the marked lack of success of all the government-sponsored prospecting—boded ill for the future. East Pool in particular was in a most precarious state. The subsidy granted to the mine after 1941 had halted the heavy losses[2] but in March 1945 the Ministry of Supply had given the mine company three months' notice that it was terminating its purchase contract[3] and, at this juncture, this was a death blow. Application for further official help was made in June and a government committee of enquiry was appointed to consider 'whether there is justification in the national interest for the granting of assistance from public funds for underground exploration at East Pool and Agar Mine for the purpose of locating additional deposits of tin ore . . .'[4] The company's loss in 1945 was £1366 and there were then over £30,000 in debentures outstanding. Abortive negotiations were continued during 1946 and early in 1947 a receiver was appointed by the Treasury, with the plant and machinery sold by auction later that year. This ended the career of what had been Cornwall's second greatest tin mine, whose life of 138 years' working had spanned five reigns, from George III to George VI.

[1] Amongst the numerous mining propositions considered and prospected were Cligga, Mount Lane (Trencrom), Wheal Bunny, Treveddoe, Ding Dong (Gunnislake), Wheal Kitty, Silver Valley, Kit Hill, Collurian, Bartinney Hill (St. Just), Wheal Bellan, and various places in east Cornwall for wolfram. Various dumps, including those of Wheal Reeth, Great Work, Basset Mines, Dolcoath, Prince of Wales, and Drakewalls, were sampled and treated, with mills renovated for the purpose at Wheal Reeth in the west, Mount Wellington in the central area and Prince of Wales in the east. Men from the Canadian Royal Engineers started mining at Collurian (adjacent to the East Wheal Margaret, Lelant, sett) in April 1942 but two months later 'old workings had been struck where virgin ground was expected' and they were transferred to Silver Valley, on Craddock Moor, adjoining the old South Phoenix. Here about sixty sappers and others were employed until July 1943 when, no output having been achieved and an expenditure of £25000 incurred, the project was terminated by order of the Ministry. The only other sizable project was Wheal Bellan at St. Just (which, by coincidence, had also been tried in 1918 and the 1930's like Silver Valley) where £9995 had been expended to little or no avail by December 1943. Chief output was secured from the re-treatment of dumps, whilst a few tons of alluvial tin were won from the St. Erth valley near Relubbus (although at a production cost of £400 per ton) using 32 Italian prisoners-of-war. By 1944 the Committee was also occupying itself with the post-war problems looming before the industry, and with a plan for the Nationalisation of Minerals. (Minutes of the Cornish Tin Mining Advisory Committee, DDX 161, CRO, Truro).

[2] In 1941, for example, the loss was £11165 whilst the profits in the ensuing three years were £3926, £2407 and £2422. (RCG 28.2.45).

[3] RCG 11.4.45. Earlier, in 1944, there had been a suggestion that the government should benefit Cornwall by forming a stock-pile of home-produced tin, on the same lines as the United States where Bolivian tin was being accumulated in enormous quantities as a strategic stock-pile.

[4] RCG 29.6.45.

Fortunately the extra water to be pumped at South Crofty as a result of East Pool's closing down did not endanger the former, although inevitably it added to working costs. South Crofty and Geevor were now the only units left in the entire industry and the county's output fell below 1000 tons of concentrate from 1946 onwards. There was at this time an acute shortage of underground workers—for which reason the closing down of East Pool had caused little or no unemployment—and Polish miners were taken on at both mines.[1]

In 1948 the Cornish Mining Development Association was formed, a body devoted primarily to furthering the possibilities of a permanent revival in the industry and in the ensuing years, with tin rising to higher prices than ever before,[2] there has been re-awakened interest in the economic potential of most of the old Cornish mining districts. Almost a score of companies are today (1966) actively prospecting in the county, from Kit Hill westward to Breage and St. Ives, including some which have the financial and technical backing of such mining giants as the Union Corporation of South Africa. Extremely expensive diamond drilling programmes have been carried out on a considerable scale in various parts of west Cornwall,[3] with almost all the old mining areas in the county held on option by one or other mining group, whilst even Dartmoor is being again prospected.

This wave of activity accompanying this latest revival in Cornish tin may be said to have depressing similarities to the earlier booms about 1906, 1926 and 1936, all of which did the image of Cornish mining more harm than good.[4] The cynic may see the current activity as yet another flutter of a new generation of moths into the old candle-flame, and the pessimist may say that despite several years of record metal prices no new mine has yet been started and output is still only about 1600 tons a year. It is true that Geevor and South Crofty remain, to date, the only producing mines[5] but the changed feature today—and perhaps the crucial one in the

[1] During the war years, all three mines had been allocated 'optants' who preferred mining to service in the armed forces but these were untrained and largely unwilling. In 1943 South Crofty and Geevor applied for Italian prisoners to supplement their available labour force—Tabb's Hotel at Redruth and Levant old count-house being suggested for use for them to be lodged in. Twenty-four were employed at East Pool in the same year but there were difficulties in fitting them to shift work due to the official restriction that they were not allowed out of doors after dark. (Minutes of the Cornish Tin Mining Advisory Committee, DDX 161, CRO, Truro).

During the war-time labour shortage, East Pool had introduced a pony for underground haulage, just as had been necessary in the old mine from about 1916 onwards.

[2] £1500 per ton in 1966, for example. In comparing this figure with former times, however, inflation and the depreciation in the present day value of money should not be forgotten. Moreover, although tin is ten times the metal price of former years, wages and other production costs in the industry have increased in similar or even greater proportion.

[3] At Wheal Vor, Tregurtha Downs, Giew, North Downs, Leedstown, near Camborne, and elsewhere. In 1964 alone, a total of sixty boreholes was put down.

[4] This is particularly true of the more peripheral ventures such as the dredging of St. Ives Bay.

[5] Output from Geevor in 1964 was 675 tons of black tin and 621 tons in 1965; from South Crofty, where the immediate reserves position is immeasurably better, 994 tons in 1964 and 1060 tons in 1965. Geevor, in conjunction with a South African mining house (Union Corporation), is at present engaged in the problems of re-opening Levant, a particularly difficult task as the sea had broken into

consideration of Cornwall's mining future—is political instability in the Far East, in Thailand, Burma, Malaysia, and Indonesia, that now supply two-thirds of the world's tin. Just as this area was the prime cause of Cornwall's decline so it may well prove an equally important one in bringing about rehabilitation. Moreover, one third of Malaya's tin dredges are expected to cease work by 1970 and no new major deposits have been found there since 1945, despite methodical prospecting.

To All Mining Companies Prospecting in Cornwall

LODE OF TIN AVAILABLE

Box P2743, West Briton, Truro.

The old spirit lives on—an advertisement in the West Briton, 2 *June* 1966

As against this, much ground there which is now regarded as exhausted could be re-worked if tin prices warranted it, and also the off-shore dredging newly instituted by Indonesia may, within a few years, open up a very great deal of new and virgin ground. Sea-going dredges capable of working in 130 feet of water and some ten miles off-shore are in commission off Indonesia and Thailand and the economic possibilities of this and other forms of submarine mining seem almost boundless. The immense resources of the Bolivian Andes are also a most important factor in the long term outlook for tin, whilst the possibility of vast quantities of alluvial deposits in remote and hitherto quite inaccessible areas in Brazil is also beginning to interest economic geologists. As a more immediate factor, it must be remembered that the high price of tin stimulates not only Cornish hopes of a revival but also those of other minor tin producing areas—Australia, for instance, having a declared production aim of 6000 tons by 1967. In this respect it is not generally appreciated that such countries as the Republic of South Africa, Burma, Australia, Brazil, France and Portugal have all rivalled Cornwall in tin output at various times in recent years.[1]

A recent survey carried out by the International Tin Council[2] contained estimates—insofar as it is ever possible to arrive at such figures in metalliferous mining—of approximately four million tons of concentrates in reserve in the world,[3] excluding Russia and China. Of this total Thailand

the latter at the notorious weak point in the back of the 40 sometime after working ended in 1930. Just prior to the closing the Levant company were filling in the 40 to render this danger point safe but this work was not complete when the mine shut down. In the past few years the same end has been achieved by covering and sealing the sea bed, using divers.

[1] In 1964 output from the smaller producing countries was: Australia 3638 tons; South Africa 1793 tons; Great Britain 1669 tons; France 469 tons; Portugal 52 tons. It is of interest to note that Portuguese output was boosted to a peak of 3000 tons a year during the Second World War.

[2] *Report on the World Tin Position, with Projections for 1965 and 1970*, The International Tin Council, London, 1965.

[3] Including measured, indicated and inferred reserves.

held an estimated 1,300,000 tons; Malaysia 900,000 tons; Indonesia 650,000 tons; Bolivia 514,000 tons; The Congo 158,000 tons; Nigeria 105,000 tons; Australia 68000 tons; United Kingdom 41000 tons; all other countries combined 62000 tons. These figures are valuable, as they put the potentialities of the various tin producing countries into world perspective. They also show how small is the remaining output if the politically unstable producers are removed from this list.

In Cornwall the basic problem today is to find areas which possess the enormous low-grade reserves of ore ground requisite to warrant the

OUTPUT OF TIN CONCENTRATES BY VARIOUS OF THE SMALLER PRODUCING COUNTRIES, 1959–64

	1959	1960	1961	1962	1963	1964
Rwanda	1124	1277	1474	1325	1271	1360 tons
Brazil	461	1556	581	731	1153	1200 tons
Japan	989	843	854	859	860	782 tons
Mexico	377	365	530	576	1055	1206 tons
Portugal	1129	644	729	679	718	676 tons
Argentina	225	238	515	571	498	715 tons
South Africa	1273	1284	1431	1422	1529	1585 tons
Australia	2351	2202	2745	2715	2860	3548 tons
Utd. Kingdom	1252	1199	1210	1181	1226	1226 tons

expense of setting up a modern mill in which an output of 500 or 600 tons of tin concentrates per annum can be produced over a period of years. It is not, as it once was, the single rich lode measured in inches that matters today and indeed it is doubtful whether any such lodes of any size exist that have not been worked at one time or another. In such a circumscribed mining field as Cornwall—unlike Bolivia, with its mountain peaks where no man has ever trodden—where miners, mine adventurers and mineral lords all had a ' nose ' for tin, there must have been very few acres that have not been critically examined and, where the slightest surface indications warranted it, tested by costeaning or an adit. In this respect it is instructive to reflect that unlike copper, where the virgin mining districts of Caradon and the Tamar were opened in the reign of Victoria, tin was found in no new district in Cornwall after 1800. The great rise in Cornish tin mining came about by vertical and not by lateral expansion and after 1875 production became concentrated more and more in the deep mines which formerly had been producers of copper. Big mines which produced tin alone, such as Wheal Vor, St. Ives Consols, Polberro, Providence, Great Work, Balleswidden, Drakewalls, Polgooth and others, disappeared after 1875—except perhaps in name only, revived from time to time by mine promoters—and virtually the only exception to this rule was West Kitty in St. Agnes. This, moreover, was despite the considerable mining and exploration programmes that were put in hand during the various booms of this century. Some of these were spurious, some indifferently managed, some badly equipped and some short of capital, but not all, and it is true to say that no lode of commercial importance has been found at shallow depth in Cornwall for a century and probably more. With a growing realisation of this, it has in consequence been the possibilities of deep tin hidden in unworked mineralized ground or left in old setts by a generation of copper miners that has occupied most attention in recent years.

The resuscitation of an old, deep mining field is a difficult task—one, moreover, that has never yet been accomplished in the history of metal mining—and it is as fraught with the possibilities of expensive failure as is the resuscitation of a single old deep mine. The more ancient, the deeper, and the more intensive has been the working in the past, the more remote are the chances of a successful re-working. It took neither money nor skill, years ago, to open a virgin mine on a lode that outcropped at surface; at a re-working, it required a little more outlay and mining ability to go down below water level, with horse-whims and a wheel for pumping; at a second re-working, at greater depth, with steam power necessary, unwatering and re-equipping involved the provision of capital and capable management; third, fourth, fifth and other successive re-workings merely increased the amount to be staked before there was even a chance of success, whilst at the same time the prize to be won had not increased and may well have been diminished. Such was, and is, the gamble of re-opening metalliferous mines.[1] Herein lies the central dilemma of Cornish tin mining revival—that there is still a great deal of tin lying scattered in the county but that even using diamond drills as a means of insurance, a great deal of money can be lost trying to locate and extract it. The history of mining in Cornwall in the twentieth century shows this only too clearly.

It is, however, interesting to speculate whether the county still possesses—unknown and untouched, perhaps under Gwennap parish or Blackwater or St. Blazey—' champion ' tin lodes as big and as rich as ever Dolcoath Main Lode once was. This is a matter for conjecture alone, for even the most experienced mining geologist can not provide any firm answer. Moreover, even the most lengthy diamond drilling programmes can only partly establish the potentialities of small areas out of the whole mining complex. The future of Cornish mining is thus problematical to a degree, hedged about not only by the inherent geological and physical difficulties found in every mining field, but affected no less vitally by sociological, economic and political factors as well. The possible spoiling of amenities, the provision of cheap money, the harsh United Kingdom system of mining taxation, the United States strategic stock-pile of tin, and the advance of Communism in south-east Asia all play a part in the hastening

[1] The alternative is a ' permanent ' mine, of which Geevor and South Crofty are now excellent examples, where the costs (and chances) of establishment have been overcome. Here working is continuous, reserves of ore and finance are maintained, and—most important of all—development is kept so far ahead of the mill that constant output is maintained.

or the delaying of any revival of Cornish mining. In this welter of possibilities, no one can foresee its future. One thing alone is certain, that if and when it returns on any large scale, Cornish mining will bear no resemblance to the days of old. There will be no grass captains and granite engine-houses, no great feats of dialling and no grand count-house dinners; equally there will be no ankylostomiosis and racking ladder-roads, no soup kitchens and no pathetic miners' choirs, for modern metal mining as an industry bears none of these scars of the past.

APPENDIX I

LIST OF CORNISH TIN SMELTING HOUSES

Name		*Established*	*Closed*	*Principal proprietors*
Angarrack	near Hayle	1704	1881	Daniell: Carne: Bolitho
Bissoe		1838	1878	Vigurs: Tregoning
Bissoe (2)		1890	1919	Ltd. Company
Calenick	near Truro	1711	1891	Daniell: Michell & Bolitho
Charlestown		1834	1884	Taylor: Enthoven: Williams Harvey, Daubuz, Bolitho
Carvedras	Truro	pre-1750	1898	Rosewarne: Daubuz: Consolidated
Chyandour	Penzance	pre-1758	1912	Oxnam: Bolitho: Consolidated
Coosebean	near Truro	pre-1771	1810	
Cornish	near Redruth	1887	1931	Lanyon: Vivian Y. & Bond
Foundry	Hayle	1816	1820	Harvey & Co.
Great Hewas	near St. Austell	1816	1822	Adventurers
Ludgvan	near Penzance	pre-1736	1835	Rogers: Gundry: Bolitho
Mellanear	Hayle	1837	1905	Williams Harvey
Newham	Truro	1703	1711	Moult
Penryn		1917	1920	London T.S.Co.
Penpoll	Restronguet (Fal)	1870	1921	James & Shakespeare: Teague & Strauss
Perran	Perranarworthal	pre-1755	1802	Reed
Portreath		1814	1825	Harvey Davey Williams
Redruth		1863	1923	Waters Lanyon: Bain
Tamar (Union)	Weir Quay (Devon)	1849	1890	Ltd. Company
Trelissick	Hayle	1820	*c.* 1850	Grenfell: various
Treloweth	St. Erth	*c.* 1715	1883	Daubuz
Trereife	Newlyn	pre-1732	1896	Batten: James: Michell: Consolidated
Trethellan	Truro	1824	1890	Grenfell: Williams Harvey
Treyew	near Truro	1712	*c.* 1765	Enys: Donnithorne
Truro		1816	1871	Magor & Turner: Daubuz: Vigurs: Gatley
Wheal Reeth	Lelant	1823	1824	Adventurers
Wheal Vor	Breage	1823	1844	Adventurers

plus others at Calstock, Gweek, Lostwithiel, Porthtowan, St. Agnes, St. Austell, and St. Blazey.

INDEX OF MINES

The various contemporary spellings of some mine names [such as Creegbroas, Creegbrawse; Wheal Caul, Wheal Call] have been retained in the main text

GENERAL INDEX